AF595079

Piezoelectric Aluminium Scandium Nitride (AlScN) Thin Films: Material Development and Applications in Microdevices

Piezoelectric Aluminium Scandium Nitride (AlScN) Thin Films: Material Development and Applications in Microdevices

Editor

Agnė Žukauskaitė

MDPI • Basel • Beijing • Wuhan • Barcelona • Belgrade • Manchester • Tokyo • Cluj • Tianjin

Editor
Agnė Žukauskaitė
S2S-Technologies and Precision Coating
Fraunhofer Institute for Organic
Electronics, Electron Beam
and Plasma Technology FEP
Dresden
Germany

Editorial Office
MDPI
St. Alban-Anlage 66
4052 Basel, Switzerland

This is a reprint of articles from the Special Issue published online in the open access journal *Micromachines* (ISSN 2072-666X) (available at: www.mdpi.com/journal/micromachines/special_issues/Piezoelectric_AlScN_Thin_Films).

For citation purposes, cite each article independently as indicated on the article page online and as indicated below:

LastName, A.A.; LastName, B.B.; LastName, C.C. Article Title. *Journal Name* **Year**, *Volume Number*, Page Range.

ISBN 978-3-0365-6367-1 (Hbk)
ISBN 978-3-0365-6366-4 (PDF)

Contents

About the Editor

Agnė Žukauskaitė

Dr. Agnė Žukauskaitė is currently splitting her time between the Fraunhofer Institute for Organic Electronics, Electron Beam and Plasma Technology FEP in Dresden and the Institute of Solid-State Electronics IFE at TU Dresden University. Her recent research activities have been in the field of application-inspired material science, especially focusing on the magnetron sputter epitaxy of thin films for sensing applications.

Preface to "Piezoelectric Aluminium Scandium Nitride (AlScN) Thin Films: Material Development and Applications in Microdevices"

Not so long ago, aluminium scandium nitride (AlScN) emerged as a material that possesses superior properties as compared to aluminium nitride (AlN). Substituting Al with Sc in AlN leads to a dramatic increase in the piezoelectric coefficient as well as in electromechanical coupling. This discovery finally allowed us to overcome the limitations of AlN thin films in various piezoelectric applications while still enabling us to benefit from all of the advantages of the parent material system, such as a high temperature stability, CMOS compatibility, and good mechanical properties.

AlScN allows for the enhanced performance of RF filters where bulk acoustic wave (BAW) or surface acoustic wave (SAW) resonators are used. Additionally, energy harvesting and sensing applications can benefit as well. Higher pyroelectric coefficients allow for new advances in, for example, infra-red (IR) detectors. Recent progress in growing this material by MOCVD and MBE has opened up new prospects in high-frequency and -power electronics, such as with high-electron-mobility transistors (HEMTs). Finally, AlScN is the first wurtzite III-nitride where ferroelectric switching was observed a few years ago, opening up another exciting research field with many possible applications in semiconductor memories; additionally, it enables the additional functionality of switching to applications where piezoelectric materials are already in use.

Recognizing the importance of this new material, internationally recognized research groups participated in this Special Issue focused on AlScN to provide the scientific community with a highly visible, multidisciplinary open access collection of the recent advances in understanding the material itself that will enable us to unlock its full potential in microdevices.

This Special Issue presents contributions addressing:

- Fundamentals, physics, theory, and modeling of AlScN material and devices;
- Growth and characterization of AlScN thin films;
- Design, fabrication, and performance of AlScN microdevices;
- Novel and niche applications for AlScN.

We hope that you will enjoy this collection!

Agnė Žukauskaitė
Editor

micromachines

MDPI

Editorial

Editorial for Special Issue "Piezoelectric Aluminium Scandium Nitride (AlScN) Thin Films: Material Development and Applications in Microdevices"

Agnė Žukauskaitė [1,2]

[1] Fraunhofer Institute for Organic Electronics, Electron Beam and Plasma Technology FEP, 01277 Dresden, Germany; agne.zukauskaite@tu-dresden.de

[2] Institute of Solid State Electronics (IFE), Faculty of Electrical and Computer Engineering, TU Dresden, 01062 Dresden, Germany

The enhanced piezoelectric properties of aluminum scandium nitride ($Al_{1-x}Sc_xN$ or AlScN) were discovered in 2009 by Morito Akiyama's team [1]. By introducing Sc into wurtzite AlN, the piezoelectric coefficient and electromechanical coupling increase remarkably due to a strong change in the response of the internal atomic coordinates to strain [2]. This had a significant and immediate impact on the 6G RF filter community as well as other fields, where AlN thin films were being developed and applied due to their CMOS compatibility, good mechanical properties, high-temperature stability, and other attractive properties. Before AlScN, the low electromechanical coupling was a barrier to commercializing AlN-based acoustic devices and having a group-III nitride that could compete with other, more established oxide-based piezoelectric materials. To this day, the number of papers on AlScN and other Al-X-N ternary nitrides grows every year, indicating a continuous interest in both academia and industry to advance the technology. In 2019, Simon Fichtner demonstrated that AlScN also has ferroelectric properties, expanding the application field for this fascinating material even further [3]. However, due to the metastability of this ternary nitride [4], there are many unanswered questions about the true limits of AlScN in material synthesis, fundamental properties, and device performance. This Special Issue features thirteen research papers that focus on recent AlScN development and explores the following aspects: growth of AlScN thin films by magnetron sputtering (three papers), fundamental investigations of material properties (four papers), fabrication and performance of AlScN-based ferroelectric and electroacoustic devices (six papers).

The first three papers [5–7] address both the growth process optimization and high rate, large area deposition on commercial and industry-relevant substrates at Sc concentration $x \approx 0.27$–0.3, which is currently regarded by industry and academia as a reasonable trade-off point, allowing enhanced device performance without suffering too much from metastability-related issues, such as phase separation, an uncontrollable amount of abnormally oriented grains (AOGs), and elemental segregation into Sc-rich and Al-rich domains. Pirro et al. [5] utilize RF bias on the substrate to better understand the relationship between stress levels and the dielectric and ferroelectric properties of the films. Tuning the RF bias resulted in 500 MPa tensile and −2 GPa compressive stress. Films with tensile stress demonstrated better ferroelectric performance and lower losses, whereas high compressive stresses were detrimental to the dielectric and ferroelectric performance of structured metal–ferroelectric–metal (MFM) structures. Su et al. [6] aim to demonstrate a substrate-independent method for depositing high-quality AlScN on various non-metallic substrates using a 20 nm thick AlN seed layer to minimize the number of AOGs. It was postulated that because AlScN is more sensitive to the substrate texture than AlN, focusing efforts on providing a good AlN seed layer rather than process optimization for AlScN on each specific substrate itself is a more applicable approach where AlScN on many different substrates is required for different applications. It was also shown that the addition of an AlN

Citation: Žukauskaitė, A. Editorial for Special Issue "Piezoelectric Aluminium Scandium Nitride (AlScN) Thin Films: Material Development and Applications in Microdevices". *Micromachines* **2023**, *14*, 1067. https://doi.org/10.3390/mi14051067

Received: 11 May 2023
Accepted: 16 May 2023
Published: 18 May 2023

seed layer has only a minor effect on the piezoelectric performance of thick AlScN films. Barth et al. [7] demonstrate AlScN sputtering at a very high deposition rate of 200 nm/min and investigate the homogeneity of structural and piezoelectric properties as well as how they are affected by different growth parameters. In particular, a hybrid unipolar-bipolar pulse mode with an optimized ratio of $S_{Unipolar}$ = 90% was shown to aid in the formation of highly uniform films with good piezoelectric and ferroelectric performance.

In the second section of the Special Issue, four papers [8–11] focus on investigating the fundamental properties of AlScN, such as the effects of temperature and Sc concentration, and the extraction of device-relevant acoustic parameters. In the in-depth Raman spectroscopy study by Solonenko et al. [8], peak broadening and various mechanisms that contribute to it are investigated. The Raman spectra of AlScN can exhibit up to eight bands, and previously unreported bands could be attributed to second-order phonon modes. In addition, temperature-dependent Raman measurements were performed to demonstrate that the temperature coefficient is a function of not only the Sc concentration but also the defect density. In the study by Wolff et al. [9], in situ annealing up to 1000 °C is combined with X-ray diffraction, and an unexpected volume expansion is observed above 550 °C in AlScN thin films prepared by magnetron sputter epitaxy. It was shown that the transition from linear to non-linear thermal expansion occurs at lower temperatures for higher Sc concentrations and that annealing induces irreversible changes in lattice parameters after cooling down. Intrinsic and extrinsic contributions related to oxygen impurities were separated and carefully explained. Continuing the theme of thermal stability and coupling it with the ferroelectric properties of AlScN, Drury et al. [10] explore the possibility of using this material in high-temperature nonvolatile memory applications. The device-relevant behavior of textured AlScN was studied up to 400 °C. While steady polarization retention was observed even at elevated temperatures, P–E loops showed a decrease in coercive field as a function of temperature, and leakage current density increased significantly at elevated temperatures. Fatigue testing indicated degraded performance above 200 °C; however, at lower temperatures, $>10^5$ cycles could be achieved before failure, indicating a high potential for AlScN-based memory applications where $\sim10^3$ cycles are expected. Next, the capabilities of the non-destructive laser ultrasonics technique are demonstrated by Meyer et al. [11] for the extraction of elastic constants of AlScN required for accurate device design. A special rotating stage was used to investigate the anisotropy of epitaxial c-plane and a-plane AlScN thin films. The latter were especially valuable due to the low amount of curvature in the dispersion curves. As a result, sensitivity analysis indicated that elastic constants c_{33} and c_{13} were the main contributors to the dispersion. Very good agreement with theoretical predictions was achieved, validating this versatile characterization method, especially for anisotropic films/substrate systems.

In the last section, six papers address the fabrication and performance of various AlScN-based devices [12–17]. It has been recognized quite early in the development of AlScN that the etching rate drops dramatically compared to pure AlN, and the existing etching approaches used for group-III nitrides must be revised, especially when vertical side-wall geometry has to be well controlled. Wet-etching with KOH is the main focus of the investigation by Tang et al. [12]. In addition to a very comprehensive review of previous dry- and wet-etching studies, both vertical and lateral etch rates are studied systematically as a function of Sc concentration as well as KOH temperature and concentration. A linearity check confirmed etch to be reaction-limited, matching AlN in KOH behavior. However, in the case of vertical etch rate, an expected gradual decrease was observed as a function of Sc. Lateral etching is a highly anisotropic process, and the rate experiences a transition after reaching the lowest point at Sc x = 0.125. The authors were able to demonstrate vertical wall formation at this Sc concentration by exposing the specific planes of AlScN to KOH, which could benefit lamb wave resonators or similar released structures in the future. In a device, it is not only the functional layer that contributes to the overall performance, as shown in the study by Nie et al. [13], where the ferroelectric properties of AlScN are investigated on different metals used for bottom electrodes. In samples with a Mo bottom electrode, a larger polarization loss was observed compared to those where Pt was used. The study indicates that while the formation of abnormally oriented grains is different and has an influence on the ferroelectric performance of AlScN, the inherent contact barrier is also a factor that needs to be considered in device fabrication. Stress control is another very important aspect of electroacoustic device fabrication, as they often contain released structures, such as membranes or cantilevers. Beaucejour et al. [14] demonstrate a stress-compensated growth process allowing low out-of-plane bending in cantilevers. First, the compressive-to-tensile stress gradient was determined as a function of AlScN film thickness, and the average film stress was determined as a function of N_2 flow. Using this information, the authors were able to produce AlScN thin films where N_2 flow varied during the growth to compensate for stress. In the best-performing compensated film, this allowed cantilever bending to be reduced from >100 μm to less than 3 μm. The theme of deflection continues in the next study, where AlN and AlScN-based micromirrors are fabricated by Stoeckel et al. [15]. Footprint MOEMS of 2×3 mm^2 and 4×6 mm^2 were designed, with geometrical parameters adjusted based on different mechanical properties of AlScN. A 10-fold increase in the deflection per electric field in AlScN micromirrors was observed compared to AlN, showing the high potential of such MOEMS for future micro-optics applications with large static scan angles. The paper by Lozano et al. [16] focuses on AlScN-on-diamond surface acoustic wave (SAW) resonators and filters. Single crystal and polycrystalline diamond were compared, and in both types of resonators, Rayleigh and Sezawa modes were excited. Although the frequency, acoustic velocity, and electromechanical coupling were similar in both, the quality factor and figure of merit (FOM) were much higher, and insertion loss improved when using the single crystal diamond as a substrate. Finally, dual-mode Lamb wave resonators are demonstrated by Rassay et al. [17]. Patterned molybdenum was used as a bottom electrode, where tapering enabled crack-free overgrowth by AlScN. Dry-etching was used for membrane release. By taking advantage of the ferroelectric properties of AlScN, the authors designed and fabricated resonators that can be periodically poled using an interdigitated transducer (IDT), thus enabling the reversible switching of frequency to cover two modes of operation—0.45–1.6 GHz and 0.8–3 GHz. This novel concept has great potential for single-chip, multi-band devices for modern wireless systems.

To summarize, this Special Issue was very successful in covering all of the main aspects of AlScN research—growth, fundamental and application-relevant properties, device fabrication and characterization. We can see that AlScN technology is mature enough to demonstrate wafer-level material development and complicated devices, but there is still much to discover in terms of deposition process control, anisotropy, and, in particular, ferroelectric behavior. AlScN research is ongoing in a number of research groups around the world, with many more discoveries to come. On behalf of the journal, the Guest Editor, Dr. Agnė Žukauskaitė, would like to thank all of the contributing authors once more and is eager to hear about future work in this field.

Conflicts of Interest: The author declares no conflict of interest.

References

1. Akiyama, M.; Kamohara, T.; Kano, K.; Teshigahara, A.; Takeuchi, Y.; Kawahara, N. Enhancement of Piezoelectric Response in Scandium Aluminum Nitride Alloy Thin Films Prepared by Dual Reactive Cosputtering. *Adv. Mater.* **2009**, *21*, 593–596. [CrossRef] [PubMed]
2. Tasnádi, F.; Alling, B.; Höglund, C.; Wingqvist, G.; Birch, J.; Hultman, L.; Abrikosov, I.A. Origin of the Anomalous Piezoelectric Response in Wurtzite $Sc_xAl_{1-x}N$Alloys. *Phys. Rev. Lett.* **2010**, *104*, 137601. [CrossRef] [PubMed]
3. Fichtner, S.; Wolff, N.; Lofink, F.; Kienle, L.; Wagner, B. AlScN: A III-V semiconductor based ferroelectric. *J. Appl. Phys.* **2019**, *125*, 114103. [CrossRef]
4. Höglund, C.; Birch, J.; Alling, B.; Bareño, J.; Czigány, Z.; Persson, P.; Wingqvist, G.; Zukauskaite, A.; Hultman, L. Wurtzite structure $Sc_{1-x}Al_xN$ solid solution films grown by reactive magnetron sputter epitaxy: Structural characterization and first-principles calculations. *J. Appl. Phys.* **2010**, *107*, 123515. [CrossRef]
5. Pirro, M.; Zhao, X.; Herrera, B.; Simeoni, P.; Rinaldi, M. Effect of Substrate-RF on Sub-200 nm $Al_{0.7}Sc_{0.3}N$ Thin Films. *Micromachines* **2022**, *13*, 877. [CrossRef] [PubMed]
6. Su, J.; Fichtner, S.; Ghori, M.U.; Wolff, N.; Islam, M.R.; Lotnyk, A.; Kaden, D.; Niekiel, F.; Kienle, L.; Wagner, B.; et al. Growth of Highly c-Axis Oriented AlScN Films on Commercial Substrates. *Micromachines* **2022**, *13*, 783. [CrossRef] [PubMed]

7. Barth, S.; Schreiber, T.; Cornelius, S.; Zywitzki, O.; Modes, T.; Bartzsch, H. High Rate Deposition of Piezoelectric AlScN Films by Reactive Magnetron Sputtering from AlSc Alloy Targets on Large Area. *Micromachines* **2022**, *13*, 1561. [CrossRef] [PubMed]
8. Solonenko, D.; Žukauskaitė, A.; Pilz, J.; Moridi, M.; Risquez, S. Raman Spectroscopy and Spectral Signatures of AlScN/Al_2O_3. *Micromachines* **2022**, *13*, 1961. [CrossRef] [PubMed]
9. Wolff, N.; Islam, M.R.; Kirste, L.; Fichtner, S.; Lofink, F.; Žukauskaitė, A.; Kienle, L. $Al_{1-x}Sc_xN$ Thin Films at High Temperatures: Sc-Dependent Instability and Anomalous Thermal Expansion. *Micromachines* **2022**, *13*, 1282. [CrossRef] [PubMed]
10. Drury, D.G.; Yazawa, K.; Zakutayev, A.; Hanrahan, B.; Brennecka, G.L. High-Temperature Ferroelectric Behavior of $Al_{0.7}Sc_{0.3}N$. *Micromachines* **2022**, *13*, 887. [CrossRef] [PubMed]
11. Mayer, E.M.R.; Rogall, O.; Ding, A.; Nair, A.; Žukauskaitė, A.; Pupyrev, P.D.; Lomonosov, A.M.; Mayer, A. Laser Ultrasound Investigations of AlScN(0001) and AlScN(11-20) Thin Films Prepared by Magnetron Sputter Epitaxy on Sapphire Substrates. *Micromachines* **2022**, *13*, 1698. [CrossRef] [PubMed]
12. Tang, Z.; Esteves, G.; Zheng, J.; Olsson, R.H. Vertical and Lateral Etch Survey of Ferroelectric AlN/$Al_{1-x}Sc_xN$ in Aqueous KOH Solutions. *Micromachines* **2022**, *13*, 1066. [CrossRef] [PubMed]
13. Nie, R.; Shamsaei, N.; Luo, Z.; Kang, X.; Wu, T. Characterization of Ferroelectric $Al_{0.7}Sc_{0.3}N$ Thin Film on Pt and Mo Electrodes. *Micromachines* **2022**, *13*, 1629. [CrossRef] [PubMed]
14. Beaucejour, R.; D'Agati, M.; Kalyan, K.; Olsson, R.H. Compensation of the Stress Gradient in Physical Vapor Deposited $Al_{1-x}Sc_xN$ Films for Microelectromechanical Systems with Low Out-of-Plane Bending. *Micromachines* **2022**, *13*, 1169. [CrossRef] [PubMed]
15. Stoeckel, C.; Meinel, K.; Melzer, M.; Žukauskaitė, A.; Zimmermann, S.; Forke, R.; Hiller, K.; Kuhn, H. Static High Voltage Actuation of Piezoelectric AlN and AlScN Based Scanning Micromirrors. *Micromachines* **2022**, *13*, 625. [CrossRef] [PubMed]
16. Lozano, M.; Fernández-García, L.; López-Romero, D.; Williams, O.A.; Iriarte, G. SAW Resonators and Filters Based on $Sc_{0.43}Al_{0.57}N$ on Single Crystal and Polycrystalline Diamond. *Micromachines* **2022**, *13*, 1061. [CrossRef] [PubMed]
17. Rassay, S.; Mo, D.; Tabrizian, R. Dual-Mode Scandium-Aluminum Nitride Lamb-Wave Resonators Using Reconfigurable Periodic Poling. *Micromachines* **2022**, *13*, 1003. [CrossRef] [PubMed]

 *micromachines*

Article

Effect of Substrate-RF on Sub-200 nm $Al_{0.7}Sc_{0.3}N$ Thin Films

Michele Pirro *, Xuanyi Zhao, Bernard Herrera, Pietro Simeoni and Matteo Rinaldi

Electrical and Computer Engineering Department, Northeastern University, Boston, MA 02115, USA; zhao.xuan@northeastern.edu (X.Z.); herrerasoukup.b@northeastern.edu (B.H.); p.simeoni@northeastern.edu (P.S.); m.rinaldi@northeastern.edu (M.R.)
* Correspondence: pirro.m@northeastern.edu

Abstract: Sc-doped aluminum nitride is emerging as a new piezoelectric material which can substitute undoped aluminum nitride (AlN) in radio-frequency MEMS applications, thanks to its demonstrated enhancement of the piezoelectric coefficients. Furthermore, the recent demonstration of the ferroelectric-switching capability of the material gives AlScN the possibility to integrate memory functionalities in RF components. However, its high-coercive field and high-leakage currents are limiting its applicability. Residual stress, growth on different substrates, and testing-temperature have already been demonstrated as possible knobs to flatten the energy barrier needed for switching, but no investigation has been reported yet on the whole impact on the dielectric and ferroelectric dynamic behavior of a single process parameter. In this context, we analyze the complete spectrum of variations induced by the applied substrate-RF, from deposition characteristics to dielectric and ferroelectric properties, proving its effect on all of the material attributes. In particular, we demonstrate the possibility of engineering the AlScN lattice cell to properly modify leakage, breakdown, and coercive fields, as well as polarization charge, without altering the crystallinity level, making substrate-RF an effective and efficient fabrication knob to ease the limitations the material is facing.

Keywords: scandium-doped aluminum nitride; ferroelectric; MEMS; substrate-RF; residual stress; coercive field; leakage current

Citation: Pirro, M.; Zhao, X.; Herrera, B.; Simeoni, P.; Rinaldi, M. Effect of Substrate-RF on Sub-200 nm $Al_{0.7}Sc_{0.3}N$ Thin Films. *Micromachines* **2022**, *13*, 877. https://doi.org/10.3390/mi13060877

Academic Editor: Aiqun Liu

Received: 9 May 2022
Accepted: 29 May 2022
Published: 31 May 2022

1. Introduction

After Akiyama [1] demonstrated the large enhancement of the piezoelectric coefficient by doping AlN with scandium, a growing number of studies have been conducted to exploit AlScN in MEMS, with particular interest in RF applications [2–8]. The augmented d_{33}, along with a reduction in the stiffness, enables AlScN-based resonators with a higher electro–mechanical coupling coefficient. This translates to filters with larger bandwidths as compared with their AlN counterparts, while maintaining a similar fabrication flow [9]. After a few years, in 2018, Fichtner [10] demonstrated a robust and repeatable ferroelectric behavior within the material, which opened new paths to multi-functional MEMS by combining the improved piezoelectric performance with the memory capability within the same BEOL-compatible process flow. Examples are the switchable FBARs [11,12], diodes [13], and ferroelectric transducers [14]. Nonetheless, the ferroelectric properties still face large limitations in their applicability. Despite the large polarization values, if compared with other ferroelectric materials [15], the coercive field and leakage current are still too high for practical integration. While a few works addressed the latter issue [16,17], several groups demonstrated how it is possible to reduce the coercive field by acting on either the AlScN structure or on the experimental set-up. A higher Sc-content [10], higher crystallinity [18], the use of different substrates [19], and testing-temperature [20] have been proven to reduce the voltage needed to switch, while still maintaining large polarization values. Another effective tuning parameter is represented by the residual bulk stress on the film, which can be controlled though different process parameters, such as pressure, N_2/Ar ratio, or substrate bias [21–23]. In 2013, while analyzing $Ga_{(1-x)}Sc_xN$, Zhang [24] suggested how substrate-induced mechanical stress can have a similar effect as the Sc-doping, both indeed leading to an increase of the internal parameter u, which reduced the coercive field. Despite bulk stress having been extensively studied in non-wurzite ferroelectrics, showing an induced variation in both the dielectric and ferroelectric behavior [25–28], only a few works show measured results from AlScN. In particular, ref. [10,21,29] demonstrate the linear dependence between coercive field and bulk stress, confirming Zhang's theory, but no general insights on the overall impact of the stress have been reported. In this context, we started from different levels of the substrate-RF applied during the deposition of sub-200 nm $Al_{0.7}Sc_{0.3}N$ within a co-planar sputtering module to compare the output films in a range of parameters from stress level to dielectric and ferroelectric properties in order to map the complete effect on the material from a single process parameter. After describing the fabrication flow and characterization method, the paper will focus on the extraction of the residual stress, breakdown fields, leakage currents, coercive fields, and polarization dynamics, demonstrating the large impact of substrate-RF on the overall AlScN behavior. We will show how substrate-RF induces stress levels on the films ranging from 500 MPa to −2 GPa, resulting in an increase of the c-axis dimensions of the AlScN cell, which is reported to degrade the breakdown–coercive field ratio and losses. Next, we demonstrate how the polarity of the films increases the impact on the static leakage current when smaller c-axis values are present, enabling different levels of current emission according to the state of the films. Last, the Nucleation-Limited-Switching (NLS) model [30,31] will be exploited to compare the ferroelectric kinetics, demonstrating the same switching mechanism among the films, with larger polarization reported for larger c-axis values.

2. Materials and Methods

2.1. Fabrication Flow

Several $Al_{0.7}Sc_{0.3}N$ (AlScN) thin films were deposited from a 12″ alloy target installed on an Evatec Clusterline-200 PVD module onto 200 mm-Si <100> 20 Ω/sq wafers coated with 20 nm titanium, with 80 nm platinum acting as a bottom electrode layer. Before the AlScN deposition, chamber and target conditioning procedures were performed to improve the base pressure and cleanliness of the target [32,33]. The thin films were deposited starting from a base pressure of 7×10^{-8} mbar and a chuck temperature set at 300 °C. The wafers sat for 5 min in the hot chuck before a 6 kW pulsed DC was applied for 250 s with 90 sccm of N_2 flow. The only variation among the films was the applied substrate-RF, which has been varied from 0 to 200 W. A final aluminum layer of 50 nm was sputtered, within the Clusterline tool, to complete the metal-ferroelectric-metal (MFM) structure. Single step photolithography and inductively coupled plasma etching were utilized to shape the top electrode features, which consisted on circular 0.144 mm^2 pads. Lastly, access to the continuous bottom electrode was created by etching the AlScN chips in phosphoric acid heated at 150 °C.

2.2. Characterization Method

The in-plane residual stress was extracted with a Flexus Tencor, which measured the wafers' bow curvature before and after the AlScN depositions. An area of 16 cm^2 from the center of the wafer was diced to locally analyze the material through X-ray diffractometry (XRD) (gonio-scan and omega-scan) and scanning electron microscopy (SEM). The electrical performance of the fabricated MFM structures was analyzed by applying different voltage waveforms, between the top and bottom electrodes, and reading the output current through a virtual ground amplifier embedded in an Aixacct TF analyzer. First, the breakdown fields were extracted from triangular waveforms at different frequencies by observing the voltage at which the MFM structures became a short circuit. A total of 25 samples were measured for each voltage–speed combination to apply a Weibull distribution, as shown in Equation (1):

$$F = 1 - exp\left(-\left(\frac{E}{E_b}\right)^{\beta}\right) \quad (1)$$

where F is the cumulative probability of the electric failure, E is the experimental breakdown, E_b is the breakdown field for which the cumulative probability is 63.2%, and β is the shape factor [34]. Next, the dielectric behavior was studied by applying 1 V_{pp} sine waves to pristine MFM structures to extract tanδ and dielectric permittivity (ε) in the kHz range. Afterwards, the DC response was evaluated though current density measurements, which allowed us to isolate the resistive behavior within the dielectric. A 50 V bi-polar, bi-directional voltage staircase with a 1 V step and 2 s holding time was applied to the samples at 25 °C, 35 °C, and 45 °C. The current response at each voltage step was extracted by averaging the output from 70% to 90% of the step length (2 s) to filter out any reactive behavior. The analysis was applied to pristine capacitors, and it started from the negative bias to induce a polarity inversion within each sweep. The measured data were then plotted as $\ln(J/T^2)$ vs. $\sqrt{E}$ and $\ln(J/E)$ vs. $\sqrt{E}$ in order to discriminate the limiting leakage mechanism, i.e., the interface-limited Schottky emission [16] or bulk-limited Pool–Frenkel [13].

Next, the ferroelectric properties were analyzed, focusing on the negative-to-positive switching since the higher leakage shown in the reverse direction overcame the current monitoring limit of the instrument. Despite this, a comparison among the wafers' response, and hence a study of the effect of the substrate-RF on the ferroelectric properties, was still possible. First, the coercive fields were studied by applying bi-polar trains of triangular pulses, with maximum voltage close to breakdown to induce a full polarization, within the films. The field values were extracted for different input voltage frequencies spanning from 10 Hz to 5 kHz. Second, the polarization extraction consisted in applying a modified PUND procedure described in [35], which allowed us to demonstrate the memory capability and the dependence on the write signal. The procedure started with a negative reset pulse (N-R-1), close to breakdown, applied to a pristine capacitor and acting as reference starting point. A second positive pulse of variable duration and intensity was subsequently applied as the write pulse (W(v,t)). After these, positive (P-R-1), negative (N-R-2), and positive (P-R-2) reset pulses were respectively applied to the MFM structure to allow two full negative-to-positive switching events. The polarization values were extracted by comparing the currents in output from the P-R-2 and P-R-1 pulses, which corresponded to a full and a variable polarization switching, respectively. The trapezoidal read voltages consisted of 50 µs rise, plateau, and fall times, with a fixed voltage intensity, which is chosen to be right above the coercive field to decrease the impact of leakage. On the other hand, the write pulse was varied in intensity (from 30 V to 60 V) and duration (from 50 µs to 1 s) to allow a complete mapping of the ferroelectric kinetics. The measured data were than fitted though the Nucleation-Limited-Switching (NLS) model, which showed better agreement than the classical Kolmogorov–Avrami–Ishibashi (KAI) approach [36,37]. In the latter model, the polarization evolution over time (i.e., $\Delta P(t)/2Ps$) follows Equation (2):

$$\frac{\Delta P(t)}{2P_S} = \left[1 - \exp(-t/\tau)^n\right] \tag{2}$$

where P_S, τ, and n are the spontaneous polarization, switching time, and dimensionality factor, respectively. On the other hand, the NLS model assumes a Lorentzian distribution over the switching time, resulting in Equation (3):

$$\frac{\Delta P(t)}{2P_S} = \int \left[1 - \exp(-t/\tau)^n\right] F(\log\tau)\, d(\log\tau) \tag{3}$$

where:

$$F(\log\tau) = \frac{A}{\pi}\left[\frac{\omega}{(\log\tau - \log\tau_1)^2 + \omega^2}\right] \tag{4}$$

where A, ω, and logτ are a normalized constant, half-width at half-maximum of the distribution, and median logarithmic value of the distribution, respectively [35].

3. Results

3.1. Material

The different substrate-RF levels induced a linear increase of the negative DC bias measured at the substrate during deposition, going from −45 V to −76 V, as reported in Figure 1a. Larger DC biases increased the bow displacement of the samples (Figure 1b), resulting in different stress levels ranging from 500 MPa to −2.2 GPa, as shown in Figure 1c.

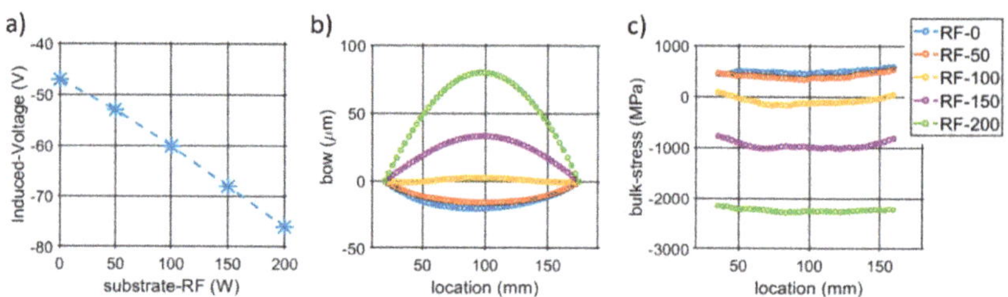

Figure 1. (**a**) Induced substrate DC biases per different RFs. (**b**) Wafers' curvatures measured with a Flexus Tencor after AlScN depositions and (**c**) corresponding bulk stresses extracted though Stoney equations.

The XRD gonio scan (Figure 2a) showed a shift in the peak corresponding to the <002> AlScN plane, suggesting a shear strain in the c-axis of the structure. No peaks corresponding to the a-plane were detected, indicating a preferred out-of-plane orientation. The full width half maximum (FWHM) of the rocking curve (RC) was 2.4° for all samples, except for RF-200, which showed RC FWHM = 3.2°, confirming the limited impact of substrate-RF on the crystallinity level of the films. SEM images of the stacks' cross-sections are shown in Figure 2b, indicating a thickness of 190 nm with no noticeable variations among different substrate-RFs.

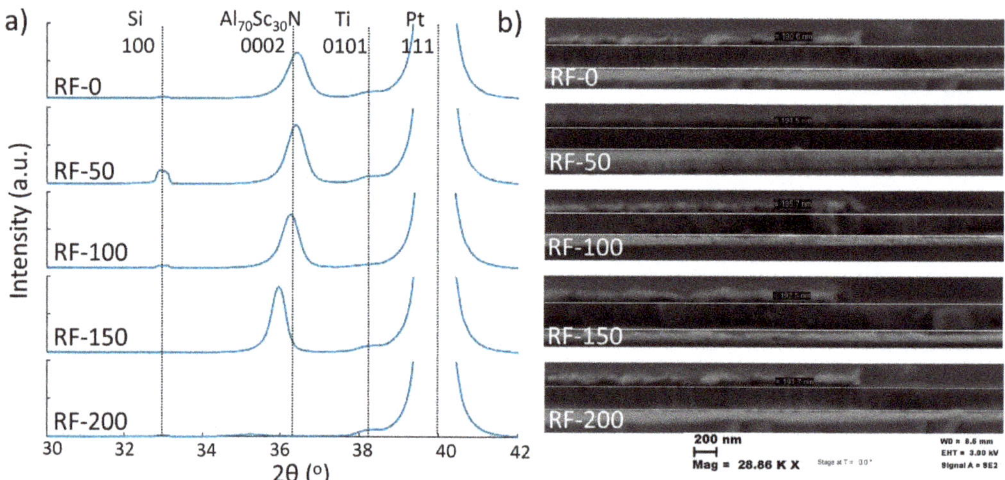

Figure 2. (**a**) Gonio-scan of the five samples showing the induced shifts of the 0002 AlScN plane. The dotted line represents the theoretical reference [18]. (**b**) SEM cross-sections of the samples with an average thickness of 190 nm.

3.2. Dielectric Properties

The breakdown fields at 5 kHz, along with the extracted parameters from the Weibull fit, are shown in Figure 3a. Comparable trends over frequency are noticed among the five samples for both average field and shape factor (Figure 2b,c). The highest breakdown is found for the RF-0 sample, which shows up to 7.5 MV/cm with a 5 kHz input frequency. A high substrate-RF, and hence a higher compressive stress, reduces the breakdown field but also increases the extracted shape factor, which indicates less scattered data values.

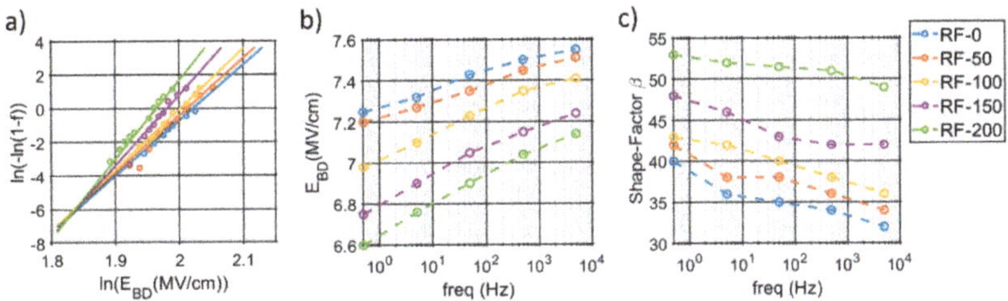

Figure 3. (**a**) Weibull distributions of the breakdown fields with a 5 kHz input voltage for the five samples. (**b**) Average breakdown fields and (**c**) shape factors extracted at different frequencies.

The summary of the electrical performance in the kHz range is presented in Figure 4a,b, which show pristine tanδ and dielectric permittivity, respectively. Low substrate-RFs are characterized by lower losses (tanδ < 0.008) and ε going from 17.2 to 18.5. On the other hand, tanδ > 0.01 and ε = 16.5 are extracted for the sample RF-150 and RF-200, indicating a larger impact on the electrical response from a GPa level of stress.

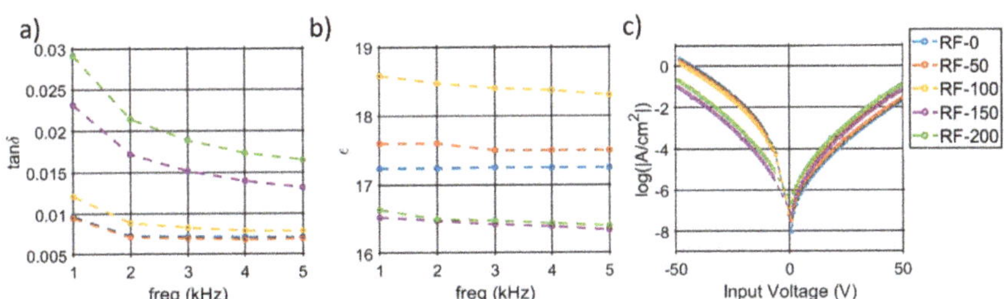

Figure 4. (**a**) Dielectric losses and (**b**) dielectric permittivity measured from 1 to 5 kHz. (**c**) Leakage current density of the non-switching sweep.

Figure 4c shows the logarithmic behavior of the non-switching leakage current density among the samples. RF-0, RF-50, and RF-100 are characterized by a large asymmetry, with higher losses when a negative bias is applied. Opposite trends are noticed according to the applied bias polarity, e.g., tensile–stress states show lower losses when a positive bias is applied and larger losses with a negative bias. In order to understand the source of the leakage, the current densities are plotted with different axis to discriminate a Schottky from Poole–Frenkel emission, as described in [38]. Despite that a linear trend is found in both $\ln(J/T^2)$ vs. $\sqrt{E}$ and $\ln(J/E)$ vs. $\sqrt{E}$, Schottky emissions result in permittivities in line with the published work in [16,17], differently than the Poole–Frenkel model, which results in unrealistic values (>30). A summary of the extracted optical permittivities is reported in the supplemental material. Examples of the Schottky emission plot at 25 °C for RF-0, RF-100, and RF-200 are reported in Figure 5, while Figure S1 shows the complete overview of the measured leakage for all samples at different temperatures. Each plot is composed of four curves, which come from the bi-polar, bi-directional sweeps, and they are labeled according to the combination of bias–polarity and polarization–state. The positive-down (blue) and the negative-up (yellow) curves show the leakage from sweeps which induce switching, while the remaining curves, positive-up (orange) and the negative-down (purple), derive from the bias-voltages with the same polarity as the films.

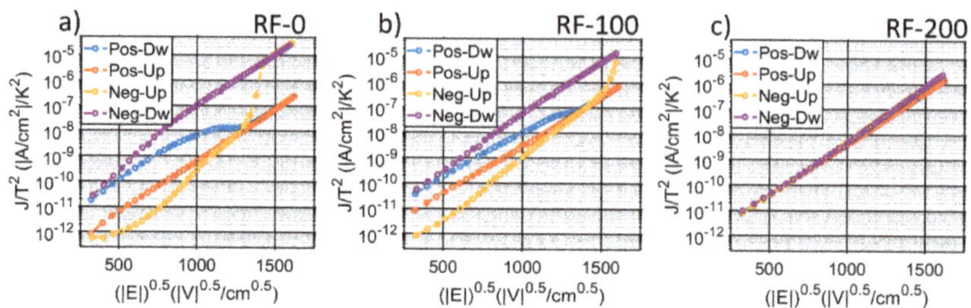

Figure 5. Example of leakage current densities for (**a**) RF-0, (**b**) RF-100, and (**c**) RF-200 measured at 25 °C and plotted versus the square root of the field to indicate the linear relationship between $\ln(J/T^2)$ and $\sqrt{E}$, typical of Schottky emissions. Each leakage measurement considers the combination of bias (positive or negative) and film polarity (up or down). Figure S1 in the supplemental material summarizes the measured leakage for all the sample at three different temperatures.

It is clear how the gap between the blue and orange curve (positive bias), as well as the gap between the purple and yellow curve (negative bias) highly changes among the samples, indicating two distinct static resistive behaviors per each polarization, which gradually converge with increased substrate-RF to have a complete un-hysteretic leakage with RF-150 and RF-200.

3.3. Ferroelectric Properties

First, the ferroelectric properties were studied from high-voltage, uni-polar triangular pulses applied at different frequencies. As shown in Figure 6a, two distinct current responses arose from the first and second pulses, i.e., the switching and non-switching pulses. A peak, in correspondence with the maximum voltage and indicating a large resistive in-phase current, is noticed for both pulses, while only the switching pulse induced a second current peak, which is distinguishable among all the tested frequencies.

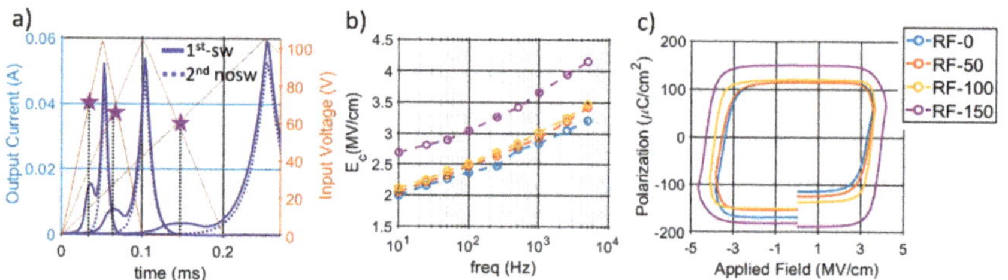

Figure 6. (**a**) Examples of switching (continuous) and non-switching (dotted) currents (blue curves) for different input frequencies, along with the corresponding coercive field extraction method. (**b**) Summary of coercive fields for all the switchable samples from 10 Hz to 5 kHz. (**c**) Extracted P-E loops with a 5 kHz input voltage.

The coercive fields were extracted from the ferroelectric current peak and plotted in a logarithmic scale in Figure 6b. Except for RF-150, whose fields were 1 MV/cm higher, comparable magnitudes are noticed among the first three wafers, with the RF-0 sample showing a switching field as low as 1.9 MV/cm at 10 Hz and up to 3.1 MV/cm at 5 kHz. No switching peaks were detected for the RF-200 sample. The corresponding P-E loops are presented in Figure 6c, which show a decrease in both coercive field and polarization with lower RFs. The study further analyzes the ferroelectric polarization of the films by measuring the electrical dynamics of the switching mechanism. After having identified the minimum voltage needed to fully reverse the films, the polarization charge is extracted from the current response in the output of a fixed-voltage reset pulse, P-R-1, which is applied 1 s after a variable positive write pulse (W(v,t)). Figure 7 shows the adopted train of pulses along with the measured output currents for the different samples. It is clear how the write pulse width, which varies from 50 µs to 1 s, with v = 60, (W(60,t)), has a clear impact on the polarization state of the films.

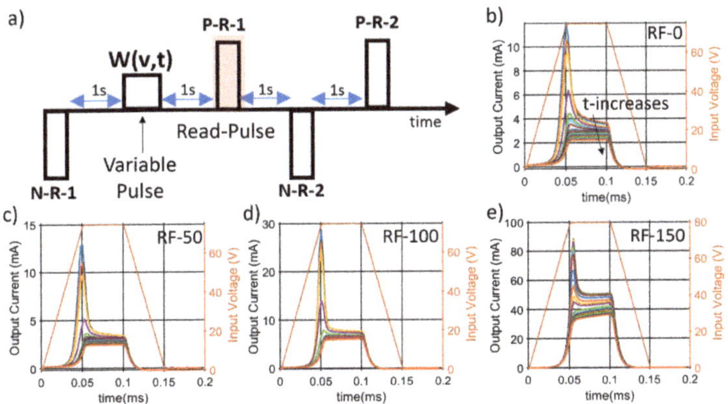

Figure 7. (**a**) Modified PUND method to analyze the polarization dynamic of the films. (**b**–**e**) Corresponding output read currents for different time-widths of the write pulse, from 50 µs to 1 s, for all the switchable samples: (**b**) RF-0, (**c**) RF-50, (**d**) RF-100, and (**e**) RF-150.

Thanks to the trapezoidal input waveform, a first analysis is possible by comparing the current peaks, which are the sum of the capacitive, resistive, and ferroelectric components, and the plateau currents, which instead filter out any time variant behavior. Similar trends are noticed among the four samples, with an increase in plateau and peak current with higher substrate-RFs, as shown in Figure S2. By varying the write amplitude along its time-width, it is possible to map the complete polarization reversal though ferroelectric switching models. As shown in Figure S3, an NLS model better described the polarization dynamic when compared with the KAI model, indicating the need to express the ferroelectric activation time through a Lorentzian distribution instead of a Delta Dirac. A summary of the extracted NLS parameters is shown in Figure 8a–c, which confirms a higher polarization for higher substrate-RFs (Figure 8a). The mean activation time and ω show similar behavior among the samples, with a linear dependence with $1/E$ and $1/E^2$, respectively, (Figure 8b,c), indicating a similar behavior to HZO [35] and poly-PZT [39].

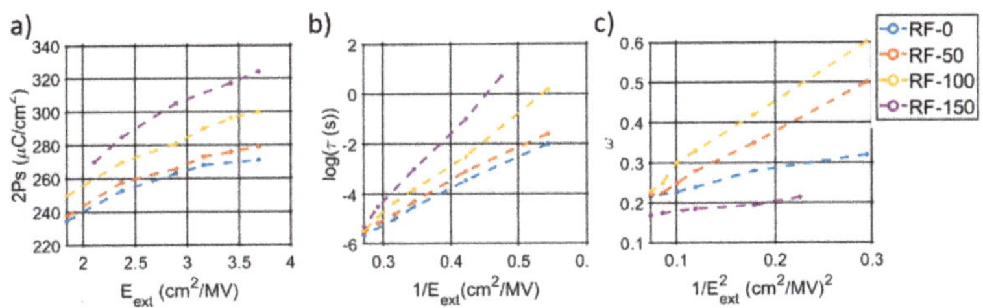

Figure 8. Summary of the extracted NSL model parameters: (**a**) switching polarization per different applied electric fields during the write signal, (**b**) average activation time versus the inverse of the applied field, and (**c**) relation between ω and the inverse of the square of the field.

4. Discussion

The substrate-RF has been demonstrated to have a direct impact on the films' structure and properties. Higher RF values increase the impinging energy of the specimen at the substrate [22], resulting in larger compressive states and a larger out-of-plane lattice dimension (Figure 9a). Even if more investigations are needed to map the whole effect on the lattice structure, i.e., the a-axis, the reported variations in the dielectric and ferroelectric domains evidence the critical role of a single process parameter in the deposition of AlScN thin films. Overall, low substrate-RFs, and hence tensile states, showed better performance, with a higher breakdown coercive field ratio and lower losses while maintaining a polarization higher than 130 $\mu C/cm^2$, as summarized in Figure 9b,c.

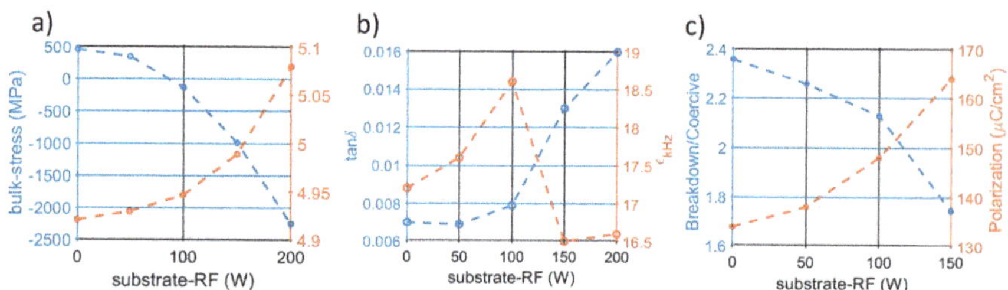

Figure 9. (**a**) Substrate-RF-induced variation of the residual stress extracted from the Flexus measurement and on the c-axis dimension extracted from the XRD scan [40]. (**b**) 5 kHz tanδ and dielectric permittivity of the pristine capacitors. (**c**) Breakdown/coercive field ratio and polarization variation per different substrate-RF.

Furthermore, the induced positive stress is reported to increase the gaps between the polarization-dependent current emissions, resulting in larger hysteresis on the $\ln(J/T^2)$ vs. $\sqrt{E}$ capacitor characteristic. As described in [13,16,17], the AlScN polarity highly affects the band alignment with the metals, which results in different barrier heights and in a clear shift in the leakage current for both positive and negative bias. Differently than [16], which was based on a symmetric capacitor structure, we report a preferred orientation state for the AlScN films, i.e., they are characterized by lower losses, which corresponds to the as-deposited polarity, unchanged among all the samples. A zoom-in of the different resistive behavior per different polarization states is presented in Figure 10a, which helps highlight the horizontal jumps corresponding to the negative-to-positive switching. As indicated by the arrows in Figure 10a, the measured shift is reported to decrease with the increase of substrate-RF until having a completely un-hysteretic leakage for RF-150 and RF-200 samples, which suggests a weaker and weaker impact of the MFM state on the contact interface. While the slopes remains invariant, i.e., with the same optical permittivities per different states, the curve intercepts, $(=\ln(A \times T_t) + q\Phi_{Bn,app}/(K_b\ T))$, change according to the MFM state, indicating a variation in the apparent Schottky barrier height ($\Phi_{Bn,app}$) and in the effective Richardson constant ($A \times T_t$). Such variations are quantifiable by plotting the extracted curve intercepts per different temperatures, as shown in Figures 10b and S4 for the positive-up and positive-down combinations, respectively [16,17]. Figure 10c summarizes how the lower the substrate-RF, the larger the impact of the polarization state on both $\Phi_{Bn,app}$ and $A \times T_t$., i.e., on the static behavior of the MFM structure. Compatible findings are obtained within the ferroelectric characterization in both coercive field and polarization extractions. As in the leakage analysis, on which low substrate-RFs induced lower fields to transition from the two resistive states, the PUND-based measurements demonstrated lower coercive fields with tensile stresses, confirming the previous works on the effect of stress on the switching field [10,29].

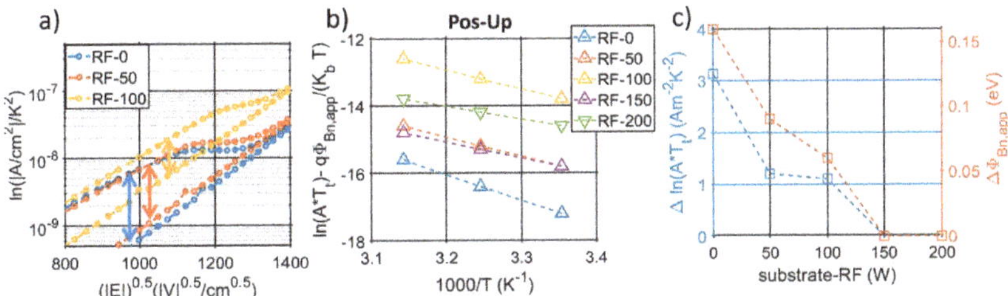

Figure 10. (**a**) Zoom-in of the $\ln(J/T^2)$ vs. $\sqrt{E}$ characteristic around the transition fields with positive bias applied per RF-0, RF-50, and RF-100. The arrows indicate the difference between the two resistive states of the films corresponding to the two polarity states of the films. (**b**) Extracted intercepts per different temperatures to evaluate the induced barrier variations with a positive bias applied to the up-polarization state. Figure S4 in supplemental material shows the intercept variations corresponding to the opposite polarity state. (**c**) Impact of the substrate-RF on the polarization-dependent resistive behavior: the left axis shows the difference between the extracted up-polarity $\ln(A \times T)$ and down-polarity $\ln(A \times T)$, while the right axis plots the same difference calculated for the apparent barrier.

Additionally, the switching currents, in output of the same input voltage, exhibit not only different voltage thresholds (i.e., voltage for which the current rises exponentially) but also different levels of saturation current (i.e., current level at the end of the trapezoidal plateau) per different polarization states, confirming the varying resistive behavior within the MFM structure. As in the leakage, such variations tend to decrease with higher substrate-RFs until they have similar voltage thresholds and plateau currents per different polarization state. It is worth mentioning how the RF-150 sample, despite an un-hysteretic leakage behavior, still shows a switching current dependent on the write signal, differently than RF-200, which does not show any sign of ferroelectric switching. The extracted NSL parameters demonstrate how the induced deformations not only vary the resistive behavior within the capacitors, but the whole switching dynamics as well. Like HZO [35] and poly-PZT [39], a linear relationship between $\log\tau$ with $1/E$ and ω vs. $1/E^2$ is noticed in the AlScN films, indicating a similar predominant switching mechanism among the materials. In particular, compressive states are reported to increase both the slopes of the $\log\tau$ vs. $1/E$ and ω vs. $1/E^2$ characteristics, suggesting slower and broader switching mechanism for intermediate polarization states, which is a sign of an increase in the pinning sites within the film. Future works will focus on the impact of electrode size, as well as on the investigation of the RF-150 sample, which does not follow the trend and shows a very small value of omega, which was constant over the different applied fields.

5. Conclusions

The paper demonstrates the effect of substrate-RF on 200 nm thin films deposited from a 12″ $Al_{0.7}Sc_{0.3}$ alloy target on a platinum substrate. From macroscopic to microscopic properties, this work compares five films deposited with different RF levels, which resulted in an induced DC bias at the substrate from −40 V to −76 V. Such a variation of energy leads to an increase of the negative bow of the wafers, which corresponded to higher compressive stresses of the films. This leads to different c-axis dimensions of the AlScN lattice structure. An overall decrease of dielectric and ferroelectric properties with an increasing c-axis was noticed. In particular, even if an undistorted lattice (and hence low stress) is preferred within the released MEMS devices to avoid cracks and peeling, we demonstrate substrate-RF as valuable knob to decrease losses and increase the breakdown to coercive field ratio, while still maintaining high polarization values. Furthermore, low substrate-RF has been demonstrated to enhance the effect of the MFM state on the contact barrier, producing two distinct resistance paths per each polarization direction. In conclusion, we demonstrate the high dependence of both dielectric and ferroelectric properties on substrate-RF, describing its potential in obtaining integrable ferroelectric AlScN thin films.

Supplementary Materials: The following supporting information can be downloaded at: https://www.mdpi.com/article/10.3390/mi13060877/s1, Figure S1. Summary of Schottky-emission plots for all the samples at 3 different temperatures. From top to bottom: RF-0, RF-50, RF-100, RF-150, RF-200, From left to right: 25 °C, 35 °C, 45 °C.; Figure S2. Summary of extracted optical permittivity per different substrate-RF. The permittivities are extracted from the slopes of the $\ln(J/T^2)$ vs. $\sqrt{E}$ characteristics [16]. Figure S3. Comparison NSL vs. KAI models, confirming a better matching with measured data when the activation time is express though a Lorentzian distribution (NSL) instead of a Delta-Dirac (KAI). Figure S4. Intercepts variations per different temperatures corresponding to the down-polarity state.

Author Contributions: Conceptualization, methodology, validation, investigation, and writing and editing, M.P.; investigation and review and editing, X.Z.; review, B.H. and P.S.; supervision and review, M.R. All authors have read and agreed to the published version of the manuscript.

Funding: This work was supported by the DARPA TUFEN Program, contact number HR00112090045.

Data Availability Statement: Data are available within the article.

Acknowledgments: The authors would like to thank the staff of the George J. Kostas Nanoscale Technology and Manufacturing Research Center for their assistance in the fabrication flow.

Conflicts of Interest: The authors declare no conflict of interest.

References

1. Akiyama, M.; Kamohara, T.; Kano, K.; Teshigahara, A.; Takeuchi, Y.; Kawahara, N. Enhancement of piezoelectric response in scandium aluminum nitride alloy thin films prepared by dual reactive cosputtering. *Adv. Mater.* **2009**, *21*, 593–596. [CrossRef] [PubMed]
2. Hashimoto, K.Y.; Sato, S.; Teshigahara, A.; Nakamura, T.; Kano, K. High-performance surface acoustic wave resonators inthe 1 to 3 GHz range using a ScAlN/6 H-SiC structure. *IEEE Trans. Ultrason. Ferroelectr. Freq. Control* **2013**, *60*, 637–642. [CrossRef] [PubMed]
3. Moreira, M.; Bjurström, J.; Katardjev, I.; Yantchev, V. Aluminum scandium nitride thin-film bulk acoustic resonators for wide band applications. *Vacuum* **2011**, *86*, 23–26. [CrossRef]
4. Konno, A.; Sumisaka, M.; Teshigahara, A.; Kano, K.; Hashimo, K.Y.; Hirano, H.; Esashi, M.; Kadota, M.; Tanaka, S. ScAlN Lamb wave resonator in GHz range released by XeF 2 etching. In Proceedings of the IEEE International Ultrasonics Symposium (IUS), Prague, Czech Republic, 21–25 July 2013; pp. 1378–1381. [CrossRef]
5. Yantchev, V.; Katardjiev, I. Thin film Lamb wave resonators in frequency control and sensing applications: A review. *J. Micromech. Microeng.* **2013**, *23*, 043001. [CrossRef]
6. Lozzi, A.; Yen, E.T.T.; Muralt, P.; Villanueva, L.G. Al0.83 Sc0.17 N contour-mode resonators with electromechanical coupling in excess of 4.5%. In *IEEE Transactions on Ultrasonics, Ferroelectrics, and Frequency Control*; IEEE: Piscataway, NJ, USA, 2018; Volume 66, pp. 146–153. [CrossRef]
7. Schneider, M.; DeMiguel-Ramos, M.; Flewitt, A.J.; Iborra, E.; Schmid, U. Scandium aluminium nitride-based film bulk acoustic resonators. *Multidiscip. Digit. Publ. Inst. Proc.* **2017**, *1*, 305. [CrossRef]
8. Ghatge, M.; Felmetsger, V.; Tabrizian, R. High kt2*Q Waveguide-Based ScAlN-on-Si UHF and SHF Resonators. In Proceedings of the IEEE International Frequency Control Symposium (IFCS), Olympic Valley, CA, USA, 21–24 May 2018; pp. 1–4. [CrossRef]
9. Ansari, A. Single crystalline scandium aluminum nitride: An emerging material for 5 G acoustic filters. In Proceedings of the IEEE MTT-S International Wireless Symposium (IWS), Guangzhou, China, 19–22 May 2019; pp. 1–3. [CrossRef]
10. Fichtner, S.; Wolff, N.; Lofink, F.; Kienle, L.; Wagner, B. AlScN: A III-V semiconductor based ferroelectric. *J. Appl. Phys.* **2019**, *125*, 114103. [CrossRef]
11. Wang, J.; Park, M.; Mertin, S.; Pensala, T.; Ayazi, F.; Ansari, A. A High-kt2 Switchable Ferroelectric $Al_{0.7}Sc_{0.3}N$ Film Bulk Acoustic Resonator. In Proceedings of the Joint Conference of the IEEE International Frequency Control Symposium and International Symposium on Applications of Ferroelectrics (IFCS-ISAF), Keystone, CO, USA, 19–23 July 2020; pp. 1–3. [CrossRef]
12. Rassay, S.; Mo, D.; Li, C.; Choudhary, N.; Forgey, C.; Tabrizian, R. Intrinsically switchable ferroelectric scandium aluminum nitride lamb-mode resonators. *IEEE Electron Device Lett.* **2021**, *42*, 1065–1068. [CrossRef]
13. Liu, X.; Zheng, J.; Wang, D.; Musavigharavi, P.; Stach, E.; Olsson, R.; Jariwala, D. Aluminum scandium nitride-based metal–ferroelectric–metal diode memory devices with high on/off ratios. *Appl. Phys. Lett.* **2021**, *118*, 202901. [CrossRef]
14. Herrera, B.; Pirro, M.; Giribaldi, G.; Colombo, L.; Rinaldi, M. AlScN Programmable Ferroelectric Micromachined Ultrasonic-Transducer (FMUT). In Proceedings of the 21st International Conference on Solid-State Sensors, Actuators and Microsystems (Transducers), Orlando, FL, USA, 20–24 June 2021; pp. 38–41. [CrossRef]
15. Olsson, R.H.; Tang, Z.; D'Agati, M. Non-Kolmogorov-Avrami switching kinetics in ferroelectric thin films. In Proceedings of the IEEE Custom Integrated Circuits Conference (CICC), Boston, MA, USA, 22–25 March 2020; pp. 1–6. [CrossRef]

16. Kataoka, J.; Tsai, S.L.; Hoshii, T.; Wakabayashi, H.; Tsutsui, K.; Kakushima, K. A possible origin of the large leakage current in ferroelectric All xScxN films. *Jpn. J. Appl. Phys.* **2021**, *60*, 030907. [CrossRef]
17. Tsai, S.L.; Hoshii, T.; Wakabayashi, H.; Tsutsui, K.; Chung, T.K.; Chang, E.Y.; Kakushima, K. Field cycling behavior and breakdown mechanism of ferroelectric $Al_{0.78}Sc_{0.22}N$ films. *Jpn. J. Appl. Phys.* **2022**. [CrossRef]
18. Yasuoka, S.; Shimizu, T.; Tateyama, A.; Uehara, M.; Yamada, H.; Akiyama, M.; Hiranaga, Y.; Cho, Y.; Funakubo, H. Effectsof deposition conditions on the ferroelectric properties of (Al1 x Sc x) N thin films. *J. Appl. Phys.* **2020**, *128*, 114103. [CrossRef]
19. Yazawa, K.; Drury, D.; Zakutayev, A.; Brennecka, G.L. Reduced coercive field in epitaxial thin film of ferroelectric wurtzite $Al_{0.7}Sc_{0.3}N$. *Appl. Phys. Lett.* **2021**, *118*, 162903. [CrossRef]
20. Zhu, W.; Hayden, J.; He, F.; Yang, J.I.; Tipsawat, P.; Hossain, M.D.; Maria, J.P.; Trolier-McKinstry, S. Strongly temperature dependent ferroelectric switching in AlN, Al1-xScxN, and Al1-xBxN thin films. *Appl. Phys. Lett.* **2021**, *119*, 062901. [CrossRef]
21. Fichtner, S.; Reimer, T.; Chemnitz, S.; Lofink, F.; Wagner, B. Stress controlled pulsed direct current co-sputtered Al1 xScxN as piezoelectric phase for micromechanical sensor applications. *APL Mater.* **2015**, *3*, 116102. [CrossRef]
22. Dubois, M.A.; Muralt, P. Stress and piezoelectric properties of aluminum nitride thin films deposited onto metal electrodes by pulsed direct current reactive sputtering. *J. Appl. Phys.* **2001**, *89*, 6389–6395. [CrossRef]
23. Iborra, E.; Olivares, J.; Clement, M.; Vergara, L.; Sanz-Hervás, A.; Sangrador, J. Piezoelectric properties and residual stress of sputtered AlN thin films for MEMS applications. *Sens. Actuators A Phys.* **2004**, *115*, 501–507. [CrossRef]
24. Zhang, S.; Holec, D.; Fu, W.Y.; Humphreys, C.J.; Moram, M.A. Tunable optoelectronic and ferroelectric properties in Sc-based III-nitrides. *J. Appl. Phys.* **2013**, *114*, 133510. [CrossRef]
25. Garino, T.J.; Harrington, M. Residual Stress in Pzt Thin Films and its Effect on Ferroelectric Properties. In *MRS Online Proceedings Library (OPL)*; Springer Nature Switzerland AG: Cham, Switzerland, 1999; Volume 243, pp. 341–347. [CrossRef]
26. Berfield, T.A.; Ong, R.J.; Payne, D.A.; Sottos, N.R. Residual stress effects on piezoelectric response of sol-gel derived lead zirconate titanate thin films. *J. Appl. Phys.* **2007**, *101*, 024102. [CrossRef]
27. Wu, Z.; Zhou, J.; Chen, W.; Shen, J.; Lv, C. Effects of residual stress on the electrical properties in $PbZr_{0.52}Ti_{0.48}O_3$ thin films. *J. Sol-Gel Sci. Technol.* **2015**, *75*, 551–556. [CrossRef]
28. Lee, J.W.; Park, C.S.; Kim, M.; Kim, H.E. Effects of residual stress on the electrical properties of PZT films. *J. Am. Ceram. Soc.* **2007**, *90*, 1077–1080. [CrossRef]
29. Rassay, S.; Hakim, F.; Li, C.; Forgey, C.; Choudhary, N.; Tabrizian, R. A Segmented-Target Sputtering Process for Growth of Sub-50 nm Ferroelectric Scandium–Aluminum–Nitride Films with Composition and Stress Tuning. *Phys. Status Solidi (RRL)-Rapid Res. Lett.* **2021**, *15*, 2100087. [CrossRef]
30. Gong, N.; Sun, X.; Jiang, H.; Chang-Liao, K.S.; Xia, Q.; Ma, T.P. Nucleation limited switching (NLS) model for HfO_2-based metal-ferroelectric-metal (MFM) capacitors: Switching kinetics and retention characteristics. *Appl. Phys. Lett.* **2018**, *112*, 262903. [CrossRef]
31. Tagantsev, A.K.; Stolichnov, I.; Setter, N.; Cross, J.S.; Tsukada, M. Non-Kolmogorov-Avrami switching kinetics in ferroelectric thin films. *Phys. Rev. B* **2002**, *66*, 214109. [CrossRef]
32. Anders, A. Physics of arcing, and implications to sputter deposition. *Thin Solid Films* **2006**, *502*, 22–28. [CrossRef]
33. Berg, S.; Nyberg, T. Fundamental understanding and modeling of reactive sputtering processes. *Thin Solid Films* **2005**, *476*, 215–230. [CrossRef]
34. Zheng, J.X.; Wang, D.; Musavigharavi, P.; Fiagbenu, M.M.A.; Jariwala, D.; Stach, E.A.; Olsson, R.H., III. Electrical breakdown strength enhancement in aluminum scandium nitride through a compositionally modulated periodic multilayer structure. *J. Appl. Phys.* **2021**, *130*, 144101. [CrossRef]
35. Wei, W.; Zhang, W.; Tai, L.; Zhao, G.; Sang, P.; Wang, Q.; Chen, F.; Tang, M.; Feng, Y.; Zhan, X.; et al. In-depth Understanding of Polarization Switching Kinetics in Polycrystalline $Hf_{0.5}Zr_{0.5}O_2$ Ferroelectric Thin Film: A Transition From NLS to KAI. In Proceedings of the IEEE International Electron Devices Meeting (IEDM), San Francisco, CA, USA, 11–16 December 2021; pp. 19.1.1–19.1.4. [CrossRef]
36. Ishibashi, Y.; Takagi, Y. Note on ferroelectric domain switching. *J. Phys. Soc. Jpn.* **1971**, *31*, 506–510. [CrossRef]
37. Dimmler, K.; Parris, M.; Butler, D.; Eaton, S.; Pouligny, B.; Scott, J.F.; Ishibashi, Y. Switching kinetics in KNO_3 ferroelectric thin-film memories. *J. Appl. Phys.* **1987**, *61*, 5467–5470. [CrossRef]
38. Pabst, G.W.; Martin, L.M.; Chu, Y.M.; Ramesh, R. Leakage mechanisms in thin $BiFeO_3$ films. *Appl. Phys. Lett.* **2007**, *90*. [CrossRef]
39. Jo, J.Y.; Han, H.S.; Yoon, J.G.; Song, T.K.; Kim, S.H.; Noh, T.W. Domain switching kinetics in disordered ferroelectric thin films. *Phys. Rev. Lett.* **2007**, *99*, 267602. [CrossRef]
40. Moram, M.A.; Vickers, M.E. X-ray diffraction of III-nitrides. *Rep. Prog. Phys.* **2009**, *72*, 036502. [CrossRef]

micromachines

MDPI

Article

Growth of Highly c-Axis Oriented AlScN Films on Commercial Substrates

Jingxiang Su [1,*], Simon Fichtner [1,2], Muhammad Zubair Ghori [1], Niklas Wolff [2], Md. Redwanul Islam [2], Andriy Lotnyk [3], Dirk Kaden [1], Florian Niekiel [1], Lorenz Kienle [2], Bernhard Wagner [1] and Fabian Lofink [1,*]

1 Fraunhofer Institute for Silicon Technology ISIT, Fraunhoferstrasse 1, 25524 Itzehoe, Germany; simon.fichtner@isit.fraunhofer.de (S.F.); muhammad.zubair.ghori@isit.fraunhofer.de (M.Z.G.); dirk.kaden@isit.fraunhofer.de (D.K.); florian.niekiel@isit.fraunhofer.de (F.N.); beha.wagner@gmail.com (B.W.)

2 Institute for Material Science, Kiel University, Kaiserstr. 2, 24143 Kiel, Germany; niwo@tf.uni-kiel.de (N.W.); mdis@tf.uni-kiel.de (M.R.I.); lk@tf.uni-kiel.de (L.K.)

3 Leibniz Institute of Surface Engineering (IOM), Permoserstr. 15, D-04318 Leipzig, Germany; andriy.lotnyk@iom-leipzig.de

* Correspondence: jxsu@mailbox.org (J.S.); fabian.lofink@isit.fraunhofer.de (F.L.)

Abstract: In this work, we present a method for growing highly *c*-axis oriented aluminum scandium nitride (AlScN) thin films on (100) silicon (Si), silicon dioxide (SiO_2) and epitaxial polysilicon (poly-Si) substrates using a substrate independent approach. The presented method offers great advantages in applications such as piezoelectric thin-film-based surface acoustic wave devices where a metallic seed layer cannot be used. The approach relies on a thin AlN layer to establish a wurtzite nucleation layer for the growth of *w*-AlScN films. Both AlScN thin film and seed layer AlN are prepared by DC reactive magnetron sputtering process where a Sc concentration of 27% is used throughout this study. The crystal quality of (0002) orientation of $Al_{0.73}Sc_{0.27}N$ films on all three substrates is significantly improved by introducing a 20 nm AlN seed layer. Although AlN has a smaller capacitance than AlScN, limiting the charge stored on the electrode plates, the combined piezoelectric coefficient $d_{33,f}$ with 500 nm AlScN is only slightly reduced by about 4.5% in the presence of the seed layer.

Keywords: aluminium scandium nitride; piezoelectric thin films; MEMS; non-metallic substrates

Citation: Su, J.; Fichtner, S.; Ghori, M.Z.; Wolff, N.; Islam, M.R.; Lotnyk, A.; Kaden, D.; Niekiel, F.; Kienle, L.; Wagner, B.; et al. Growth of Highly c-Axis Oriented AlScN Films on Commercial Substrates. *Micromachines* **2022**, *13*, 783. https://doi.org/10.3390/mi13050783

Academic Editor: Nam-Trung Nguyen

Received: 24 April 2022
Accepted: 13 May 2022
Published: 17 May 2022

Publisher's Note: MDPI stays neutral with regard to jurisdictional claims in published maps and institutional affiliations.

1. Introduction

Already for decades, piezoelectric thin film AlN has been of interest for its excellent dielectric properties as well as its chemical and temperature stability and has been widely used in piezoelectric MEMS (microelectromechanical systems) sensors and actuator elements [1–4]. In 2009, Akiyama et al. first reported that the piezoelectric coefficient of AlN could be significantly increased by doping with Sc [5,6]. Since then, AlScN has attracted great attention and has become a promising piezoelectric material for MEMS applications [7–11]. Multiple studies on AlScN-based MEMS magnetoelectric sensors [12], MEMS energy harvesters [13], MEMS quasistatic mirrors [14], and acoustic wave resonators [15–21] have been reported.

In most piezoelectric MEMS devices, AlN or AlScN thin films are grown on metallic seed layers such as molybdenum (Mo) or platinum (Pt) to ensure good *c*-axis orientation, where full width at half maximum (FWHM) values of the rocking curve measurements are typically larger than 1.3° for AlN and larger than 1.6° for $Al_{0.73}Sc_{0.27}N$ [22–26]. However, metallic seed layers cannot be used for, e.g., piezoelectric thin-film-based surface acoustic wave (SAW) devices [21], optical waveguides [27], or MEMS actuators with doped silicon used as bottom electrode [28]. For these applications, the piezoelectric layer (AlN or AlScN) has to be grown directly on substrates such as (100) silicon (Si), silicon dioxide (SiO_2) or epitaxial polysilicon (poly-Si), but still a high degree of *c*-axis orientation of AlN or AlScN is required. For AlN the deposition on various Si-based substrates has been studied by

Jiao et al. in [29] and SiO_2 was found to be most suitable substrate for AlN *c*-axis growth. Due to its higher piezoelectric response compared to AlN, AlScN films have attracted more interest, thus becoming a focus in piezoelectric MEMS research. Although there is success in growing AlScN films with low Sc concentrations on different nonmetallic substrates (on high-resistivity (100) Si [21], low-resistivity boron-doped (001) Si [17,30], SiC [31]), higher Sc concentrations present an increasingly difficult challenge for the growth of AlScN films with exclusive *c*-axis orientation [10].

In this work, we present a largely substrate-independent method to grow wurtzite-type AlScN films with exclusive *c*-axis orientation even for high Sc concentrations. Throughout this paper, AlScN with 27% Sc is chosen as a balance between high piezoelectric coefficient and robust deposition process. Hereafter, AlScN is used synonymously with $Al_{0.73}Sc_{0.27}N$. AlN and AlScN films are deposited directly on (100) Si, SiO_2 and poly-Si substrates, using process parameters established for the growth on metallic nucleation layers. The microstructure and *c*-axis texture quality of AlN and AlScN films are investigated using scanning electron microscopy (SEM), transmission electron microscopy (TEM), and high-resolution X-ray diffraction (XRD). The evaluation of the surface morphology and rocking curve XRD scans reveals that AlN films show a high degree of *c*-axis orientation on all investigated substrates. This confirms that sputter deposited AlN is able to realize a good texture with low distortions on many substrates, largely independent of the underlying texture [32–36]. In contrast, AlScN films grown directly on various nonmetallic substrates exhibit a high density of misaligned grains. The superior crystalline quality of AlN films motivates the approach reported herein to grow high quality AlScN on nonmetallic substrates using AlN as the nucleation layer.

2. Experimental

2.1. Sample Preparation

AlN and AlScN thin films are prepared by DC reactive magnetron sputtering in an Evatec Clusterline multichamber sputtering system. AlN films are sputter deposited in a gas plasma mixture of argon (Ar) and nitrogen (N_2) at a temperature of 300 °C, while AlScN films are prepared by cosputtering of Sc and Al targets in a pure N_2 plasma system. The detailed process parameters are listed in Table 1. In this work, (100) Si, oxidized Si and poly-Si are used as substrates to grow AlN and AlScN films. All three substrates have a smooth surface on the front side (<2 nm RMS (root mean square)). Prior to the deposition of piezoelectric layers, all substrates are heated to 200 °C for 30 s for degassing and then cleaned by a soft Ar plasma surface etching (300 W for 40 s) in vacuum. In the first part of this study, AlN as well as AlScN films with a thickness of 1 µm are deposited directly on three different substrates, respectively. In the second part, in order to investigate the effect of AlN seed layer on the crystalline orientation and piezoelectric response of AlScN thin films, a 20 nm AlN layer is first deposited on substrates, followed by a 500 nm AlScN layer after a vacuum break. For the characterization of the piezoelectric coefficient, the AlScN film has to be sandwiched between bottom and top electrodes [37]. Therefore, a stack of four layers consisting of bottom electrode, 20 nm AlN, 500 nm AlScN, and top electrode is deposited on an oxidized Si wafer. Here, 20 nm Ti/100 nm Pt and 200 nm Mo are used as bottom and top electrodes, respectively. All samples in this work are 8-inch wafer-scale in size.

Table 1. Sputter deposition parameters of AlN and $Al_{0.73}Sc_{0.27}N$ thin films.

	AlN	**$Al_{0.73}Sc_{0.27}N$**
Power on Al target (kW)	7.5	4.5
Power on Sc target (kW)	/	3.5
Temperature (°C)	300	300
Ar flow (sccm)	28	/
N_2 flow (sccm)	84	70

2.2. Characterization Methods

The surface structure of AlN and AlScN thin films is investigated using a critical dimension scanning electron microscope (CDSEM, Hitachi), which allows loading and imaging of 8-inch wafers. The crystal structures of AlN and AlScN films are examined by performing XRD θ-2θ scans and ω scans in a Rigaku SmartLab diffractometer (9 kW, Hypix detector) with Cu Kα radiation, a Ge(220)x2 monochromator and soller slit of 5°.

Transmission electron microscopy analysis of the Si/AlN/AlScN stack is conducted on a cross-section specimen, which is prepared using the focused ion beam (FIB) method and milled down to electron-transparency (FEI DualBeam Helios600 FIB-SEM). The nanoscopic structural and chemical analyses are performed on a probe C_S-corrected Titan^3 G2 60-300 microscope operating at 300 kV and a JEOL JEM-2100 transmission electron microscope (thermionic source LaB_6, acceleration voltage 200 kV) for selected area electron diffraction (SAED). The elemental distribution across the Si/AlN/AlScN interfaces is probed by energy dispersive X-ray spectroscopy (EDS) mapping using a Super-X EDS detector on the Titan^3 microscope.

The piezoelectric coefficient $d_{33,f}$ is characterized using a double beam laser interferometer (DBLI) from aixACCT systems, which allows automatic measurement with an 8-inch wafer. As the piezoelectric coefficient depends on the ratio of the electrode size to the substrate thickness [38], the $d_{33,f}$ shown in this work has been calibrated to its geometry independent value.

3. Results and Discussion

3.1. Microstructure Investigations of AlN and AlScN Thin Films

Figure 1 shows the surface structures of AlN and AlScN films grown directly on three different substrates. All AlN samples (Figure 1a–c) show a homogeneous surface with small round grains, which indicates the successful growth of columnar grains with c-axis orientation as already reported in multiple studies [11,25,30]. On the other hand, the surface of AlScN on all three substrates (Figure 1d–f) is dominated by crystallites with wedge-shaped structure, implying a poor c-axis orientation [25]. In addition, AlScN films seem to be grown slightly better on Si and poly-Si substrates compared to SiO_2 because few areas without misorientated grains can be observed.

To further examine the crystal phase and quality of the samples presented in Figure 1, XRD θ-2θ scans and ω scans are performed and shown in Figure 2. AlN and AlScN 0002 reflections at 2θ of around 36° [39] are detected. In addition, the reflections of crystalline Si (100) orientation, poly-Si (111) and (220) planes are recorded in samples with corresponding substrates. The full width at half maximum (FWHM) of the AlN 0002 reflection rocking curve for all samples is less than 1.5°. This confirms that the AlN films on all investigated substrates are indeed well c-axis oriented. In contrast, for all AlScN samples, the measured FWHM values of 0002 reflection are larger than 2.2°, which is slightly higher than the reported FWHM of AlScN 0002 reflection (approx. 1.6°) reported in [25,39], which use Ti/Pt and Si as substrates, respectively. This indicates a lower quality of c-axis orientation of AlScN. The [0001] crystallographic direction of the out-of-the-plane misoriented grains has been shown to be tilted between 60° and 90° [40].

By using the given process parameters, the growth of highly c-axis oriented AlN directly on smooth surfaces (RMS $<$ 2 nm) of amorphous SiO_2, (100) Si and poly-Si wafers can be achieved, despite the different crystallographic texture of the substrate materials, in agreement with several studies [29,33,34,41]. However, a smooth surface alone is not sufficient to grow high quality AlScN films on these substrates as the misaligned grains and high FWHM values are measured. Our preliminary investigations show that the growth of AlScN on Ti/Pt bottom electrode for identical process parameters is stable and highly c-axis oriented (FWHM of 1.43°). Consequently, AlScN films are more sensitive to the substrate texture and irregularities. This fits with the observation that the c-axis orientation of AlScN even on metallic electrodes decreases significantly with increasing Sc concentration [25].

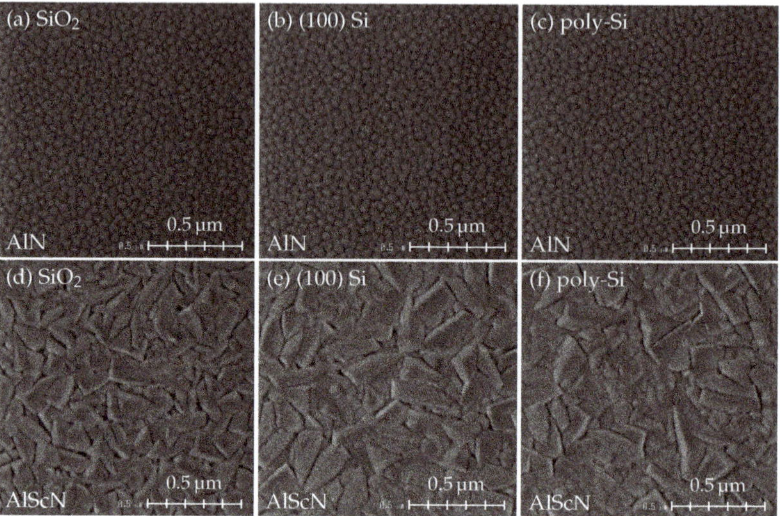

Figure 1. SEM surface view of 1 µm AlN deposited directly on (**a**) SiO_2, (**b**) (100) Si, (**c**) poly-Si, and 1 µm AlScN deposited directly on (**d**) SiO_2, (**e**) (100) Si, (**f**) poly-Si without a seed layer.without a seed layer.

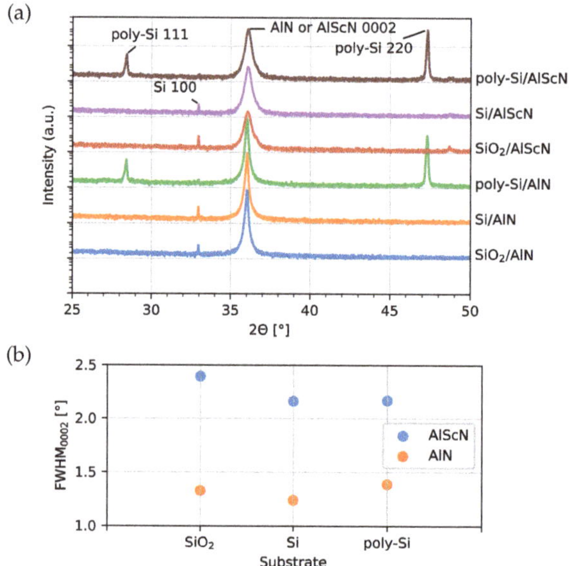

Figure 2. (**a**) θ-2θ scans of 1 µm AlN and AlScN grown directly on SiO_2, Si and poly-Si substrates without a seed layer; (**b**) Results of rocking curve measurements of AlN and AlScN 0002 reflections. The FWHM is determined by fitting a pseudo-Voight profile using the XRD fit module (Python based open source tool for XRD peak fitting [42]).

3.2. Microstructure and Piezoelectric Response of AlScN Films with a Thin AlN Seed Layer

To improve the *c*-axis orientation of AlScN on the investigated substrates, one option is to optimize the deposition parameters. In our previous work [43], we showed that the quality of $Al_{0.73}Sc_{0.27}N$ films on SiO_2 can be significantly improved by increasing the cathode–substrate distance offset. In this work, instead of optimizing the process parameters, an ultrathin AlN seed layer is introduced to improve the growth of *c*-axis oriented AlScN films on these substrates. We consider this approach to be more generally applicable and easier to transfer to different substrates.

The SEM images of 500 nm AlScN with 20 nm AlN seed layer on the investigated substrates are shown in Figure 3. Since the misoriented grains originate close to the substrate surface [25,30], there is no major difference in the number of misoriented grains between 500 nm and 1 µm AlScN films. A homogeneous surface with small grains is observed for all three samples, on which only a small number of misoriented grains is visible. Compared to the samples grown without the seed layer (Figure 1d–f), the *c*-axis orientation of AlScN films is significantly improved. The structural quality of AlN/AlScN films grown on SiO_2, Si and poly-Si is characterized using XRD (Figure 4). The FWHM values of AlScN 0002 reflection are slightly below 2° for all samples, which demonstrates only a moderate improvement in respect to Figure 1d–f. Although there is a small difference in 2θ values of AlN and AlScN 0002 orientations [10,11,39], a broadening of the 0002 FWHM due to the reflection from the 20 nm thin AlN seed layer can be expected to be negligible.

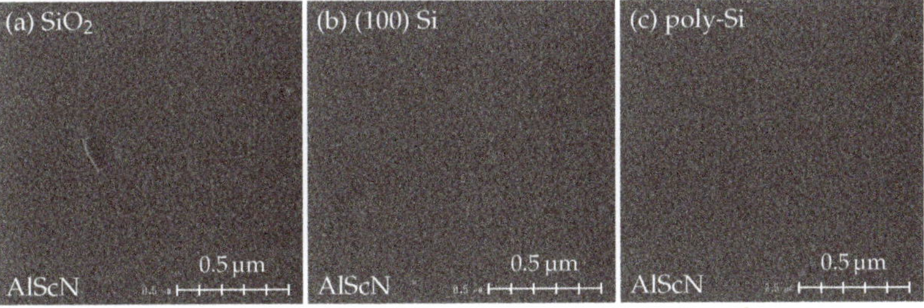

Figure 3. SEM surface view of 500 nm AlScN grown on (**a**) SiO_2, (**b**) (100) Si and (**c**) poly-Si with a 20 nm AlN seed layer.

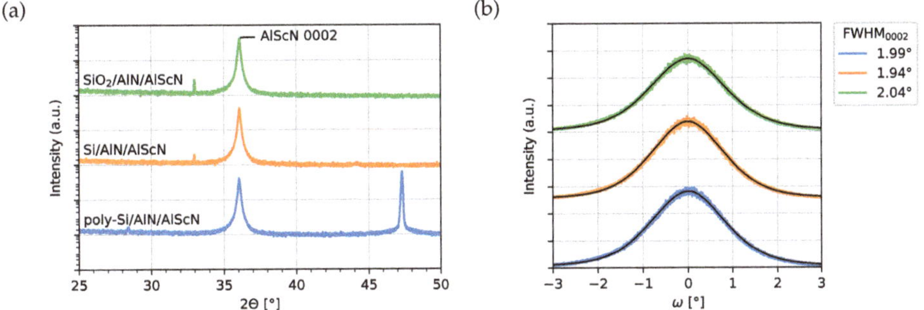

Figure 4. (**a**) θ-2θ scans of 500 nm AlScN grown on SiO_2, Si and poly-Si substrates with the AlN seed layer; (**b**) Results of rocking curve measurements of AlScN 0002 reflections. The FWHM is determined by fitting a pseudo-Voight profile using the XRD fit module (Python-based open source tool for XRD peak fitting [42]).

For further investigation of the local chemical composition and nanostructure at the interfaces, the sample with Si/AlN/AlScN is selected for TEM analysis. The scanning TEM annular bright-field (ABF) micrograph in Figure 5a provides an overview of the film cross-section, showing columnar grain structures of AlN and AlScN layers. The quality of the AlN/AlScN interface is further investigated by high-resolution STEM imaging and elemental analysis. A magnified HRSTEM ABF image of the interface is given in Figure 5b. By using the ABF detector, contrast-rich images with atoms shown by black dots are recorded. The columnar grains with diameters <5 nm growing along the *c*-axis on both sides of the interface are well displayed. However, obtaining a clear image of the interface is limited by the in-plane rotational disorder of the columnar grains and their three dimensional superposition along the finite sample thickness, as well as the patchy contrast spanning 2–3 nm in vertical direction across the interface region. The mosaic tilt along the *c*-axis is additionally visualized in the electron diffraction pattern recorded on the Si/AlN/AlScN multi-layers (see Figure 5c). The displayed intensity distribution can be explained by the superposition of the individual [110] Si, [2$\bar{1}\bar{1}$0] and [1$\bar{1}$00] AlN and AlScN zone axis patterns. The high coherency of the Si lattice causes electrons to scatter into sharp and bright reflections, whereas the different lattice constants of AlN and AlScN, as well as the small out-of-plane mosaic tilts of individual columns and the in-plane rotation of the fiber textured microstructure, result into diffuse and elongated intensities. The chemical composition is examined by elemental maps and profiles of the averaged intensity, as shown in Figure 5. Here, a peak in the oxygen signal is detected directly at the AlN/AlScN interface indicating a partial oxidation of the AlN surface during the vacuum break. Such partially oxidized interface has been reported before on a similar system and could not be avoided even after applying an RF etch cleaning step [44]. However, the oxide interface does not impede high-quality *c*-axis-oriented growth.

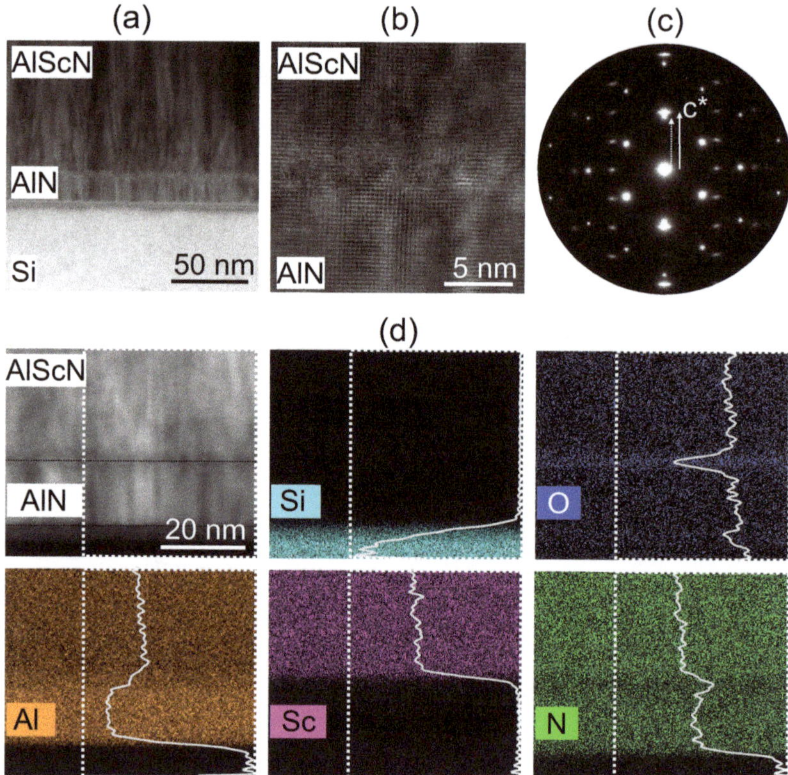

Figure 5. TEM study of the sample Si/AlN/AlScN. (**a**) STEM ABF overview image showing the columnar grain structures of AlN and AlScN layers on a natively passivated Si substrate; (**b**) HRSTEM ABF image showing structural disorder at the AlN/AlScN interface; (**c**) SAED pattern containing reflections of all layers corresponding to the [110] Si, [2$\bar{1}\bar{1}$0] and [1$\bar{1}$00] zone axes of AlN and AlScN; (**d**) STEM EDS elemental maps with integrated intensity profiles over the region of interest (dashed frame). The O-K map demonstrates the formation of an interfacial oxide layer between AlN and AlScN as well as the native oxide on the Si substrate.

To investigate the effect of 20 nm AlN seed layer on the piezoelectric response, the piezoelectric coefficient $d_{33,f}$ of 500 nm AlScN on sputtered Ti/Pt without and with the seed layer are measured and shown in Figure 6. The measured average $d_{33,f}$ of AlScN layer with the seed layer is 8.91 ± 0.03 pm/V, which is slightly lower (4.5%) compared to the one without the seed layer (9.33 ± 0.02 pm/V). However, the homogeneity of the distribution is not affected. The slightly lower piezoelectric coefficient is due to the lower dielectric permittivity of AlN which limits the electric charge storage on electrode plates.

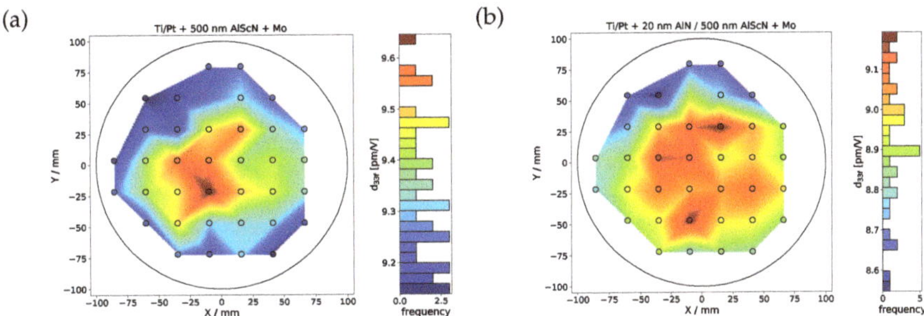

Figure 6. Measured $d_{33,f}$ of (**a**) 500 nm AlScN and of (**b**) 500 nm AlScN with 20 nm AlN seed layer on Ti/Pt_{sput} on a wafer level.

4. Conclusions

In this paper, we demonstrate a method to grow AlScN films with a high degree of *c*-axis orientation using same process parameters on various types of substrates, e.g., SiO_2, (100) Si and poly-Si. This approach is to introduce a 20 nm thin AlN seed layer, which itself grows with good textural properties on most smooth substrates. By using an AlN seed layer, the wurtzite-type structure is established in the AlScN layer, resulting in a good *c*-axis orientation. The lattice mismatch between AlN and AlScN films seems to be of secondary importance in this context. In addition, there is only a small reduction (4.5%) from 9.33 pm/V to 8.91 pm/V in the piezoelectric coefficient $d_{33,f}$ of AlScN layers in the presence of a 20 nm AlN seed layer with lower dielectric permittivity and piezoelectric coefficient.

Author Contributions: Conceptualization, J.S., S.F., B.W., and F.L.; methodology, J.S., S.F., D.K., and F.N.; software, J.S. and N.W.; validation, J.S.; formal analysis, J.S. and N.W.; investigation, J.S., M.Z.G., N.W., M.R.I., and A.L.; resources, D.K., L.K., B.W., and F.L.; data curation, J.S. and N.W.; writing original draft preparation, J.S.; writing review and editing, J.S., S.F., N.W., F.N., L.K., and F.L.; visualization, J.S.; supervision, S.F., D.K., and F.L.; project administration, J.S., S.F., and F.L.; funding acquisition, F.L., B.W. and L.K. All authors have read and agreed to the published version of the manuscript.

Funding: This work was funded by the German Research Foundation (Deutsche Forschungsgemeinschaft, DFG) under the scheme of the collaborative research center (CRC) 1261 'Magnetoelectric Sensors: From Composite Materials to Biomagnetic Diagnostics' and the project "ForMikro-SALSA" (grant no. 16ES1053) from the Federal Ministry of Education and Research (BMBF).

Institutional Review Board Statement: Not applicable.

Informed Consent Statement: Not applicable.

Data Availability Statement: The data presented in this study are available on reasonable request from the corresponding author.

Conflicts of Interest: The authors declare no conflict of interest. The funders had no role in the design of the study; in the collection, analyses, or interpretation of data; in the writing of the manuscript; or in the decision to publish the results.

References

1. Dubois, M.A.; Muralt, P. Properties of aluminum nitride thin films for piezoelectric transducers and microwave filter applications. *Appl. Phys. Lett.* **1999**, *74*, 3032–3034. [CrossRef]
2. Kim, T.; Kim, J.; Dalmau, R.; Schlesser, R.; Preble, E.; Jiang, X. High-temperature electromechanical characterization of AlN single crystals. *IEEE Trans. Ultrason. Ferroelectr. Freq. Control* **2015**, *62*, 1880–1887. [CrossRef] [PubMed]
3. Jackson, N.; Keeney, L.; Mathewson, A. Flexible-CMOS and biocompatible piezoelectric AlN material for MEMS applications. *Smart Mater. Struct.* **2013**, *22*, 115033. [CrossRef]
4. Olivares, J.; Iborra, E.; Clement, M.; Vergara, L.; Sangrador, J.; Sanz-Hervás, A. Piezoelectric actuation of microbridges using AlN. *Sens. Actuators A Phys.* **2005**, *123–124*, 590–595. [CrossRef]
5. Akiyama, M.; Kamohara, T.; Kano, K.; Teshigahara, A.; Takeuchi, Y.; Kawahara, N. Enhancement of Piezoelectric Response in Scandium Aluminum Nitride Alloy Thin Films Prepared by Dual Reactive Cosputtering. *Adv. Mater.* **2009**, *21*, 593–596. [CrossRef]
6. Akiyama, M.; Kano, K.; Teshigahara, A. Influence of growth temperature and scandium concentration on piezoelectric response of scandium aluminum nitride alloy thin films. *Appl. Phys. Lett.* **2009**, *95*, 162107. [CrossRef]
7. Zywitzki, O.; Modes, T.; Barth, S.; Bartzsch, H.; Frach, P. Effect of scandium content on structure and piezoelectric properties of AlScN films deposited by reactive pulse magnetron sputtering. *Surf. Coat. Technol.* **2017**, *309*, 417–422. [CrossRef]
8. Zukauskaite, A.; Wingqvist, G.; Palisaitis, J.; Jensen, J.; Persson, P.O.A.; Matloub, R.; Muralt, P.; Kim, Y.; Birch, J.; Hultman, L. Microstructure and dielectric properties of piezoelectric magnetron sputtered w-$Sc_xAl_{1-x}N$ thin films. *J. Appl. Phys.* **2012**, *111*, 093527. [CrossRef]
9. Matloub, R.; Hadad, M.; Mazzalai, A.; Chidambaram, N.; Moulard, G.; Sandu, C.S.; Metzger, T.; Muralt, P. Piezoelectric $Al_{1-x}Sc_xN$ thin films: A semiconductor compatible solution for mechanical energy harvesting and sensors. *Appl. Phys. Lett.* **2013**, *102*, 152903. [CrossRef]
10. Mertin, S.; Heinz, B.; Rattunde, O.; Christmann, G.; Dubois, M.A.; Nicolay, S.; Muralt, P. Piezoelectric and structural properties of c-axis textured aluminium scandium nitride thin films up to high scandium content. *Surf. Coat. Technol.* **2018**, *343*, 2–6. [CrossRef]
11. Fichtner, S.; Reimer, T.; Chemnitz, S.; Lofink, F.; Wagner, B. Stress controlled pulsed direct current co-sputtered $Al_{1-x}Sc_xN$ as piezoelectric phase for micromechanical sensor applications. *APL Mater.* **2015**, *3*, 116102. [CrossRef]
12. Su, J.; Niekiel, F.; Fichtner, S.; Thormaehlen, L.; Kirchhof, C.; Meyners, D.; Quandt, E.; Wagner, B.; Lofink, F. AlScN-based MEMS magnetoelectric sensor. *Appl. Phys. Lett.* **2020**, *117*, 132903. [CrossRef]
13. Yarar, E.; Fichtner, S.; Hayes, P.; Piorra, A.; Reimer, T.; Lisec, T.; Frank, P.; Wagner, B.; Lofink, F.; Meyners, D.; et al. MEMS-Based AlScN Resonating Energy Harvester With Solidified Powder Magnet. *J. Microelectromech. Syst.* **2019**, *28*, 1019–1031. [CrossRef]
14. Gu-Stoppel, S.; Lisec, T.; Fichtner, S.; Funck, N.; Claus, M.; Wagner, B.; Lofink, F. AlScN based MEMS quasi-static mirror matrix with large tilting angle and high linearity. *Sens. Actuators A Phys.* **2020**, *312*, 112107. [CrossRef]
15. Matloub, R.; Artieda, A.; Sandu, C.; Milyutin, E.; Muralt, P. Electromechanical properties of $Al_{0.9}Sc_{0.1}N$ thin films evaluated at 2.5 GHz film bulk acoustic resonators. *Appl. Phys. Lett.* **2011**, *99*, 092903. [CrossRef]
16. Park, M.; Hao, Z.; Dargis, R.; Clark, A.; Ansari, A. Epitaxial Aluminum Scandium Nitride Super High Frequency Acoustic Resonators. *J. Microelectromech. Syst.* **2020**, *29*, 490–498. [CrossRef]
17. Kurz, N.; Ding, A.; Urban, D.F.; Lu, Y.; Kirste, L.; Feil, N.M.; Žukauskaitė, A.; Ambacher, O. Experimental determination of the electro-acoustic properties of thin film AlScN using surface acoustic wave resonators. *J. Appl. Phys.* **2019**, *126*, 075106. [CrossRef]
18. Gillinger, M.; Shaposhnikov, K.; Knobloch, T.; Schneider, M.; Kaltenbacher, M.; Schmid, U. Impact of layer and substrate properties on the surface acoustic wave velocity in scandium doped aluminum nitride based SAW devices on sapphire. *Appl. Phys. Lett.* **2016**, *108*, 231601. [CrossRef]
19. Schneider, M.; DeMiguel-Ramos, M.; Flewitt, A.J.; Iborra, E.; Schmid, U. Scandium Aluminium Nitride-Based Film Bulk Acoustic Resonators. *Proceedings* **2017**, *1*, 305. [CrossRef]
20. Wang, J.; Park, M.; Mertin, S.; Pensala, T.; Ayazi, F.; Ansari, A. A Film Bulk Acoustic Resonator Based on Ferroelectric Aluminum Scandium Nitride Films. *J. Microelectromech. Syst.* **2020**, *29*, 741–747. [CrossRef]
21. Meyer, J.M.; Schell, V.; Su, J.; Fichtner, S.; Yarar, E.; Niekiel, F.; Giese, T.; Kittmann, A.; Thormählen, L.; Lebedev, V.; et al. Thin-Film-Based SAW Magnetic Field Sensors. *Sensors* **2021**, *21*, 8166. [CrossRef] [PubMed]
22. Jackson, N. Influence of silicon crystal orientation on piezoelectric textured aluminium nitride deposited on metal electrodes. *Vacuum* **2016**, *132*, 47–52. [CrossRef]
23. Felmetsger, V.V.; Laptev, P.N.; Tanner, S.M. Crystal orientation and stress in AC reactively sputtered AlN films on Mo electrodes for electro-acoustic devices. In Proceedings of the 2008 IEEE Ultrasonics Symposium, Beijing, China, 2–5 November 2008; pp. 2146–2149. [CrossRef]
24. Lee, J.B.; Jung, J.P.; Lee, M.H.; Park, J.S. Effects of bottom electrodes on the orientation of AlN films and the frequency responses of resonators in AlN-based FBARs. *Thin Solid Film.* **2004**, *447–448*, 610–614. [CrossRef]
25. Fichtner, S.; Wolff, N.; Krishnamurthy, G.; Petraru, A.; Bohse, S.; Lofink, F.; Chemnitz, S.; Kohlstedt, H.; Kienle, L.; Wagner, B. Identifying and overcoming the interface originating c-axis instability in highly Sc enhanced AlN for piezoelectric micro-electromechanical systems. *J. Appl. Phys.* **2017**, *122*, 035301. [CrossRef]
26. Kamohara, T.; Akiyama, M.; Ueno, N.; Kuwano, N. Improvement in crystal orientation of AlN thin films prepared on Mo electrodes using AlN interlayers. *Ceram. Int.* **2008**, *34*, 985–989. [CrossRef]

27. Alvarado, M.; Pelegrini, M.; Pereyra, I.; Assumpção, T.d.; Kassab, L.; Alayo, M. Fabrication and characterization of aluminum nitride pedestal-type optical waveguide. *Can. J. Phys.* **2014**, *92*, 951–954. [CrossRef]
28. Meinel, K.; Melzer, M.; Stoeckel, C.; Shaporin, A.; Forke, R.; Zimmermann, S.; Hiller, K.; Otto, T.; Kuhn, H. 2D Scanning Micromirror with Large Scan Angle and Monolithically Integrated Angle Sensors Based on Piezoelectric Thin Film Aluminum Nitride. *Sensors* **2020**, *20*, 6599. [CrossRef]
29. Jiao, X.; Shi, Y.; Zhong, H.; Zhang, R.; Yang, J. AlN thin films deposited on different Si-based substrates through RF magnetron sputtering. *J. Mater. Sci. Mater. Electron.* **2015**, *26*, 801–808. [CrossRef]
30. Lu, Y.; Reusch, M.; Kurz, N.; Ding, A.; Christoph, T.; Kirste, L.; Lebedev, V.; Žukauskaitė, A. Surface Morphology and Microstructure of Pulsed DC Magnetron Sputtered Piezoelectric AlN and AlScN Thin Films. *Phys. Status Solidi A* **2018**, *215*, 1700559. [CrossRef]
31. Teshigahara, A.; Hashimoto, K.Y.; Akiyama, M. Scandium aluminum nitride: Highly piezoelectric thin film for RF SAW devices in multi GHz range. In Proceedings of the 2012 IEEE International Ultrasonics Symposium, Dresden, Germany, 7–10 October 2012; IEEE: Piscataway, NJ, USA, 2012; pp. 1–5. [CrossRef]
32. Gillinger, M.; Shaposhnikov, K.; Knobloch, T.; Stöger-Pollach, M.; Artner, W.; Hradil, K.; Schneider, M.; Kaltenbacher, M.; Schmid, U. Enhanced c-axis orientation of aluminum nitride thin films by plasma-based pre-conditioning of sapphire substrates for SAW applications. *Appl. Surf. Sci.* **2018**, *435*, 432–437. [CrossRef]
33. Iriarte, G.; Rodríguez, J.; Calle, F. Synthesis of c-axis oriented AlN thin films on different substrates: A review. *Mater. Res. Bull.* **2010**, *45*, 1039–1045. [CrossRef]
34. Artieda, A.; Sandu, C.; Muralt, P. Highly piezoelectric AlN thin films grown on amorphous, insulating substrates. *J. Vac. Sci. Technol. A Vac. Surfaces Film.* **2010**, *28*, 390–393. [CrossRef]
35. Clement, M.; Vergara, L.; Sangrador, J.; Iborra, E.; Sanz-Hervás, A. SAW characteristics of AlN films sputtered on silicon substrates. *Ultrasonics* **2004**, *42*, 403–407. [CrossRef]
36. Xiong, J.; Gu, H.; Hu, K.I.; Hu, M.Z. Influence of substrate metals on the crystal growth of AlN films. *Int. J. Miner. Metall. Mater.* **2010**, *17*, 98–103. [CrossRef]
37. Liu, J.M.; Pan, B.; Chan, H.; Zhu, S.; Zhu, Y.; Liu, Z. Piezoelectric coefficient measurement of piezoelectric thin films: An overview. *Mater. Chem. Phys.* **2002**, *75*, 12–18. [CrossRef]
38. Sivaramakrishnan, S.; Mardilovich, P.; Schmitz-Kempen, T.; Tiedke, S. Concurrent wafer-level measurement of longitudinal and transverse effective piezoelectric coefficients ($d_{33,f}$ and $e_{31,f}$) by double beam laser interferometry. *J. Appl. Phys.* **2018**, *123*, 014103. [CrossRef]
39. Lu, Y.; Reusch, M.; Kurz, N.; Ding, A.; Christoph, T.; Prescher, M.; Kirste, L.; Ambacher, O.; Zukauskaite, A. Elastic modulus and coefficient of thermal expansion of piezoelectric $Al_{1-x}Sc_xN$ (up to x = 0.41) thin films. *APL Mater.* **2018**, *6*, 076105. [CrossRef]
40. Sandu, C.S.; Parsapour, F.; Mertin, S.; Pashchenko, V.; Matloub, R.; LaGrange, T.; Heinz, B.; Muralt, P. Abnormal Grain Growth in AlScN Thin Films Induced by Complexion Formation at Crystallite Interfaces. *Phys. Status Solidi A* **2019**, *216*, 1800569. [CrossRef]
41. Tamariz, S.; Martin, D.; Grandjean, N. AlN grown on Si (111) by ammonia-molecular beam epitaxy in the 900–1200 °C temperature range. *J. Cryst. Growth* **2017**, *476*, 58–63. [CrossRef]
42. Crowther, P.; Daniel, C.S. xrdfit: A Python package for fitting synchrotron X-ray diffraction spectra. *J. Open Source Softw.* **2020**, *5*, 2381. [CrossRef]
43. Fichtner, S. *Development of High Performance Piezoelectric AlScN for Microelectromechanical Systems: Towards a Ferroelectric Wurtzite Structure*; Books on Demand: Norderstedt, Germany 2020.
44. Parsapour, F.; Pashchenko, V.; Mertin, S.; Sandu, C.; Kurz, N.; Nicolay, P.; Muralt, P. Ex-situ AlN seed layer for (0001)-textured $Al_{0.84}Sc_{0.16}N$ thin films grown on SiO_2 substrates. In Proceedings of the 2017 IEEE International Ultrasonics Symposium (IUS), Washington, DC, USA, 6–9 September 2017; pp. 1–4. [CrossRef]

Article

High Rate Deposition of Piezoelectric AlScN Films by Reactive Magnetron Sputtering from AlSc Alloy Targets on Large Area

Stephan Barth *, Tom Schreiber, Steffen Cornelius, Olaf Zywitzki, Thomas Modes and Hagen Bartzsch

Fraunhofer Institute for Organic Electronics, Electron Beam and Plasma Technology FEP, 01277 Dresden, Germany
* Correspondence: stephan.barth@fep.fraunhofer.de

Abstract: This paper reports on the deposition and characterization of piezoelectric $Al_XSc_{1-X}N$ (further: AlScN) films on Si substrates using AlSc alloy targets with 30 at.% Sc. Films were deposited on a Ø200 mm area with deposition rates of 200 nm/min using a reactive magnetron sputtering process with a unipolar–bipolar hybrid pulse mode of FEP. The homogeneity of film composition, structural properties and piezoelectric properties were investigated depending on process parameters, especially the pulse mode of powering in unipolar–bipolar hybrid pulse mode operation. Characterization methods include energy-dispersive spectrometry of X-ray (EDS), X-ray diffraction (XRD), piezoresponse force microscopy (PFM) and double-beam laser interferometry (DBLI). The film composition was $Al_{0.695}Sc_{0.295}N$. The films showed good homogeneity of film structure with full width at half maximum (FWHM) of AlScN(002) rocking curves at $2.2 \pm 0.1°$ over the whole coating area when deposited with higher share of unipolar pulse mode during film growth. For a higher share of bipolar pulse mode, the films showed a much larger c-lattice parameter in the center of the coating area, indicating high in-plane compressive stress in the films. Rocking curve FWHM also showed similar values of 1.5° at the center to 3° at outer edge. The piezoelectric characterization method revealed homogenous $d_{33,f}$ of 11–12 pm/V for films deposited at a high share of unipolar pulse mode and distribution of 7–10 pm/V for a lower share of unipolar pulse mode. The films exhibited ferroelectric switching behavior with coercive fields of around 3–3.5 MV/cm and polarization of 80–120 μC/cm^2.

Keywords: AlScN; aluminum scandium nitride; piezoelectric thin films; piezoelectric; ferroelectric

Citation: Barth, S.; Schreiber, T.; Cornelius, S.; Zywitzki, O.; Modes, T.; Bartzsch, H. High Rate Deposition of Piezoelectric AlScN Films by Reactive Magnetron Sputtering from AlSc Alloy Targets on Large Area. *Micromachines* **2022**, *13*, 1561. https://doi.org/10.3390/mi13101561

Academic Editor: Agnė Žukauskaitė

Received: 29 July 2022
Accepted: 13 September 2022
Published: 21 September 2022

Publisher's Note: MDPI stays neutral with regard to jurisdictional claims in published maps and institutional affiliations.

1. Introduction

In 2009, Akiyama et al. first reported on the Sc doping of AlN films by the cosputtering process [1]. These $Al_XSc_{1-X}N$ films exhibited significantly increased piezoelectric properties depending on the Sc concentration with a maximum at 43 at.% Sc (Al + Sc = 100 at.%). Since then, research and development on AlScN films has increasingly gained attention from research groups as well as industry worldwide. The focus is mostly on MEMS [2,3] and acoustic wave applications [4,5], but energy harvesting applications [6–8] are also gaining traction.

Furthermore, Fichtner et al. demonstrated in 2019 the possibility of ferroelectric switching in AlScN films with scandium concentrations between 27 at.% and 43 at.% [9]. The ferroelectric switching occurs at coercive fields of 2–5 MV/cm, with higher Sc concentrations showing reduced necessary fields for switching but in turn also lower polarization values from above 100 μC/cm^2 to around 80 μC/cm^2. Further research by different groups worldwide is ongoing [10,11].

One aspect that still proves very challenging for the development of AlScN deposition processes is the increasing probability of the formation of abnormally oriented grains (AOG) at the grain boundaries, especially at higher Sc concentrations [12], tensile film stress [5] and higher film thickness [12,13]. This limits the process window in which films with good piezoelectric properties can be fabricated as well as the possible applications. Therefore, to further extend the material and process range and enable new applications, additional process parameters and adjustment options are necessary. One such process parameter is the pulse mode of powering in reactive pulse magnetron sputter processes. Usually, the configuration can either be unipolar or bipolar pulse mode. This results in different plasma conditions and consecutively film properties. By developing and using a unipolar–bipolar hybrid pulse mode, Barth et al. could freely influence film properties of cosputtered AlScN films from Al and Sc targets in a much wider range, realizing highly oriented films with thicknesses of up to several 10 μm [14]. This paper investigates if such pulse mode variation can be used to deposit highly oriented $Al_{0.7}Sc_{0.3}N$ films from $Al_{0.7}Sc_{0.3}$ alloy targets with a high deposition rate and good homogeneity on a large area with adjustable film properties in a wider range. This would offer new application opportunities, e.g., in thicker films of several μm for energy harvesting or ultrasonic devices or as stress-optimized thin films in multilayer stacks.

2. Materials and Methods

2.1. Film Deposition

The films were deposited using a reactive pulse magnetron sputter process described in [14]. The double-ring magnetron DRM 400 uses two concentric targets, whose discharges overlap to deposit homogenously on a wide area. Thus, films can be deposited with a very high film growth rate of 200 nm/min with film thickness homogeneity of up to 0.5% on a Ø200 mm area. Film thicknesses can be several tens of microns. By using the pulse unit UBS-C2 developed by FEP and standard DC power supplies, the pulse mode of the pulse magnetron sputtering can be adjusted as unipolar, bipolar or as a unipolar–bipolar hybrid pulse mode. In unipolar pulse mode, a pulsed dc voltage is applied between each of the two targets acting as cathodes and the separate hidden anode. In bipolar pulse mode, the two targets are alternately the cathode and the anode. In hybrid pulse mode, different shares of unipolar (S_u) or bipolar (S_b) pulse modes can be applied in a period of 1 ms to influence plasma parameters. The share of unipolar pulse mode is defined as the ratio between the time fraction of unipolar mode t_u with respect to the total time of one cycle of unipolar–bipolar hybrid pulse mode, i.e., $Su = t_u/(t_u + t_b)$. The plasma parameters can be adjusted in a wide range between both pure pulse modes, as shown in the plasma density measurements of an example process of Si sputtering in Figure 1.

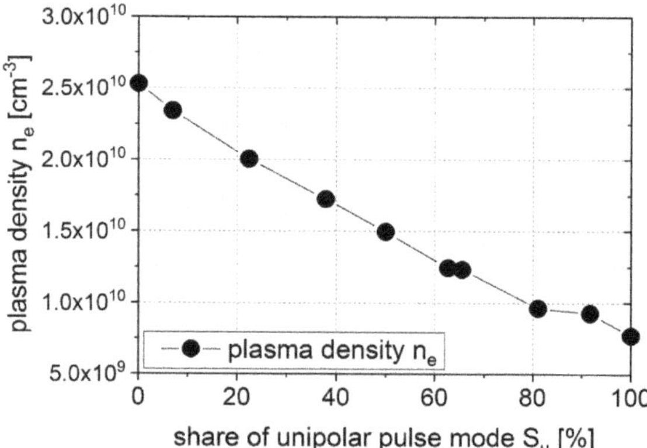

Figure 1. Plasma density in dependence of share of unipolar pulse mode in unipolar–bipolar hybrid pulse mode operation (example of Si process, adapted from [14]).

In the case of AlScN film depositions, AlSc targets (3N5) with Sc contents of 30 at.% were used. Argon and nitrogen (5N) were used as inert and reactive gas, respectively. The gas flow was between 20 and 60 sccm, and the pressure was between 0.3 Pa and 0.8 Pa. A closed-loop reactive gas control was applied to stabilize the process in the transition mode [15]. The share of unipolar pulse mode in hybrid pulse mode operation was investigated between 60% and 90% of the period. The chamber base pressure was 2×10^{-7} mbar. Before coating each sample, a precleaning of the substrate surface was performed using rf bias etching in an Ar atmosphere. There was no additional substrate heating or cooling beyond water cooling of the substrate platform applied.

2.2. Characterization

Film composition was determined by energy-dispersive spectrometry (EDS) of X-ray (Octane Elect Plus, EDAX, Pleasanton, CA, USA) using 10 kV accelerating voltage and the aluminum and scandium Kα lines for quantification.

Film stress measurements were performed using the wafer curvature method based on Stoney's equation. The measurements were carried out with a surface profiler, P15-Ls (KLA Tencor, Milpitas, CA, USA).

The XRD characterizations were carried out in a Bruker D8 Discover diffractometer equipped with a Göbel mirror for Cu-Ka parallel beam geometry and a 1D LynxEye^{TM} XET semiconductor detector (Bruker AXS, Karlsruhe, Germany). θ–2θ scans in 1D detector mode were used to characterize lattice parameters of the wurtzite structure. Film quality in terms of crystallographic orientation was investigated by evaluating the (002) rocking curve FWHM in 0D detector mode. For calculation of the wurtzite lattice parameter c, the (002) peaks in θ–2θ scans were used. Peak fitting including a Cu-$K_{\alpha 1/\alpha 2}$ doublet correction followed by a correction for instrumental 2θ alignment errors was carried out to extract the exact 2θ peak positions from the raw data.

The piezoresponse force microscopy (PFM) measurements were carried out using an AFM NX 20 from Park Systems (Suwon, Republic of Korea). Silicon wafers coated with a Ti/Pt seed layer and bottom electrode for grounding connection were used as the substrate for AlScN layers. An ac voltage with an amplitude of 10 V and 17 kHz frequency was applied to a platinum-coated AFM tip (Spark 350 Pt) to measure PFM amplitude and phase of inverse piezoelectric effect in off-resonance mode.

The piezoelectric and ferroelectric characterization was performed using a double-beam laser interferometer (DBLI) of aixACCT systems GmbH (Aachen, Germany). This method uses laser interferometry to measure surface displacement of the sample depending on applied electrical voltage. By using a double-beam configuration, substrate bending can be considered. The effective piezoelectric coefficient $d_{33,f}$ of the film material can be calculated by measuring at different electrode pad sizes and correcting using a geometric factor f(r), which is a function of the ratio of the pad size to substrate thickness and substrate Poisson's ratio. Besides $d_{33,f}$, this also allows calculation of transverse effective piezoelectric coefficient $e_{31,f}$. The method is described in further detail in [16].

3. Results and Discussion

3.1. Composition and Structure of AlScN Films

The first coatings were performed directly on Si substrates, and the film thickness distribution was ±1.5% at Ø180 mm and ±3% at Ø200 mm. The film composition was determined by EDS. The results are shown in Figure 2. The Sc content of the $Al_XSc_{1-X}N$ films was slightly below that of the nominal target composition as given by the manufacturer, especially at a longer distance from the symmetry axis of the magnetron, going as low as 28 at.% Sc at the radial position 100 mm, while being around 29.5 at.% Sc in the inner coating area. These (minor) variations can be attributed to geometric effects such as shadowing of particle flux by the outer plasma shields and the difference in angular distribution of sputtered Al and Sc atoms. The latter effect is expected to be more pronounced at low target–substrate distances (TSD 90 mm) and low pressure, i.e., a high mean free path of atoms.

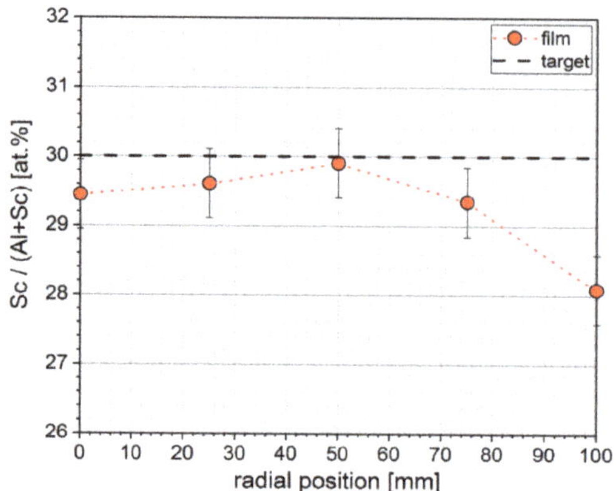

Figure 2. EDS measurement of Sc content in AlScN and comparison to the $Al_{0.7}Sc_{0.3}$ target composition.

Figure 3 shows the θ–2θ scans of AlScN with 2.5 μm and 0.9 μm thicknesses on Si/Ti/Pt with a 5 nm Ti and 40 nm Pt seed layer for different pulse modes and radial positions. All films were highly c-axis-oriented, with (002) peaks being the main intensity by far. The 2.5 μm-thick films deposited with a more unipolar pulse mode (S_u = 90%), revealing that there were also weak peaks of (102) and (103) orientation present, although those had a very small intensity of 1:3500 relative to the (002) peak. On the other hand, the films deposited with a lower share of unipolar pulse mode (S_u = 60%) exhibited no (102) or (103) peaks in the inner radial positions, while they appeared only at the edge of the coating area (r = 90 mm), with intensity ratios to the (002) peak being 1/1000 and 1/2000, respectively. The thinner films with 900 nm thickness, on the other hand, exhibited no visible (102) or (103) peaks, regardless of pulse mode or radial position, indicating a high crystalline quality over the whole coating area for the thinner films, similar to results for thickness-dependent occurrences of AOG reported in [12].

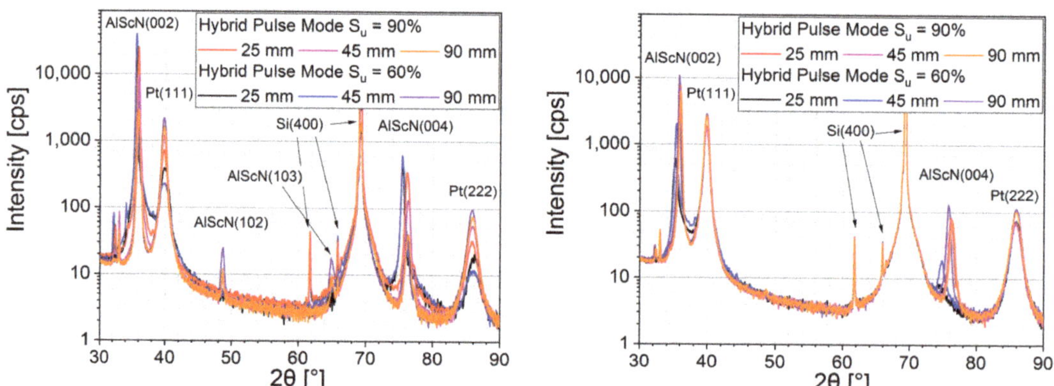

Figure 3. θ–2θ scan of 2.5μm (**left**) and 0.9 μm (**right**) AlScN films at different radial positions (25 mm, 45 mm, 90 mm); peaks labeled as Si(400) are the Si substrate peaks by Cu $K_{\alpha1/2}$ (69.2°), Cu $K_{\beta1}$ (61.7°) and Cu W $L_{\alpha1}$ (65.9°); the peaks at 32.2° and 34.2° are the AlScN(002) Peaks by Cu $K_{\beta1}$ and Cu $WL_{\alpha1}$, as well as Si(200) by Cu $K_{\alpha1}$ at 33.0°, respectively.

From the 2θ position of AlScN (002) peaks in Figure 3, the c-lattice constant was calculated as described in the Methods section. Film stress, e.g., due to energetic particle bombardment during layer deposition, results in a shift of the c-lattice parameter compared to stress free values [17]. For c-axis-oriented films, the lattice parameter c increased with increasing in-plane compressive film stress. Additionally, besides the effect due to film stress, the lattice parameters a and c of AlScN increased with the incorporation of scandium into the wurtzite structure. The lattice parameters of AlN are a = 0.31114 nm and c = 0.49792 nm according to ICDD PDF-25-1133. According to calculations by other groups, the a-axis lattice parameter of $Al_{0.7}Sc_{0.3}cN$ should be increased compared to that of AlN by ca. 4.4% to 0.3248–0.3250 nm, and the c-axis lattice parameter should increase by ca. 0.5–1% to 0.5003–0.5027 nm [18,19]. The calculated c-axis values from XRD measurements for different pulse modes and radial positions are shown in Figure 4. As can be seen, the mostly unipolar pulse mode (S_u = 90%) was almost homogenous along the whole radius. Film stress, as calculated with the wafer curvature method on a reference sample, i.e., an AlScN-coated 8 inch Si wafer without any electrode layers, was homogenous over the deposition area and around −700 MPa, therefore compressive. On the other hand, the pulse mode with a lower share of unipolar (S_u = 60%) was more inhomogeneous, with a 2% bigger c axis in the wafer middle compared to the edge and thus a much higher compressive stress of above 1.5 GPa. This is in line with previous results, revealing higher plasma density near the substrate and higher energetic bombardment of the substrate at a higher share of bipolar pulse mode (see Figure 1, [14]).

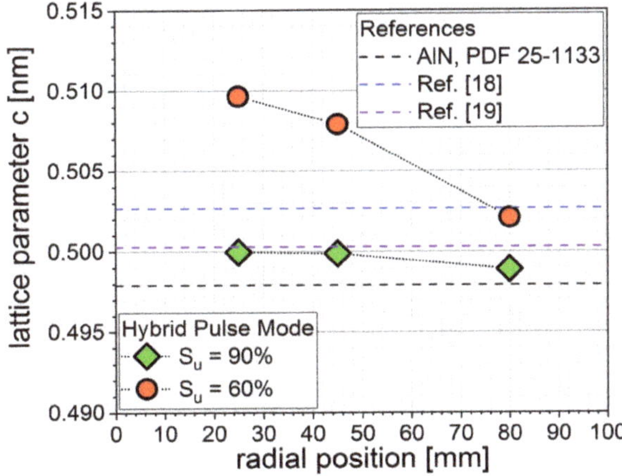

Figure 4. Calculated c-lattice parameter of 0.9 µm films deposited with different hybrid pulse modes depending on radial position (PDF 25-1133; theoretical AlN lattice parameter c = 0.49792 nm, as well as calculated values for $Al_{0.7}Sc_{0.3}N$ from [18,19] for comparison).

The rocking curves of (002) peaks shown in Figure 5 exhibited a similar behavior of radial dependencies. The films deposited at pulse mode S_u = 90% had homogenous values of around 2.1–2.3° over the whole deposition range. In contrast to this, films deposited with a higher share of bipolar pulse mode showed a significant increase in FWHM from 1.5° in the middle, where energetic particle bombardment during film growth was highest, to 3° at the outer position, where energy bombardment was much lower. This is also in line with Figure 3, where the 2.5 µm-thick film deposited at pulse mode S_u = 90% showed a homogenous occurrence of, although barely visible, (102) and (103) peaks, whereas the film deposited at pulse mode S_u = 60% exhibited them only at the outermost position.

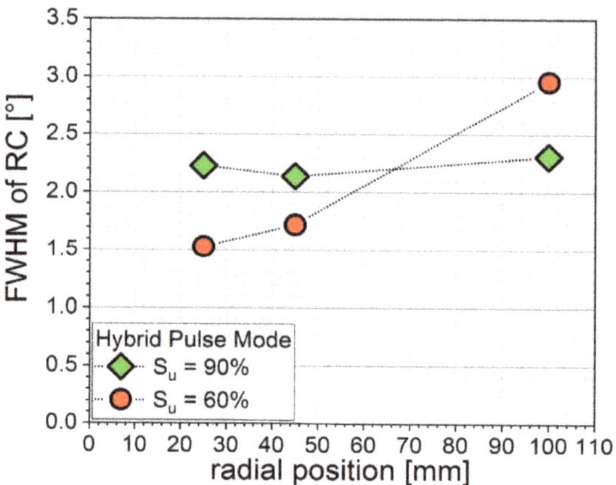

Figure 5. FWHM of (002) rocking curve of 2.5 μm films depending on radial position and pulse mode.

3.2. Piezoelectric and Ferroelectric Properties

The piezoelectric properties were characterized on 1.5 mm-thick 4 inch Si wafers to take substrate deformation into account [16]. The bottom electrodes were 5 nm Ti and 40 nm Pt, and the top electrodes were 100 nm Al with different electrode diameters from Ø0.5 mm to Ø2 mm for calculation of $d_{33,f}$ to take pad size effect into account.

The material parameter $d_{33,f}$ of 0.9μm AlScN films is shown in Figure 6. Every data point refers to a least 20 single electrodes. These electrodes are within < ±5 mm distance to the noted radial position in the figure. As can be seen, the films deposited at a more unipolar pulse mode S_u = 90% showed a homogeneous radial distribution of around 11–12 pm/V with only a slight increase towards the outer coating area. This conforms with the XRD data, which show homogenous rocking curve FWHM. On the other hand, the more bipolar pulse mode resulted in films, which showed a maximum of $d_{33,f}$ = 10 pm/V in the middle of the deposition area and a decreasing value until reaching 7 pm/V at radial position around ca. 70 mm from the center with a slight increase towards higher radial positions again. The maximum value of $d_{33,f}$ was lower compared to films grown at higher unipolar pulse mode. This is somewhat congruent with the XRD data, as it shows an increase in FWHM of (002) rocking curve towards the edge, although FWHM started at a lower value compared to films deposited with a more unipolar pulse mode.

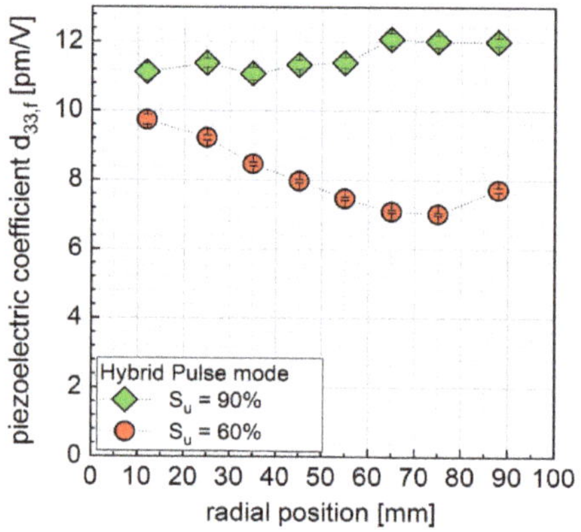

Figure 6. Dependency of $d_{33,f}$ of 1 μm AlScN films on radial position and share of unipolar pulse mode S_u.

Figure 7 shows the piezoresponse force microscopy measurements of AlScN deposited with pulse mode S_u = 60% at two radial positions. The film surface at the inner radial position of 25 mm was homogenous with an arithmetic average roughness Ra of 1.4 nm, whereas the films at outer position of 100 nm showed a high density of abnormal-oriented grains (AOG) and thus a much higher roughness Ra of 9.7 nm. The measured piezoelectric amplitude showed the same behavior, being homogenous in the inner 25 mm position and having many areas without any piezoelectric activity at the 100 mm edge position of the deposition area. Additionally, the piezoelectric phase of films was unipolar (N-polar) at the inner position and bipolar at the outer position, meaning the film had grown both Al- and N-polar. The cause for this is still under investigation. The most probable cause may be some local distortions during film growth, supported by the higher ratio of AOG on the surface.

Figure 7. PFM results (area 5 µm × 5 µm) of 2.5 µm AlScN, deposited with S_u = 60% pulse mode, at radial positions 25 mm (**a–c**) and 100 mm (**d–f**): topography (**a**,**d**), piezoelectric amplitude (**b**,**e**) and piezoelectric phase (**c**,**f**).

The ferroelectric switching of AlScN films deposited with pulse mode S_u = 60% and 90% are exemplarily shown in Figure 8 for Ø0.75 mm top electrodes (area of 0.44 mm^2). The polarization after switching was ca. 80 µC/cm^2 on the positive axis for both pulse modes. On the negative axis of applied electrical field, the polarization was 120 µC/cm^2 for S_u = 60% and around 80 µC/cm^2 with higher leakage current for S_u = 90% at negative electrical field. The coercive fields were at ca. 3.1 MV/cm and −3.5 MV/cm for S_u = 60% and ca. 3 MV/cm and −3.3 MV/cm for S_u = 90%, respectively. This is in line with the values reported by Fichtner et al., where AlScN with 27 at.% or 32 at.% Sc was reported to have a polarization of above 100 µC/cm^2 with coercive fields of around 4 to 5 MV/cm or 3 to 4 MV/cm, depending on the sign of electrical field, residual stress and Sc concentration [9]. Furthermore, Giribaldi et al. reported an effect of the electrode material on the Pt-AlScN-Al layer stack [20], a layout similar to this paper. They reported the effect of asymmetry of electrical properties as resulting from the combination of a Schottky junction at the Pt–AlScN interface and an ohmic contact at the AlScN–Al interface. As an additional factor, since the film stress of our sample was highly compressive, the resulting P-E loop and strain response on the applied electrical field shifted accordingly. Maximum strain at ±4 MV/cm for S_u = 60% and 90% was 0.25% and 0.35% at positive strain (elongation of c axis) and 0.13%/0.18% and 0.23%/0.29% at negative strain (compression of c axis), respectively. This is in line with the higher piezoelectric activity (Figure 6) and better crystalline quality (Figure 3).

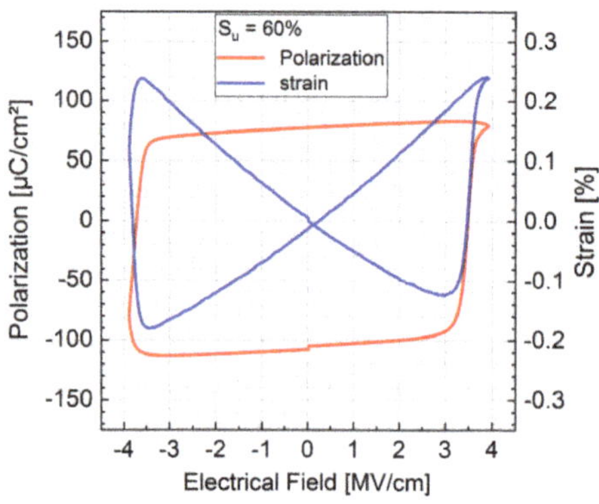

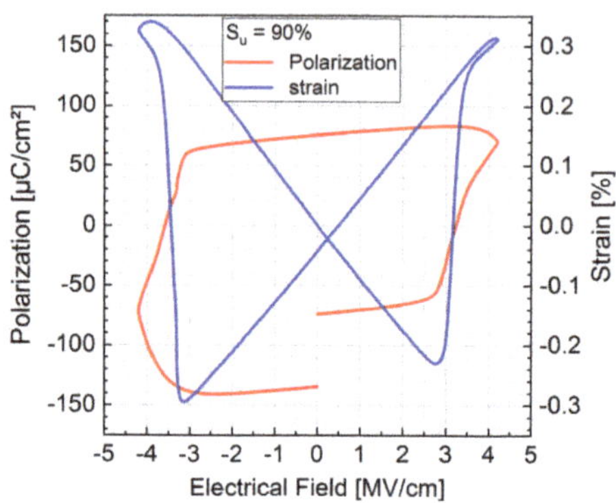

Figure 8. DBLI measurement of polarization and strain of 1µm $Al_{0.695}Sc_{0.295}N$ deposited with pulse mode S_u = 60% (**left**) and S_u = 90% (**right**) on Ti/Pt seed layer with Ø0.75 mm Al top electrode (= 0.44 mm^2) depending on electrical field, measured at 1 kHz, average of 100 measurements.

4. Conclusions

This paper reports the deposition and characterization of ferroelectric AlScN films with a Sc content of 30 at.%. Film composition was largely homogenous at 29.5 ± 0.5 at.% Sc, with a decrease to 28 at.% at the edge due to geometric effects.

XRD θ–2θ scans revealed a highly oriented structure with high c-axis (002) orientation at all parameters, although thick films with a large share of unipolar pulse mode (S_u = 90%) exhibited a small occurrence of (102) and (103) orientations with peak intensities of 1/3500th of (002) peak intensity. Calculation of c-lattice parameters revealed a homogenous distribution for more unipolar process conditions, with more bipolar pulse modes (S_u = 60%) resulting in films with a higher c axis, i.e., more compressive stress, at the inner coating area. This is in line with known plasma parameter measurements, meaning higher plasma density in the center for bipolar process conditions. Rocking curve measurements of AlScN (002) peaks confirm this, as they revealed homogenous distribution of FWHM values for the 90% share of unipolar pulse mode of around 2.2 ± 0.1° and a radial distribution for the 60% share of unipolar pulse mode, with 1.5° at the center of the deposition area and 3° at the edge.

Piezoelectric characterization by double-beam laser interferometry showed $d_{33,f}$ values of 11–12 pm/V for films deposited by hybrid pulse mode S_u = 90% process conditions and 10–7 pm/V for S_u = 60% process conditions with a decrease towards outer coating areas. PFM measurements attribute the overall lower piezoelectric behavior of the films deposited by S_u = 60% to a stronger occurrence of AOG and a bipolar phase distribution.

The films showed ferroelectric switching behavior with coercive fields of 3 MV/cm to 3.5 MV/cm for 0.9 µm AlScN films. Polarization of the films reached 80–120 µC/cm^2, depending on switching direction and pulse mode. The maximum strain at the highest field of 4 MV/cm was around 0.35%, also depending on switching direction and pulse mode.

In short, the influence of pulse powering variation by special unipolar–bipolar hybrid pulse mode on resulting film properties of AlScN was investigated. It was shown that good structural and piezoelectric properties can be achieved with very high deposition rates of up to 200 nm/min on large areas using this hybrid pulse mode operation.

Author Contributions: Conceptualization, S.B. and H.B.; Formal analysis, S.B., S.C., O.Z. and T.M.; Funding acquisition, S.B. and H.B.; Investigation, S.B., T.S., O.Z. and T.M.; Methodology, S.B. and S.C.; Project administration, S.B. and H.B.; Supervision, S.B. and H.B.; Writing—original draft, S.B.; Writing—review and editing, S.B., T.S., S.C., O.Z., T.M. and H.B. All authors have read and agreed to the published version of the manuscript.

Funding: This research was partly funded by the ECSEL Joint Undertaking (JU), for the project "Pilot Integration 3nm Semiconductor technology (Pin3S)", grant number 826422. The JU receives support from the European Union's Horizon 2020 research and innovation program and Netherlands, Belgium, Germany, France, Romania and Israel. This research was also partly funded under the project "Tragbare, Autarke und Kompakte Strom Generatoren (TASG)" by the European Regional Development Fund (ERDF) and the Free State of Saxony under grant number 100347675.

Data Availability Statement: The data presented in this study are available on reasonable request from the corresponding author.

Conflicts of Interest: The authors declare no conflict of interest. The funders had no role in the design of the study, in the collection, analyses, or interpretation of data, in the writing of the manuscript or in the decision to publish the results.

References

1. Akiyama, M.; Kamohara, T.; Kano, K.; Teshigahara, A.; Takeuchi, Y.; Kawahara, N. Enhancement of Piezoelectric Response in Scandium Aluminum Nitride Alloy Thin Films Prepared by Dual Reactive Cosputtering. *Adv. Mater.* **2009**, *21*, 593–596. [CrossRef] [PubMed]
2. Ng, D.K.T.G.; Ho, C.-P.; Zhang, T.; Xu, L.; Siow, L.-Y.; Chung, W.-W.; Cai, H.; Lee, L.Y.T.; Zhang, Q.; Singh, N. CMOS compatible MEMS pyroelectric infrared detectors: From AlN to AlScN. In Proceedings of the SPIE 11697, MOEMS and Miniaturized Systems XX, Online. 6–12 March 2021; Volume 116970. [CrossRef]
3. Su, J.; Niekiel, F.; Fichtner, S.; Thormaehlen, L.; Kirchhof, C.; Meyners, D.; Quandt, E.; Wagner, B.; Lofink, F. AlScN-based-MEMS magnetoelectric sensor. *Appl. Phys. Lett.* **2020**, *117*, 132903. [CrossRef]
4. Wang, J.; Zheng, Y.; Ansari, A. Ferroelectric Aluminum Scandium Nitride Thin Film Bulk Acoustic Resonators with Polarization-Dependent Operating States. *Phys. Status Solidi RRL* **2021**, *15*, 2100034. [CrossRef]
5. Liu, C.; Li, M.; Chen, B.; Zhang, Y.; Zhu, Y. Evaluation of the impact of abnormal grains on the performance of Sc0.15Al0.85N-based BAW resonators and filters. *J. Micromech. Microeng.* **2022**, *32*, 034002. [CrossRef]
6. Barth, S.; Bartzsch, H.; Glöß, D.; Frach, P.; Modes, T.; Zywitzki, O.; Suchaneck, G.; Gerlach, G. Magnetron sputtering of piezoelectric AlN and AlScN thin films and their use in energy harvesting applications. *Microsyst. Technol.* **2016**, *22*, 1613–1617. [CrossRef]
7. Barth, S.; Frach, P.; Bartzsch, H.; Glöß, D.; Zywitzki, O.; Modes, T. Sputter deposition of AlN and AlScN films for actuator, transducer and energy harvesting applications. In Proceedings of the Society of Vacuum Coaters 61st Annual Technical Conference Proceedings 2018, Orlando, FL, USA, 5–10 May 2018. [CrossRef]
8. Yarar, E.; Fichtner, S.; Hayes, P.; Piorra, A.; Reimer, T.; Lisec, T.; Frank, P.; Wagner, B.; Lofink, F.; Meyners, D.; et al. MEMS-Based AlScN Resonating Energy Harvester with Solidified Powder Magnet. *J. Microelectromech. Syst.* **2019**, *28*, 1019–1031. [CrossRef]
9. Fichtner, S.; Wolff, N.; Lofink, F.; Kienle, L.; Wagner, B. AlScN: A III-V semiconductor based ferroelectric. *J. Appl. Phys.* **2019**, *125*, 114103. [CrossRef]
10. Tsai, S.L.; Hoshii, T.; Wakabayashi, H.; Tsutsui, K.; Chung, T.-K.; Chang, E.Y.; Kakushima, K. Room-temperature deposition of a poling-free ferroelectric AlScN film by reactive sputtering. *Appl. Phys. Lett.* **2021**, *118*, 082902. [CrossRef]
11. Pirro, M.; Herrera, B.; Assylbekova, M.; Giribaldi, G.; Colombo, L.; Rinaldi, M. Characterization of dielectric and piezoelectric properties of ferroelectric AlScN films. In Proceedings of the 2021 IEEE 34th International Conference on Micro Electro Mechanical Systems (MEMS), Gainesville, FL, USA, 25–29 January 2021; p. 646. [CrossRef]
12. Sandu, C.S.; Parsapour, F.; Mertin, S.; Pashchenko, V.; Matloub, R.; LaGrange, T.; Heinz, B.; Muralt, P. Abnormal Grain Growth in AlScN Thin Films Induced by Complexion Formation at Crystallite Interfaces. *Phys. Status Solidi A* **2019**, *216*, 1800569. [CrossRef]
13. Sandu, C.S.; Parsapour, F.; Xiao, D.; Nigon, R.; Riemer, L.M.; LaGrange, T.; Muralt, P. Impact of negative bias on the piezoelectric properties through the incidence of abnormal oriented grains in Al0.62Sc0.38N thin films. *Thin Solid Film.* **2020**, *697*, 137819. [CrossRef]
14. Barth, S.; Bartzsch, H.; Glöß, D.; Frach, P.; Gittner, M.; Labitzke, R. Adjustment of plasma properties in magnetron sputtering by pulsed powering in unipolar/bipolar hybrid pulse mode. *Surf. Coat. Technol.* **2016**, *290*, 73–76. [CrossRef]
15. Frach, P.; Gottfried, C.; Bartzsch, H.; Goedicke, K. The double ring process module—A tool for stationary deposition of metals, insulators and reactive sputtered compounds. *Surf. Coat. Technol.* **1997**, *90*, 75–81. [CrossRef]
16. Sivaramakrishnan, S.; Mardilovich, P.; Schmitz-Kempen, T.; Tiedke, S. Concurrent wafer-level measurement of longitu-dinal and transverse effective piezoelectric coefficients ($d_{33,f}$ and $e_{31,f}$) by double beam laser interferometry. *J. Appl. Phys.* **2018**, *123*, 014103. [CrossRef]
17. Tasnadi, F.; Alling, B.; Höglund, C.; Winqvist, G.; Birch, J.; Hultmann, L.; Abrikosov, I.A. Origin of the Anomalous Piezoelectric Response inWurtzite ScxAl1-xN Alloys. *Phys. Rev. Lett.* **2021**, *104*, 137601. [CrossRef] [PubMed]
18. Österlund, E.; Ross, G.; Caro, M.A.; Paulasto-Kröckel, M.; Hollmann, A.; Klaus, M.; Meixner, M.; Genzel, C.; Koppinen, P.; Pensala, T.; et al. Stability and residual stresses of sputtered wurtzite AlScN thin films. *Phys. Rev. Mater.* **2021**, *5*, 035001. [CrossRef]

19. Ambacher, O.; Christian, B.; Feil, N.; Urban, D.F.; Elsässer, C.; Preschner, M.; Kirste, L. Wurtzite ScAlN, InAlN, and GaAlN crystals, a comparison of structural, elastic, dielectric, and piezoelectric properties. *J. Appl. Phys.* **2021**, *130*, 045102. [CrossRef]
20. Giribaldi, G.; Pirro, M.; Soukup, B.H.; Assylbekova, M.; Colombo, L.; Rinaldi, M. Compensation of contact nature-dependent assymetry in the leakage current of ferroelectric ScXAl1-XN thin film capacitors. In Proceedings of the 2021 IEEE 34th International Conference on Micro Electro Mechanical Systems (MEMS), Online, 25–29 January 2021; p. 650. [CrossRef]

Article

Raman Spectroscopy and Spectral Signatures of $AlScN/Al_2O_3$

Dmytro Solonenko [1,*], Agnė Žukauskaitė [2,†], Julian Pilz [1], Mohssen Moridi [1] and Sarah Risquez [1]

1 Microsystems Division, Silicon Austria Labs, 9524 Villach, Austria
2 Fraunhofer Institute for Applied Solid State Physics, IAF, Tullastr. 72, D-79108 Freiburg, Germany
* Correspondence: dmytro.solonenko@silicon-austria.com
† Current affiliation: Fraunhofer Institute for Organic Electronics Electron Beam and Plasma Technology FEP, Winterbergstr. 28, D-01277 Dresden, Germany.

Abstract: III-V solid solutions are sensitive to growth conditions due to their stochastic nature. The highly crystalline thin films require a profound understanding of the material properties and reliable means of their determination. In this work, we have investigated the Raman spectral fingerprint of $Al_{1-x}Sc_xN$ thin films with Sc concentrations x = 0, 0.14, 0.17, 0.23, 0.32, and 0.41, grown on Al_2O_3(0001) substrates. The spectra show softening and broadening of the modes related to the dominant wurtzite phase with increasing Sc content, in agreement with the corresponding XRD results. We investigated the primary scattering mechanism responsible for the immense modes' linewidths by comparing the average grain sizes to the phonon correlation length, indicating that alloying augments the point defect density. The low-frequency Raman bands were attributed to the confined spherical acoustic modes in the co-forming ScN nanoparticles. Temperature-dependent Raman measurements enabled the temperature coefficient of the E_2(high) mode to be determined for all Sc concentrations for the precise temperature monitoring in AlScN-based devices.

Keywords: aluminium scandium nitride; piezoelectric films; Raman spectroscopy; alloy scattering; temperature coefficient

Citation: Solonenko, D.; Žukauskaite, A.; Pilz, J.; Moridi, M.; Risquez, S. Raman Spectroscopy and Spectral Signatures of $AlScN/Al_2O_3$. *Micromachines* **2022**, *13*, 1961. https://doi.org/10.3390/mi13111961

Academic Editor: Alberto Tagliaferro

Received: 20 September 2022
Accepted: 8 November 2022
Published: 11 November 2022

1. Introduction

Aluminium nitride (AlN) thin films, being a staple for commercial acoustic wave resonators, have been well investigated over the last two decades to improve understanding of the growth–performance relation [1–3]. The advent of the $Al_{1-x}Sc_xN$ (AlScN) pseudobinary alloys [4] instigated further studies on the enhancement of the piezoelectric properties, yielding the interplay of the wurtzite and hexagonal lattice configurations, which led to the overall lattice softening [5] and the increased electromagnetic coupling [6]. The phase formation diagram is complicated by the fact that pure ScN crystallises in the rock-salt structure [7]. It was initially shown that only up to $x = 0.22$ could be introduced into the AlN lattice before it converts to the cubic system [8]. However, the maximum amount of Sc within the wurtzite lattice was later optimised: $x = 0.41$ [9]. The amount of Sc atoms dispersed in the AlN wurtzite lattice has a dominant role on the physical properties of the AlScN, which can, for instance, show even ferroelectric behaviour for $x > 0.28$ [10].

The phonon properties of these pseudobinary alloys are of great interest not only to enable insights into the structural integrity and thermal properties of their crystals, but also to provide reliable and non-destructive characterisation of the thin films in terms of their dielectric and optical properties. The first results on the infrared-active phonon modes were shown by Mayrhofer et al. for AlScN films of varied Sc composition in a narrow range up to 15% [11]. The redshift of the E_1(TO) band of AlN was shown to be proportional to the Sc content, which was interpreted as the indication of the elongating M-N (M = Al, Sc) bonds. Similar behaviour was also observed for the E_2(high) and A_1(TO) Raman bands and compared to the evolution of the rest IR-active bands, thereby confirming a decrease in the bond length as a result of increased bond ionicity [12].

Despite the narrow range of the Sc concentration, the main trends in phonon properties of the AlScN lattice have been unravelled so far while largely omitting the obvious impact of the film's microstructure on the spectral data. While the phonon frequency shift is evident and expected in the framework of the lattice softening, the rest of the peak characteristics are intricately entangled with the quality of the AlScN films. Although the influence of the microstructure can be observed using both IR and Raman spectra, it can be approached more accurately in the latter case [13,14]. Indeed, the large peak broadening observed in the Raman spectra of AlScN alloys remains poorly understood, and so does their overall spectral signature, which includes a few Raman bands addressed earlier [12,15–19]. It is necessary for understanding the contributions of various mechanisms responsible for the dramatic broadening of the Raman-active spectral bands, such as the possibility of the rock-salt phase formation, alloy- or disorder-based phonon scattering, and size effect.

In this work, we investigated in detail the vibrational signatures of $Al_{1-x}Sc_xN$ pseudobinary alloys in a broad range of the Sc concentrations (up to $x = 0.41$). Our results confirm the previous observations of the frequency shift and peak broadening for the Raman bands observed in the c–axis–oriented films [12,15]. We show that the Raman spectra of the alloys actually exhibit up to eight bands. Unreported bands are attributed to the second-order phonon modes of the wurtzite AlScN lattice. Moreover, we investigated the low–frequency spectral region, the investigation of which is usually limited by technical difficulties. We found the presence of Raman bands which are proposed to be related to the confined acoustic modes. In contrast to pure AlN, the second–order spectral features stemming from two-phonon modes and phonon density of states are greatly enhanced as a result of the bandgap shrinkage and a high density of the midgap states. The presence of these electronic states is attributed to the point defects, the density of which is two orders of magnitude higher than in pure wurtzite, as assessed via the phonon correlation length. This was corroborated by temperature–dependent Raman measurements showing that the phonon-defect scattering dominates over the phonon–phonon mechanism in AlScN regardless of the Sc concentration. The determined temperature coefficients can be used for the precise temperature determination in AlScN films via Raman spectroscopy.

2. Materials and Methods

First 1 µm thick $Al_{1-x}Sc_xN$ layers with Sc concentrations $x = 0, 0.14, 0.17, 0.23, 0.32$, and 0.41 were deposited on single-sided, polished, 100 mm diameter $Al_2O_3(0001)$ substrates at chuck temperatures of 300–400 °C by reactive pulsed DC magnetron co-sputtering (Evatec cluster sputter tool). In addition 99.9995% pure Al and 99.99% pure Sc targets were sputtered in a pure N_2 atmosphere; a constant total power applied to the Al and Sc targets to achieve different Sc concentrations; all other growth parameters were kept constant. The deposition parameters are described in detail in [9,20]. The Sc content with an accuracy of $x \pm 0.02$ was determined by energy dispersive X-ray (EDX) spectroscopy (Bruker Quantax, Bruker Corporation, Billerica, MA, USA) in a scanning electron microscope (Zeiss Auriga Crossbeam FIB-SEM, Carl Zeiss, Oberkochen, Germany). The compositional analysis of the samples is described elsewhere [21]. The $\theta/2\theta$ scans were obtained using a X-ray diffractometer equipped with a 4-bounce Ge 220 monochromator, a parallel beam X-ray mirror on the incident side, and a PiXcel3D detector (X'Pert^3 MRD XL, PANalytical, Almelo, The Netherlands). An atomic force microscope (NX20, Park Systems, Suwon, Korea) with the AC160TS tip (radius: 2 nm) was used to investigate the surface morphology and determine the roughness using Gwyddion software [22]. Raman spectra were collected using the micro-Raman spectrometer (inVia Qontor, Renishaw, London, UK). The excitation wavelength of the DPSS laser (Renishaw RL532) was 532 nm when used in combination with the 2400 L/mm grating, yielding the spectral resolution of about 0.1 cm^{-1}. The power density of about 10 $mW \cdot \mu m^{-2}$ estimated for the 100× objective (NA = 0.75) was used to avoid the heating of samples. The near-excitation tunable (NExT, Renishaw) filter was used, enabling the collection of the anti-Stokes spectra. The baseline stemming from the photoluminescence was subtracted from all obtained spectra using a polynomial function, describing the background signal increasing towards longer wavelengths. The temperature-dependent Raman measurements were recorded using a thermostat, a hot plate-like stage (T96-P, Linkam Scientific Instruments, Tadworth, UK). Due to the heating of the samples from backside of the sapphire substrate, it was important to realise that the heating of the AlScN films corresponded to the temperature setpoints. Provided the outstanding thermal conductivity of sapphire, the temperature in the films was assumed to be equal to that of the sapphire substrate. The calibration and tracking of the sample's temperature was performed using the A_{1g} Raman-active mode of the sapphire substrate (417.4 cm^{-1} [23]). The thermal coefficient of the Raman band was found to be -0.015 K·cm^{-1}, which is in an excellent agreement with previous studies [24].

3. Results and Discussion

3.1. Film Microstructure

The XRD scans of the $Al_{1-x}Sc_xN$ ($x = 0, 0.14, 0.23, 0.32, 0.41$) films grown on the c-plane Al_2O_3 substrates show the 000l ($l = 2, 4$) reflections of the nitride and 000l ($l = 6, 9$) reflections of the oxide compounds (Figure 1). The lack of additional peaks assigned to the AlScN and the reflections stemming from the c-plane suggest that the films are highly c-axis-oriented. Pole figure measurements confirm in-plane oriented growth of AlScN with the epitaxial relationship defined as [10-10]AlScN//[11-20]sapphire and (0001)AlScN//(0001)sapphire [21]. Provided that the film thickness values are similar, the variation in the peak intensity indicates different amounts of the diffracting domains, and the variation in the peak linewidths suggests their diverging size distributions (Figure 1, inset). The trends for the peak position, which are dependent on the Sc concentration and the thermal strain, are in agreement with the previously reported ones [20,25,26]. The alterations in the peak positions and linewidths of the (0004) reflection peak mirror the behaviour observed for the (0002) one scaled due to the higher 2θ angles, which suggests good uniformity for the long-range order. The peak intensity is continuously reduced for higher amounts of Sc, which can be related to the size reduction of crystalline domains. No peak solely related to the rock-salt ScN or AlScN phase, which might have been expected in the alloy phase diagram [8], was observed in the diffractograms, confirming the dominating wurtzite phase in the pseudobinary AlScN alloy.

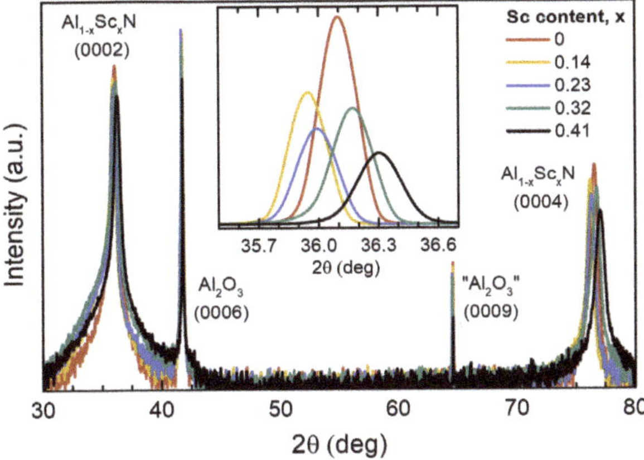

Figure 1. XRD symmetric $\theta/2\theta$ scans of AlScN films with various Sc amounts. Inset: the 2θ range in the vicinity of the (0002) peak of the AlScN alloy.

Figure 2 shows AFM micrographs of the films, revealing the pebble-like morphology characteristic of the AlN films grown via magnetron sputtering [27,28]. The AFM results revealed a less random arrangement of the grains being seemingly clustered into short chains of hemispherical droplets, which resembled the surface of AlN after the annealing process [29] and thus indicated the relation to the alteration of the adatom surface mobility with adding more Sc. When applying the structure zone model [30], the worm-like surface texture is characteristic for the low adatom mobility [31], which is caused by the different kinetic energies of co-sputtered species. The surface of the $Al_{0.59}Sc_{0.41}N$ film reveals the presence of larger grains located at the nodes of the worm-like surface. These specific grains may be attributed to the abnormally oriented grains of AlScN observed for the cases of high Sc concentration [21,32]. The surface roughness of the films decreased when alloying to more Sc atoms and reached its minimum value for $x = 0.32$ of the scandium composition. The following increase in the roughness values for the film alloyed with the highest Sc amount was related to the bright protrusions visible in the image. Nevertheless, the roughness values below 2 nm revealed that AlScN films were largely smooth and exhibited no surface structures to be assigned to other crystalline phases. Thus, the investigation of the film microstructure showed that the films of the pseudobinary AlScN alloys on Al_2O_3 consisted of one crystalline wurtzite phase, which was oriented along the c-axis, indicating columnar growth.

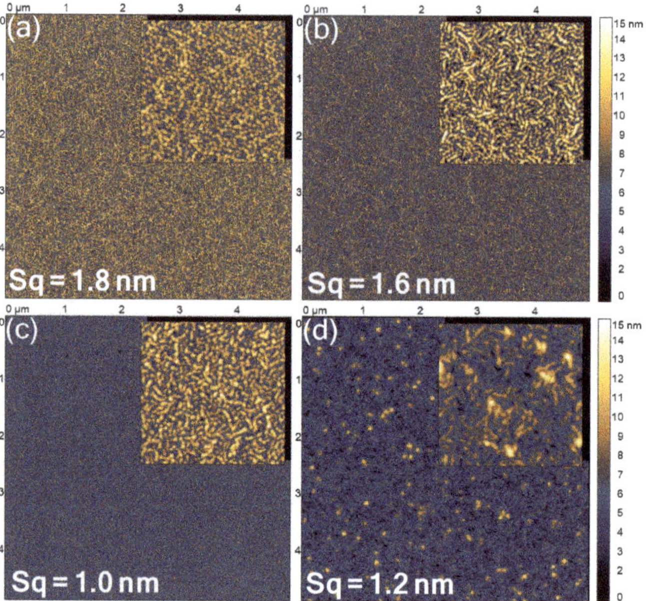

Figure 2. Surface morphology of the thin films with the following Sc content, x = (**a**) 0 (AlN), (**b**) 0.14, (**c**) 0.32, (**d**) 0.41 obtained via AFM. The sample with $x = 0.23$ is not shown. The false colour scale is common for all images. The rms roughness, Sq, values is indicated on the corresponding images. Insets: complimentary 1×1 μm^2 AFM images.

3.2. Vibrational Properties

The Raman spectra of the $AlScN/Al_2O_3$ system, shown in Figure 3, are complex because of the multiple spectral bands of different origins. In order to properly assign them, the spectra have to be approached by inspecting the film and the substrate separately. The spectrum of the aluminium oxide substrate indicates the single-crystalline c-plane oriented corundum by the numerous sharp peaks. The most pronounced ones are at about 416.7, 576.3, and 748.8 cm^{-1}, corresponding to the A_{1g} and 2 E_g modes, respectively [23]. These spectral bands are present in all Raman spectra shown, indicating the penetration depth of the laserline, which spanned through the entire depth of a nitride film. The spectrum of pure AlN exhibits the first-order Raman-active modes, such as E_2(low), E_2(high), and quasi A_1(LO) (labelled as "QLO") [33]; and the second-order modes—optic overtone [K_3, M] and

A_1-symmetry optic combination and overtone [34]. The spectral band at about 1189.4 cm^{-1} is assigned to the overtone of the A_1(TO) mode. The peak parameters, such as the spectral position and linewidth, are given in Table 1, to which we will refer from here on. According to the symmetry selection rules for the wurtzite-type crystals [35], the present combination of the visible spectral bands confirms the c-axis texture of the nitride films.

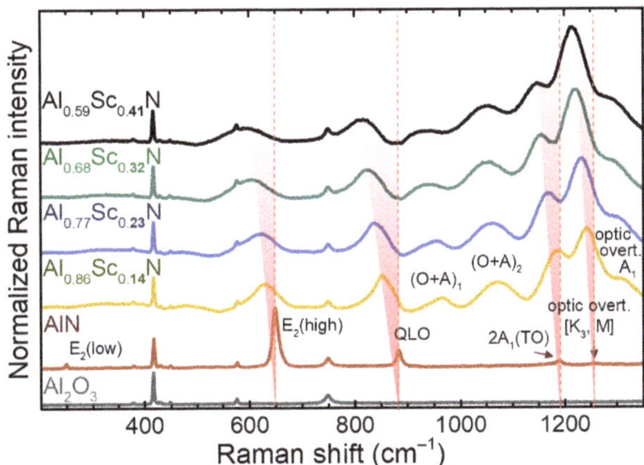

Figure 3. Raman spectra of AlScN films with various Sc concentrations. The spectra are stacked for clarity. The spectral bands of pure AlN are labelled by the dash lines to emphasise their low-frequency shifts in $Al_{1-x}Sc_xN$ as a guide to the eye. The Raman spectrum of the Al_2O_3 substrate revealed the bands unrelated to the nitride films.

The spectral position of the E_2(high) mode in the spectrum of pure AlN (cf. Table 1) suggests a significant level of the tensile residual stress, which can be estimated to reach up to ca. 2 GPa, according to the Raman biaxial stress coefficient of −3.8 cm^{-1}·GPa^{-1} and stress-free phonon frequency of 656.7 cm^{-1} [36]. Such a high value of the residual stress, beyond the yield strength of AlN (ca. 0.3 GPa), however, suggests that the biaxial stress is not the only contribution to the peak position. Moreover, wafer bow measurements revealed the residual stress of ca. 1.1 GPa [37]. As we show below, another important factor is the hydrostatic stress caused by diverse point defects. This is corroborated by the large linewidth values of all AlN bands, being inversely proportional to the phonon lifetime limited by phonon scattering. The phonons can be scattered by other quasiparticles, such as phonons and electrons, or the lattice irregularities. Considering the columnar microsctructures of the films, we expect significant contributions to the phonon scattering in AlN by the grain boundaries and overall point defects.

Table 1. Peak position and linewdith (FWHM) of the Raman-active bands collected from the AlScN films with the various Sc concentrations.

	E_2(high)	QLO	(O + A)$_1$	(O + A)$_2$	2A_1(TO)	Overtone [K_3, M]	A_1 Optic Comb & over
	Position (FWHM) [cm^{-1}]	Position (FWHM) [cm^{-1}]	Position (FWHM) [cm^{-1}]	Position (FWHM) [cm^{-1}]	Position (FWHM) [cm^{-1}]	Position (FWHM) [cm^{-1}]	Position (FWHM) [cm^{-1}]
AlN	649.1 (12)	883.7 (10)			1189.4 (16)	1258.1 (23)	1347.3 (35)
$Al_{0.86}Sc_{0.14}N$	630.2 (53)	852.1 (36)	963.5 (38)	1071.8 (77)	1185.6 (57)	1241.8 (45)	1303.5 (79)
$Al_{0.73}Sc_{0.23}N$	622.6 (55)	839.0 (57)	952.9 (85)	1061.6 (87)	1169.9 (55)	1232.0 (47)	1292.3 (94)
$Al_{0.68}Sc_{0.32}N$	603.8 (72)	824.5 (65)	942.0 (123)	1054.3 (89)	1156.4 (53)	1221.3 (52)	1282.6 (84)
$Al_{0.59}Sc_{0.41}N$	592.2 (94)	814.6 (69)	930.0 (147)	1052.3 (85)	1149.5 (54)	1215.9 (53)	1279.0 (95)

The Raman spectra of the AlScN films evolve with the Sc concentration (Figure 3) in perfect agreement with the previous reports [12,15]. Namely, the peaks, which correspond to the E_2(high) and QLO bands, shift towards lower frequencies, making the trend inversely proportional to the Sc concentration. We also observed a drastic enhancement of the bands above 900 cm^{-1}, corresponding to the two-phonon modes, which may be related to the resonance enhancement in AlScN, the dielectric function of which drastically changes upon alloying [9]. Moreover, the expected high defect density promotes the midgap electronic states. The defect-state electron transitions in the visible light range can thus facilitate the resonance enhancement of the Raman scattering effect [38]. Apart from the three modes observed in the spectrum of the AlN film, two more bands at around 963.5 and 1071.8 cm^{-1} are evident for AlScN. Their assignment to the combination of the acoustic and optic modes, for example, TA + A_1(TO) or E_2(high), requires further investigations. The assignment of the two-phonon modes to the "interference fringes" [15] can be ruled out by the fact that the peak positions (cf. Table 1) are also redshifted, though not as drastically as in the case of the first-order modes. Another observation concerns the overall amplification of the spectral background, which may be attributed to the relaxation of the Raman selection rules as a result of the lattice disorder, enabling the visibility of the phonon density of states (pDOS). This is the reason for the absence of the E_2(low) mode in the spectra of AlScN. The relaxation of the momentum (q) conservation, observed as the enhancement of the background signal and typical broadening of the spectral features, is well known from the studies of other III-V solid solutions [33,39–42]. The origin of the relaxation is the substitutional disorder, which leads to the activation of the non-Γ-point ($q \neq 0$) phonon modes [41], which may either be seen as separate peaks or as peak tails contributing to the asymmetry of the pristine first-order modes [40]. Their actual spectral appearance is governed by the curvature of the phonon dispersion relations in the Brilloun zone. Given that AlScN pseudobinary alloys represent the amalgamation type of optical phonon behaviour [42], and since rock-salt ScN exhibits no first-order modes, no localisation of disorder structure is expected [41]. This implies the typical effects of the compositional fluctuation [43], and thus the asymmetric peak broadening and the non-linear linewidth variation with the Sc concentration, which is obvious from the retrieved peak characteristics (cf. Table 1). Due to the absent phonon dispersions for AlScN, we can approximate the behaviour of spectral modes using the ones of AlN [33,34], focusing on the two main first-order modes, E_2(high) and QLO, visible in all spectra (Figure 3). One can clearly see that the dispersions of these two modes are dramatically different: while the E_2(high) mode almost does not disperse from the zone-centre towards the edges, the QLO one spreads in the range of 200 cm^{-1} in Γ-H direction. This difference explains the stark asymmetry of the QLO mode in the spectra of AlScN, while E_2(high) remains symmetric even at the highest Sc concentration. The further theoretical studies would be necessary to verify this likely interpretation. We can hence conclude that the disorder-activated spectral features can be understood in the framework of typical compositional disorder, which originates from alloying, and the high-degree long-range ordering in the films is validated by the corresponding XRD patterns.

Thus, the evolution of the Raman bands due to Sc alloying can be traced only for two one-phonon bands and two-phonon bands (observed in AlN). Using the knowledge of the residual film stress in the AlScN films [37], we can estimate the shift of E_2(high) mode as a function of Sc concentration, x, assuming that the Raman biaxial stress coefficient is invariable:

$$\Delta\omega = -126.14 \cdot x$$

The slope factor of -126.14 is obtained via a linear fit of the peak position values (cf. Table 1) with a subtracted contribution from the residual film stress. Comparing the slope to the one reported by Deng et al. [12], our slope factor is almost two times lower, which stems from the fact the Raman peak is visible at much higher Sc concentrations, confirming the high quality of these epitaxial films. Additionally, we surmise that the biaxial stress coefficient of the E_2(high) mode in AlScN films may drastically differ from the one of AlN, which further increases the uncertainty for the application of Raman spectra for the Sc concentration determination.

Apart from the one-phonon Raman bands, the Raman spectra of AlScN films also feature bands in the low-frequency region (Figure 4a). The suppression of the intense Rayleigh peak via the notch filter also enables us to observe the anti-Stokes side of the spectra. In the spectrum of AlN, we can clearly observe the E_2(low) mode at 249.5 cm^{-1} (on the Stokes side) and its anti-Stokes counterpart at −249.5 cm^{-1}. The intensity of the Stokes-side Raman band is substantially higher than the one of the anti-Stokes counterpart, in accordance with the Bose–Einstein distribution of phonons [44]. The alloying of AlN with Sc leads to the emergence of the intense and broad band between 100 and 250 cm^{-1} in the spectra of AlScN films. Moreover, its spectral position is seemingly proportional to the Sc content (Figure 4b). The redshifted Stokes and anti-Stokes maxima and the peak intensities differ negligibly. Although their relation to Sc alloying is clear, the underlying mechanism of their Raman activity can only be controversially discussed. As discussed earlier, the alloying practically increases the overall defect density, which might lead to amorphisation in its extreme case. The latter is known to relax the momentum conservation law of the photon–phonon scattering process, allowing the detection of the phonon density of state in Raman spectra [45,46]. According to the phonon structure calculated for pure AlN [34] and ScN [47] lattices, no considerable density is expected to be observed in the spectral range below 200 cm^{-1}. This fact rules out the assignment of the bands to the pDOS and a possible resonant enhancement of the acoustic phonons by the bandgap narrowing or alloy-induced absorption in AlScN. Surmising no relation to a particular crystalline phase, their origin may have a purely geometry nature. This is corroborated by the similar peak intensity, which contradicts the ratio of the intensity values of the Stokes and anti-Stokes bands dictated by the Bose–Einstein statistics. The similar low-frequency bands were observed in the Raman spectra of various semiconductor nanoparticles [48,49] emerging due to the confinement of the spheroidal acoustic phonons [50,51]. The obvious linear dependence of the peak position on the Sc content (Figure 4b) resembles the $\omega_{max} \propto d^{-1}$ relation, where d is a particle diameter [48] suggesting that alloying AlN with more Sc leads to an increase in the particles' average size. The existence of AlN nanoparticles can also be ruled out by the comparably large crystallites detected via the XRD study. Thus, the actual origin of the nanoparticles can only be assumed to stem from the rock-salt ScN phase. In addition, it is more energetically favourable for rock-salt phases to form the spherical nanoparticles due to the isotropy of their crystal structure.

While the fundamental softening mechanism of the one-phonon bands due to Sc alloying is understood [5,12], the means of their broadening are not sufficiently explained in the discussion of the phonon lifetime reduction [17]. From Table 1, it is clear that broadening of the one-phonon bands depends on the Sc concentration more then that of the two-phonon bands, which may be expected taking into account their probabilistic origin. We thus focus on the linewdiths, Γ, of the E_2(high) and QLO modes, for which the correlation length (i.e., mean free path) can be estimated as follows

$$l = \frac{s}{\sqrt{\Gamma \omega_0}}$$

where s is the dispersion parameter, having the same order as the acoustic velocity and ω_0 is the mode position [52]. The longitudinal acoustic velocity was estimated via the relation between the elastic constant, c_{33}, and the film density, ρ [53], using the data obtained via the DFT simulation of AlScN pseudobinary alloys [54]:

$$V_L = \sqrt{\frac{c_{33}}{\rho}}$$

The theoretically obtained values of the film density agree well with the experimental ones, corroborating their use in our analysis [55]. The derived values of the acoustic velocity and the phonon correlation length can eventually be used to assess the point defect density, l^{-3}, provided in Table 2.

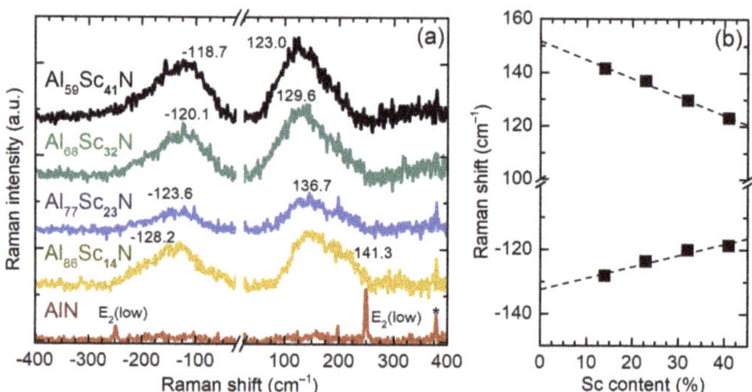

Figure 4. (**a**) Raman spectra of AlScN films at the low frequency spectral region. The spectra are stacked for clarity. The spectral bands are labelled by the positions of their maxima. The asterisk marks the band related to the Al_2O_3 substrate. (**b**) The positions of the low-frequency bands plotted as a function of the Sc concentration. The dashed lines are the linear fits.

Table 2. Material properties of the AlScN pseudobinary alloys, such as the elastic constant, c_{33}, and the mass density, ρ [54] necessary to estimate the longitudinal sound velocity, V_l, and the point defect density, $(l)^{-3}$. The mode position and linewidth are taken from Table 1.

	c_{33} [GPa]	ρ [g·cm^{-3}]	V_l [m s^{-1}]	$(l)^{-3}$ [cm^{-3}]
AlN	351.7	3.194	10,493.5	1.65×10^{19}
$Al_{0.86}Sc_{0.14}N$	292.8	3.240	9506.0	1.91×10^{20}
$Al_{0.73}Sc_{0.23}N$	253.1	3.268	8800.2	2.54×10^{20}
$Al_{0.68}Sc_{0.32}N$	211.9	3.295	8020.4	4.73×10^{20}
$Al_{0.59}Sc_{0.41}N$	169.3	3.319	7141.7	9.75×10^{20}

The defect density value is two orders of magnitude larger in the case of AlScN with $x = 0.42$ of Sc when compared to pure AlN, which underlines the contribution of the point defects in the phonon scattering. This means that every 100th lattice position can present a certain defect, which, however, is still way too low to account for all substituting Sc atoms, comprising almost one half of Al atoms in the lattice of the sample with the highest Sc concentration. This example shows a very high sensitivity of the Raman scattering process, whereby even a minute presence of defects is reflected by the width of Raman peaks. Despite our assumption of the point defects due to alloying [56], another contribution to the defect density can be considered originating from the grain boundaries. The contributions from the phonon–phonon and phonon–electron scattering are discussed in the next section. To estimate the share of the scattering on grain boundaries, it is compelling to compare the phonon coherence length values to the average grain size in the AlScN films. The average size of the crystallites (i.e., coherently diffracting crystalline domains) can be estimated using XRD symmetric scans (Figure 1) via the well-known Scherrer Equation [57] assuming the shape factor of 0.9. We employed the AlScN (0002) reflection accessing the crystallite size in the out-of-plane direction, i.e., perpendicular to the film surface. Provided the anisotropy of the film growth due to the columnar structure of the nitride films, the in-plane grain size can be retrieved from the AFM topography images via the grain analysis algorithms. The average grain size values are plotted together with the phonon coherence lengths as a function of the Sc concentration (Figure 5).

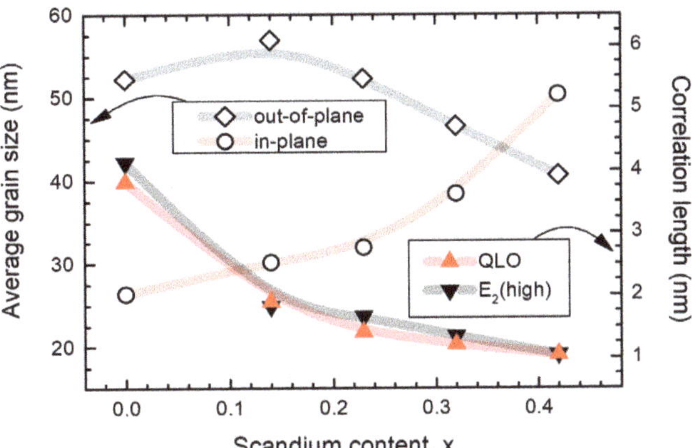

Figure 5. Average grain size and phonon correlation length as a function of the Sc content in the AlScN films. The continuous lines are given as guides to the eye.

The out-of-plane grain size of AlN grains was determined to be twice as large compared to the in-plane size, which agrees with the c-axis texture and the columnar growth of the nitride. The difference in the grain size values becomes smaller with the higher Sc content, reaching the equal magnitudes for the samples with $x > 0.3$ due to the larger grains seen via AFM (Figure 2) and the reduction of the grains in the orthogonal direction. The decrease in the out-of-plane grain dimension was observed by a similar trend in the phonon correlation length values, determined for the E_2(high) and QLO modes separately. Despite the different polarisation of the Raman modes, the phonon correlation length of the double-degenerate E_2(high) mode follows the one of the out-of-plane QLO modes, suggesting the phonons also travel less along the basal plane of the AlScN lattice. Even though this interpretation contradicts the positive trend of the in-plane size growth shown in Figure 5, the discrepancy may arise due to the worm-like surface morphology facilitated by clustering of individual grains. Thus, the similar trends in the correlation length values estimated using both Raman modes indicated the isotropic distribution of the scattering centres, suggesting point defects to be responsible for the low correlation length values. The phonon correlation length was estimated to be one order of magnitude lower than the average grain size in the out-of-plane direction overall, indicating that the lattice irregularities are ubiquitous and they become even more pronounced with the addition of Sc. We can hence speculate that the grain boundaries play a lesser role in the scattering of the optical phonons. Provided that the point defects also contribute to the hydrostatic stress in the lattice, high point defect density influences the spectral position of the mode itself, which explains the "impossibly high" residual biaxial stress in the AlN sample.

3.3. Temperature-Dependent Raman Measurements

Figure 6 shows the position and the linewidth of the E_2(high) mode obtained from the Raman spectra recorded at the elevated temperatures. The softening of the mode with the film heating was observed for all the AlScN films with different Sc contents (Figure 6a), which indicates the thermal expansion of the lattice due to the increase in the anharmonic phonon–phonon interactions, as expected. The phonon softening in this narrow temperature range is linear [58], conveniently enabling the temperature monitoring of the thin films. The linear approximations of the temperature coefficient [58] for the E_2(high) mode, shown in Figure 6a, varied with the Sc content. Compared to the one in AlN (0.0222 $K^{-1}{\cdot}cm^{-1}$ [59]), the increase in the coefficient values can be seen reaching up to 0.1157 $K^{-1}{\cdot}cm^{-1}$ for the highest Sc content. The five-fold difference suggests the dramatic changes in the anharmonic potential in AlScN corresponding to the stronger interatomic interaction [60], which contradicts the widely observed softening of the one-phonon bands as a result of bond length increase [12,15]. Additionally, the softening of the material should be accompanied by the lowering of the coefficient of the thermal expansion (CTE), but the experimental results showed the opposite behaviour [20]. The mechanism behind the ascending temperature coefficient should be related to the increase in the bound charge carrier density, realised through the alloying with Sc and the defect formation. It is instructive to see that the film with the least Sc added resulted in a temperature coefficient value (0.0143 $K^{-1}{\cdot}cm^{-1}$) lower than the one in AlN. The similarly low temperature coefficient values were found for CVD-grown AlScN with 20% of Sc [19]. Thus, we can conclude that the temperature coefficient is not only a function of the Sc concentration, but it also depends on the defect density, which may significantly alter the temperature monitoring.

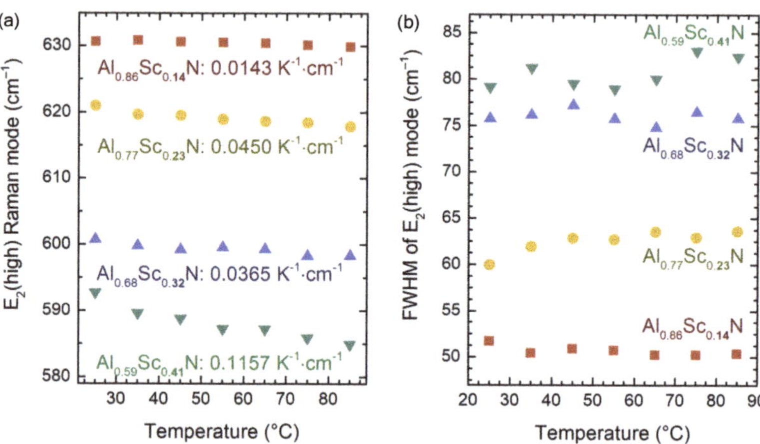

Figure 6. (**a**) E_2(high) mode position and (**b**) FWHM of AlScN films with various Sc concentration plotted as a function of temperature. The values given in (**a**) are the Raman temperature coefficients determined from the linear fit of the data.

In contrast to the clear trends in the peak position, the peak linewidth (FWHM), also known as the damping constant, weakly increases with the temperature increase (Figure 6b). Moreover, it may seem inconsistent between the AlScN films with different Sc concentrations. Despite the large peak widths, the determination error increases proportionally to the alloying degree owing to the possible contributions from the enhanced signals of the pDOS. Assuming a considerable uncertainty in the FWHM values, we conclude that the damping constant values failed to show any temperature dependence, and the fluctuation of the values was caused by a fitting uncertainty exclusively. This interpretation leads to the conclusion that the low phonon lifetimes as a result of the considerable point defect density cloak the anharmonic phonon–phonon scattering contribution usually responsible for the temperature-related peak broadening [61].

4. Conclusions

The vibrational structure of AlScN pseudobinary alloys in the thin-film form with various Sc content grown on sapphire substrates was investigated in this work by means of Raman spectroscopy. The films were confirmed to exhibit the wurtzite structure with the c-axis orientation facilitated by the columnar growth and the pebble-like surface, regardless of Sc content. Apart from the modes expected for the AlN-based wurtzite lattice, the Raman spectra of the AlScN films exhibited an enhanced phonon density of states and the long-time neglected second-order phonon modes. The one-phonon modes were used to analyse the defect density, which rose by almost two orders of magnitude in the case of AlScN with the highest Sc content examined. The comparison between the phonon correlation length and the average grain size showed the dominant role of the point defects originating largely due to the alloying. The temperature-dependent Raman measurements enabled the temperature coefficients for the E_2(high) mode to be determined, which can be used for the temperature monitoring in the AlScN-based devices. Using the near-excitation tunable notch filter, the low-frequency Raman spectra were recorded for the first time, demonstrating the Raman-active bands between 100 and 150 cm^{-1} attributed to the acoustic phonons confined in the spherical nanoparticles. We showed that Raman spectroscopy allows multifaceted material characterisation in the field of III-V compounds and their alloys, owing to the robust vibrational properties of the condensed matter. This work enables further studies of the vibrational properties of $Al_{1-x}Sc_xN$ alloys concerning the dependence of pressure and temperature in broader ranges.

Author Contributions: Conceptualisation, methodology, D.S.; resources, A.Z.; validation, formal analysis, D.S.; writing—review and editing, D.S., J.P., S.R. and A.Z.; project administration; funding acquisition, S.R. and M.M. All authors read and agreed to the published version of the manuscript.

Funding: This project was performed within the COMET Centre ASSIC Austrian Smart Systems Integration Research Center, which is funded by BMK, BMDW, and the Austrian provinces of Carinthia and Styria, within the framework of COMET—Competence Centers for Excellent Technologies. The COMET programme is run by FFG.

Data Availability Statement: The data that support the findings of this study are available from the corresponding author upon reasonable request.

Acknowledgments: The authors would like to thank Vladimir Pashchenko, Yuriy Azhniuk, and Volodymyr Dzhagan for fruitful discussions on the presented topics of this paper.

Conflicts of Interest: The authors declare no conflict of interest.

References

1. Karmann, S.; Schenk, H.P.D.; Kaiser, U.; Fissel, A.; Richter, W. Growth of columnar aluminum nitride layers on Si(111) by molecular beam epitaxy. *Mater. Sci. Eng. B* **1997**, *50*, 228–232. [CrossRef]
2. Iriarte, G.; Bjurstrom, J.; Westlinder, J.; Engelmark, F.; Katardjiev, I. Synthesis of c-axis-oriented AlN thin films on high-conducting layers: Al, Mo, Ti, TiN, and Ni. *IEEE Trans. Ultrason. Ferroelectr. Freq. Control* **2005**, *52*, 1170–1174. [CrossRef] [PubMed]
3. Akiyama, M.; Kamohara, T.; Ueno, N.; Sakamoto, M.; Kano, K.; Teshigahara, A.; Kawahara, N. Polarity inversion in aluminum nitride thin films under high sputtering power. *Appl. Phys. Lett.* **2007**, *90*, 151910. [CrossRef]
4. Akiyama, M.; Kano, K.; Teshigahara, A. Influence of growth temperature and scandium concentration on piezoelectric response of scandium aluminum nitride alloy thin films. *Appl. Phys. Lett.* **2009**, *95*, 162107. [CrossRef]
5. Tasnádi, F.; Alling, B.; Höglund, C.; Wingqvist, G.; Birch, J.; Hultman, L.; Abrikosov, I.A. Origin of the Anomalous Piezoelectric Response in Wurtzite $Sc_xAl_{1-x}N$ Alloys. *Phys. Rev. Lett.* **2010**, *104*, 137601. [CrossRef]
6. Wingqvist, G.; Tasnadi, F.; Zukauskaite, A.; Birch, J.; Arwin, H.; Hultman, L. Increased electromechanical coupling in w-$Sc_xAl_{1-x}N$. *Appl. Phys. Lett.* **2010**, *97*, 112902. [CrossRef]
7. Travaglini, G.; Marabelli, F.; Monnier, R.; Kaldis, E.; Wachter, P. Electronic structure of ScN. *Phys. Rev. B* **1986**, *34*, 3876–3882. [CrossRef]
8. Höglund, C.; Birch, J.; Alling, B.; Bareno, J.; Czigany, Z.; Persson, P.O.A.; Wingqvist, G.; Zukauskaite, A.; Hultman, L. Wurtzite structure $Sc_xAl_{1-x}N$ solid solution films grown by reactive magnetron sputter epitaxy: Structural characterization and first-principles calculations. *J. Appl. Phys.* **2010**, *107*, 123515. [CrossRef]
9. Baeumler, M.; Lu, Y.; Kurz, N.; Kirste, L.; Prescher, M.; Christoph, T.; Wagner, J.; Žukauskaitė, A.; Ambacher, O. Optical constants and band gap of wurtzite $Al_{1-x}Sc_xN/Al_2O_3$ prepared by magnetron sputter epitaxy for scandium concentrations up to $x = 0.41$. *J. Appl. Phys.* **2019**, *126*, 045715. [CrossRef]
10. Fichtner, S.; Wolff, N.; Lofink, F.; Kienle, L.; Wagner, B. AlScN: A III-V semiconductor based ferroelectric. *J. Appl. Phys.* **2019**, *125*, 114103. [CrossRef]
11. Mayrhofer, P.M.; Eisenmenger-Sittner, C.; Euchner, H.; Bittner, A.; Schmid, U. Influence of c-axis orientation and scandium concentration on infrared active modes of magnetron sputtered $Sc_xAl_{1-x}N$ thin films. *Appl. Phys. Lett.* **2013**, *103*, 251903. [CrossRef]
12. Deng, R.; Jiang, K.; Gall, D. Optical phonon modes in $Al_{1-x}Sc_xN$. *J. Appl. Phys.* **2014**, *115*, 013506. [CrossRef]
13. Zi, J.; Büscher, H.; Falter, C.; Ludwig, W.; Zhang, K.; Xie, X. Raman shifts in Si nanocrystals. *Appl. Phys. Lett.* **1996**, *69*, 200–202. [CrossRef]
14. Dzhagan, V.M.; Ya Valakh, M.; Raevskaya, A.E.; Stroyuk, A.L.; Kuchmiy, S.Y.; Zahn, D.R.T. Size effects on Raman spectra of small CdSe nanoparticles in polymer films. *Nanotechnology* **2008**, *19*, 305707. [CrossRef]

15. Hähnlein, B.; Hofmann, T.; Tonisch, K.; Pezoldt, J.; Kovac, J.; Krischok, S. Structural Analysis of Sputtered $Sc_xAl_{1-x}N$ Layers for Sensor Applications. *Key Eng. Mater.* **2020**, *865*, 13–18. [CrossRef]
16. Solonenko, D.; Lan, C.; Schmidt, C.; Stoeckel, C.; Hiller, K.; Zahn, D. Co-sputtering of $Al_{1-x}Sc_xN$ thin films on Pt(111): A characterization by Raman and IR spectroscopies. *J. Mater. Sci.* **2020**, *55*, 1–11. [CrossRef]
17. Song, Y.; Perez, C.; Esteves, G.; Lundh, J.S.; Saltonstall, C.B.; Beechem, T.E.; Yang, J.I.; Ferri, K.; Brown, J.E.; Tang, Z.; et al. Thermal Conductivity of Aluminum Scandium Nitride for 5G Mobile Applications and Beyond. *ACS Appl. Mater. Interfaces* **2021**, *13*, 19031–19041. [CrossRef]
18. Lin, W.; Cheng, W.; Wang, Y.; Sun, Y.; Zha, Q.; Zeng, C.; Cui, Q.; Zhang, B. Fabrication and characterization of Al0.8Sc0.2N piezoelectric thin films. In Proceedings of the 2021 5th IEEE Electron Devices Technology & Manufacturing Conference (EDTM), Chengdu, China, 8–11 April 2021; pp. 1–3. [CrossRef]
19. Sui, W.; Wang, H.; Lee, J.; Qamar, A.; Rais-Zadeh, M.; Feng, P.X.L. AlScN-on-SiC Thin Film Micromachined Resonant Transducers Operating in High-Temperature Environment up to 600 °C. *Adv. Funct. Mater.* **2022**, *32*, 2202204. [CrossRef]
20. Lu, Y.; Reusch, M.; Kurz, N.; Ding, A.; Christoph, T.; Prescher, M.; Kirste, L.; Ambacher, O.; Žukauskaitė, A. Elastic modulus and coefficient of thermal expansion of piezoelectric $Al_{1-x}Sc_xN$ (up to $x = 0.41$) thin films. *APL Mater.* **2018**, *6*, 076105. [CrossRef]
21. Lu, Y.; Reusch, M.; Kurz, N.; Ding, A.; Christoph, T.; Kirste, L.; Lebedev, V.; Žukauskaitė, A. Surface Morphology and Microstructure of Pulsed DC Magnetron Sputtered Piezoelectric AlN and AlScN Thin Films. *Phys. Status Solidi A* **2018**, *215*, 1700559. [CrossRef]
22. Nečas, D.; Klapetek, P. Gwyddion: An open-source software for SPM data analysis. *Open Phys.* **2012**, *10*, 181–188. [CrossRef]
23. Watson, G.H.; Daniels, W.B.; Wang, C.S. Measurements of Raman intensities and pressure dependence of phonon frequencies in sapphire. *J. Appl. Phys.* **1981**, *52*, 956–958. [CrossRef]
24. Thapa, J.; Liu, B.; Woodruff, S.D.; Chorpening, B.T.; Buric, M.P. Raman scattering in single-crystal sapphire at elevated temperatures. *Appl. Opt.* **2017**, *56*, 8598. [CrossRef] [PubMed]
25. Zukauskaite, A.; Wingqvist, G.; Palisaitis, J.; Jensen, J.; Persson, P.O.A.; Matloub, R.; Muralt, P.; Kim, Y.; Birch, J.; Hultman, L. Microstructure and dielectric properties of piezoelectric magnetron sputtered w-$Sc_xAl_{1-x}N$ thin films. *J. Appl. Phys.* **2012**, *111*, 093527. [CrossRef]
26. Moram, M.A.; Zhang, S. ScGaN and ScAlN: Emerging nitride materials. *J. Mater. Chem. A* **2014**, *2*, 6042–6050. [CrossRef]
27. Artieda, A.; Barbieri, M.; Sandu, C.S.; Muralt, P. Effect of substrate roughness on c-oriented AlN thin films. *J. Appl. Phys.* **2009**, *105*, 024504. [CrossRef]
28. Aissa, K.A.; Achour, A.; Elmazria, O.; Simon, Q.; Elhosni, M.; Boulet, P.; Robert, S.; Djouadi, M.A. AlN films deposited by dc magnetron sputtering and high power impulse magnetron sputtering for SAW applications. *J. Phys. Appl. Phys.* **2015**, *48*, 145307. [CrossRef]
29. Prakash, J.; Bose, G. Aluminum Nitride (AlN) Film Based Acoustic Devices: Material Synthesis and Device Fabrication. In *Acoustic Waves—From Microdevices to Helioseismology*; Beghi, M.G., Ed.; InTech: Vienna, Austria, 2011.
30. Thornton, J.A.; Hoffman, D. Stress-related effects in thin films. *Thin Solid Films* **1989**, *171*, 5–31. [CrossRef]
31. Tanner, S.M.; Felmetsger, V.V. Microstructure of piezoelectric AlN films deposited by AC reactive sputtering. In Proceedings of the 2009 IEEE International Ultrasonics Symposium, Rome, Italy, 20–23 September 2009; pp. 1691–1694. [CrossRef]
32. Sandu, C.S.; Parsapour, F.; Mertin, S.; Pashchenko, V.; Matloub, R.; LaGrange, T.; Heinz, B.; Muralt, P. Abnormal Grain Growth in AlScN Thin Films Induced by Complexion Formation at Crystallite Interfaces. *Phys. Status Solidi A* **2019**, *216*, 1800569. [CrossRef]
33. Haboeck, U.; Siegle, H.; Hoffmann, A.; Thomsen, C. Lattice dynamics in GaN and AlN probed with first- and second-order Raman spectroscopy. *Phys. Status Solidi C* **2003**, *0*, 1710–1731. [CrossRef]
34. Davydov, V.Y.; Kitaev, Y.E.; Goncharuk, I.N.; Smirnov, A.N.; Graul, J.; Semchinova, O.; Uffmann, D.; Smirnov, M.B.; Mirgorodsky, A.P.; Evarestov, R.A. Phonon dispersion and Raman scattering in hexagonal GaN and AlN. *Phys. Rev. B* **1998**, *58*, 12899–12907. [CrossRef]
35. Arguello, C.A.; Rousseau, D.L.; Porto, S.P.S. First-Order Raman Effect in Wurtzite-Type Crystals. *Phys. Rev.* **1969**, *181*, 1351–1363. [CrossRef]
36. Lundh, J.S.; Coleman, K.; Song, Y.; Griffin, B.A.; Esteves, G.; Douglas, E.A.; Edstrand, A.; Badescu, S.C.; Moore, E.A.; Leach, J.H.; et al. Residual stress analysis of aluminum nitride piezoelectric micromachined ultrasonic transducers using Raman spectroscopy. *J. Appl. Phys.* **2021**, *130*, 044501. [CrossRef]
37. Lu, Y. Development and Characterization of Piezoelectric AlScN-Based Alloys for Electroacoustic Applications. Ph.D. Thesis, Albert-Ludwigs-Universität Freiburg, Freiburg im Breisgau, Germany, 2019. [CrossRef]
38. Richter, W. Resonant Raman scattering in semiconductors. In *Solid-State Physics*; Dornhaus, R., Nimtz, G., Richter, W., Eds.; Springer Tracts in Modern Physics; Springer: Berlin/Heidelberg, Germany, 1976; pp. 121–272.
39. Beserman, R.; Hirlimann, C.; Balkanski, M.; Chevallier, J. Raman detection of phonon-phonon coupling in $Ga_xIn_{1-x}P$. *Solid State Commun.* **1976**, *20*, 485–488. [CrossRef]
40. Jusserand, B.; Slempkes, S. Evidence by Raman scattering on $In_{1-x}Ga_xAs_yP_{1-y}$ of the two-mode behaviour of $In_{1-x}Ga_xP$. *Solid State Commun.* **1984**, *49*, 95–98. [CrossRef]
41. Wang, X.-J.; Zhang, X.-Y. Disorder effects in $Ga_{1-x}Al_xAs$. *Solid State Commun.* **1986**, *59*, 869–872. [CrossRef]
42. Hayashi, K.; Itoh, K.; Sawaki, N.; Akasaki, I. Raman scattering in $Al_xGa_{1-x}N$ alloys. *Solid State Commun.* **1991**, *77*, 115–118. [CrossRef]

43. Ramkumar, C.; Jain, K.P.; Abbi, S.C. Raman-scattering probe of anharmonic effects due to temperature and compositional disorder in III-V binary and ternary alloy semiconductors. *Phys. Rev. B* **1996**, *53*, 13672–13681. [CrossRef]
44. Pinczuk, A.; Burstein, E. Fundamentals of Inelastic Light Scattering in Semiconductors and Insulators. In *Light Scattering in Solids*; Cardona, M., Ed.; Topics in Applied Physics; Springer: Berlin/Heidelberg, Germany, 1975; pp. 23–78. [CrossRef]
45. Beeman, D.; Tsu, R.; Thorpe, M.F. Structural information from the Raman spectrum of amorphous silicon. *Phys. Rev. B* **1985**, *32*, 874–878. [CrossRef]
46. Ferrari, A.C.; Robertson, J. Interpretation of Raman spectra of disordered and amorphous carbon. *Phys. Rev. B* **2000**, *61*, 14095–14107. [CrossRef]
47. Paudel, T.R.; Lambrecht, W.R.L. Calculated phonon band structure and density of states and interpretation of the Raman spectrum in rocksalt ScN. *Phys. Rev. B* **2009**, *79*, 085205. [CrossRef]
48. Fujii, M.; Kanzawa, Y.; Hayashi, S.; Yamamoto, K. Raman scattering from acoustic phonons confined in Si nanocrystals. *Phys. Rev. B* **1996**, *54*, R8373–R8376. [CrossRef]
49. Ikezawa, M.; Okuno, T.; Masumoto, Y.; Lipovskii, A.A. Complementary detection of confined acoustic phonons in quantum dots by coherent phonon measurement and Raman scattering. *Phys. Rev. B* **2001**, *64*, 201315. [CrossRef]
50. Lamb, H. On the Vibrations of an Elastic Sphere. *Proc. Lond. Math. Soc.* **1881**, *s1-13*, 189–212. [CrossRef]
51. Montagna, M.; Dusi, R. Raman scattering from small spherical particles. *Phys. Rev. B* **1995**, *52*, 10080–10089. [CrossRef]
52. Lughi, V.; Clarke, D.R. Defect and stress characterization of AlN films by Raman spectroscopy. *Appl. Phys. Lett.* **2006**, *89*, 241911. [CrossRef]
53. Parsapour, F.; Pashchenko, V.; Kurz, N.; Sandu, C.S.; LaGrange, T.; Yamashita, K.; Lebedev, V.; Muralt, P. Material Parameter Extraction for Complex AlScN Thin Film Using Dual Mode Resonators in Combination with Advanced Microstructural Analysis and Finite Element Modeling. *Adv. Electron. Mater.* **2019**, *5*, 1800776. [CrossRef]
54. Urban, D.F.; Ambacher, O.; Elsässer, C. First-principles calculation of electroacoustic properties of wurtzite (Al,Sc)N. *Phys. Rev. B* **2021**, *103*, 115204. [CrossRef]
55. Kurz, N.; Ding, A.; Urban, D.F.; Lu, Y.; Kirste, L.; Feil, N.M.; Žukauskaitė, A.; Ambacher, O. Experimental determination of the electro-acoustic properties of thin film AlScN using surface acoustic wave resonators. *J. Appl. Phys.* **2019**, *126*, 075106. [CrossRef]
56. Gurunathan, R.; Hanus, R.; Jeffrey Snyder, G. Alloy scattering of phonons. *Mater. Horizons* **2020**, *7*, 1452–1456. [CrossRef]
57. Scherrer, P. Nachr Ges wiss goettingen. *Math. Phys.* **1918**, *2*, 98–100.
58. Cui, J.B.; Amtmann, K.; Ristein, J.; Ley, L. Noncontact temperature measurements of diamond by Raman scattering spectroscopy. *J. Appl. Phys.* **1998**, *83*, 7929–7933. [CrossRef]
59. Hayes, J.M.; Kuball, M.; Shi, Y.; Edgar, J.H. Temperature Dependence of the Phonons of Bulk AlN. *Jpn. J. Appl. Phys.* **2000**, *39*, L710–L712. [CrossRef]
60. Solonenko, D.; Gordan, O.D.; Lay, G.L.; Şahin, H.; Cahangirov, S.; Zahn, D.R.T.; Vogt, P. 2D vibrational properties of epitaxial silicene on Ag(111). *2D Mater.* **2016**, *4*, 015008. [CrossRef]
61. Balkanski, M.; Wallis, R.F.; Haro, E. Anharmonic effects in light scattering due to optical phonons in silicon. *Phys. Rev. B* **1983**, *28*, 1928–1934. [CrossRef]

 *micromachines*

Article

$Al_{1-x}Sc_xN$ Thin Films at High Temperatures: Sc-Dependent Instability and Anomalous Thermal Expansion

Niklas Wolff [1,*], Md Redwanul Islam [1], Lutz Kirste [2], Simon Fichtner [1,3], Fabian Lofink [3], Agnė Žukauskaitė [2,*,†] and Lorenz Kienle [1,*]

1 Department of Material Science, Faculty of Engineering, Kiel University, Kaiserstr. 2, D-24143 Kiel, Germany
2 Fraunhofer Institute for Applied Solid State Physics, IAF, Tullastr. 72, D-79108 Freiburg, Germany
3 Fraunhofer Institute for Silicon Technology ISIT, Fraunhoferstr. 1, D-25524 Itzehoe, Germany
* Correspondence: niwo@tf.uni-kiel.de (N.W.); agne.zukauskaite@fep.fraunhofer.de (A.Ž.); lk@tf.uni-kiel.de (L.K.); Tel.: +49-431-880-6179 (N.W.); +49-351-2586-167 (A.Ž.); +49-431-880-6176 (L.K.)
† Current address: Fraunhofer Institute for Organic Electronics, Electron Beam and Plasma Technology FEP, Winterbergstr. 28, D-01277 Dresden, Germany.

Abstract: Ferroelectric thin films of wurtzite-type aluminum scandium nitride ($Al_{1-x}Sc_xN$) are promising candidates for non-volatile memory applications and high-temperature sensors due to their outstanding functional and thermal stability exceeding most other ferroelectric thin film materials. In this work, the thermal expansion along with the temperature stability and its interrelated effects have been investigated for $Al_{1-x}Sc_xN$ thin films on sapphire Al_2O_3(0001) with Sc concentrations x (x = 0, 0.09, 0.23, 0.32, 0.40) using in situ X-ray diffraction analyses up to 1100 °C. The selected $Al_{1-x}Sc_xN$ thin films were grown with epitaxial and fiber textured microstructures of high crystal quality, dependent on the choice of growth template, e.g., epitaxial on Al_2O_3(0001) and fiber texture on Mo(110)/AlN(0001)/Si(100). The presented studies expose an anomalous regime of thermal expansion at high temperatures >~600 °C, which is described as an isotropic expansion of a and c lattice parameters during annealing. The collected high-temperature data suggest differentiation of the observed thermal expansion behavior into defect-coupled intrinsic and oxygen-impurity-coupled extrinsic contributions. In our hypothesis, intrinsic effects are denoted to the thermal activation, migration and curing of defect structures in the material, whereas extrinsic effects describe the interaction of available oxygen species with these activated defect structures. Their interaction is the dominant process at high temperatures >800 °C resulting in the stabilization of larger modifications of the unit cell parameters than under exclusion of oxygen. The described phenomena are relevant for manufacturing and operation of new $Al_{1-x}Sc_xN$-based devices, e.g., in the fields of high-temperature resistant memory or power electronic applications.

Keywords: aluminum scandium nitride; thermal stability; structure analysis; X-ray diffraction; ferroelectrics

Citation: Wolff, N.; Islam, M.R.; Kirste, L.; Fichtner, S.; Lofink, F.; Žukauskaitė, A.; Kienle, L. $Al_{1-x}Sc_xN$ Thin Films at High Temperatures: Sc-Dependent Instability and Anomalous Thermal Expansion. *Micromachines* **2022**, *13*, 1282. https://doi.org/10.3390/mi13081282

Academic Editor: Niall Tait

Received: 22 June 2022
Accepted: 4 August 2022
Published: 8 August 2022

1. Introduction

Thin films of scandium-substituted aluminum nitride ($Al_{1-x}Sc_xN$ or 'AlScN') are a pioneering new class of ferroelectric materials with wurtzite-type structure and unidirectional polarization reversal resulting in square-like hysteresis loops [1–5]. Their ultra-stable remnant polarization and high coercive fields promise ferroelectric field-effect transistor-based non-volatile memory devices maintaining large memory windows combined with high access speed, high endurance and low energy consumption [6]. In this respect, the integration of AlScN on Si has been demonstrated recently for ferroelectric field-effect transistors as well as GaN technology-based structures, e.g., for high electron mobility transistors [7–10] with integrated memory functions. Further, thin films of AlScN are potential candidates for high-temperature actuation and sensing applications in harsh environments (T > 500 °C) or for high-temperature non-volatile memory due to remarkable temperature stability of the wurtzite-type structure and its piezo/ferroelectric properties at least up to 1100 °C [11].

In contrast, many conventional oxide-based piezoelectric and especially ferroelectric thin film materials such as $Pb_{1-x}Zr_xTiO_3$ or $BaTiO_3$ are limited in their usable temperature range either by phase transitions (relatively low Curie temperatures), which describe the transition from the pyroelectric and thus ferroelectric state to a paraelectric state, or chemical decomposition [12,13]. Numerous detailed studies on temperature-dependent physical properties, e.g., the change in coercive fields [14,15], the pyroelectric coefficient [16], thermo-electro-acoustic coupling and thermal expansion [17–19], have been conducted to date, but only moderate temperatures of max. 400 °C were investigated. Recently, we also reported on in situ and ex situ high-temperature investigations, addressing the structural stability of $Al_{1-x}Sc_xN$ (x~0.3) thin films [11,20]. However, depending on the annealing conditions, thin film stack and deposition conditions, both degradation of crystal quality and improvement after annealing were observed. Thereafter, the overall picture of the thermal stability of AlScN thin films in the high-temperature regime remains diffuse and a discussion of the influence of the Sc content is still missing.

Structurally, the substitution of Al atoms by Sc atoms into the AlN_4 tetrahedra distorts the (Al, Sc)-N bond lengths and bond angles of the three bonds forming the basal plane of the new tetrahedral MN_4 (M = Al, Sc) unit. As a result, the in-plane lattice parameter a and out-of-plane lattice parameter c change anisotropically with increasing concentration of Sc [19]. Statistical calculations on supercell models support this relationship showing that Sc increases disorder and symmetry breaking, establishing a tilt of the MN_4 tetrahedra [21]. In consequence, a pronounced elastic softening along the c-axis is observed [17] which couples to the evolution of the internal displacement parameter u of the cation–anion distance described by the product uc. The compromise between the ideal tetrahedral coordination of Al ($u = 3/8$) and the favored octahedral coordination of Sc ($u = 1/2$) results in the displacement of Sc in the tetrahedra. With increasing Sc concentration, this displacement approaches a theoretical layered hexagonal phase with $u = 0.5$ promoting phase transition [21]. The flattened energy landscape with increasing Sc concentration [22] also improves the materials' piezoelectric response up to d_{33}~27 pC/N at $x = 0.43$ [23] and promotes ferroelectric switching in high electric fields [1]. In conclusion, the substitution of Al by Sc in AlN in tetrahedral positions destabilizes the wurtzite-type crystal structure, resulting in a phase transformation above $x > 0.46$ to a non-polar cubic AlScN exhibiting rocksalt-type structure with octahedral coordination [24,25].

With this contribution, we extend our previous work on $Al_{0.73}Sc_{0.27}N$ thin films on templates of Mo(110)/AlN(0001)/Si(001) [11] and match our new results to high-temperature data of $Al_{1-x}Sc_xN(0001)/Al_2O_3(0001)$ to provide a joint perspective on AlScN films subjected to the high-temperature regime. In the latter systems, the Sc content was varied from Sc $x = 0.0$ to 0.40 to study Sc content-dependent thermal expansion behavior and high-temperature stability. The discussed results are based on the combination of X-ray diffraction techniques and in situ annealing experiments. Our measurements reveal the increasing structural destabilization of AlScN with increasing Sc content. Further, the demonstration of thermally activated effects is described, which drive an unexpected volume expansion at intermediate and high temperatures exceeding 550 °C. The high-temperature branch of thermal expansion is separated into intrinsic and extrinsic contributions related to oxygen impurities resulting in an irreversible change in the lattice parameters after annealing. Our results could have implications for the integration of AlScN thin films into metal–ferroelectric–metal capacitors for actuator, sensor or computing structures designed for high-temperature operation exceeding 500 °C up to the materials' Curie temperature of >1100 °C.

2. Materials and Methods

2.1. Thin Film Samples

Three different sets of $Al_{1-x}Sc_xN$ thin film samples summarized in Table 1 were observed:

Table 1. Summary of all thin film samples investigated in this study.

$Al_{1-x}Sc_xN(0001)$	#1	#2	#3
Template	Mo(110)/AlN(0001)/Si(001)	$Al_2O_3(0001)$	Mo(110)/AlN(0001)/$Al_2O_3(0001)$
Sc concentration x	0.27	0, 0.09, 0.23, 0.32, 0.40	0.40
Thickness	400 nm	1000 nm	400 nm
Microstructure	Fiber textured	Epitaxial with columnar growth	Epitaxial with columnar growth
XRC (FWHM)	1.6°	0.9–1.6°	0.7
SiN_x passivation	As specified	No	Yes

(#1) Thin films of $Al_{0.73}Sc_{0.27}N$ (thickness 400 nm) were deposited on high-temperature-stable metal electrode Mo(110)/AlN(0001)/Si(001) templates using an Oerlikon MSQ200 multi-source pulsed-DC sputter chamber and previously reported processes [26,27]. These films show a typical highly *c*-axis-oriented fiber texture (the 0002 XRD rocking curve (XRC) full-width at half-maxima (FWHM) is ~1.6°) with a small number of misoriented grains [26]. Surface capping of the films with a 100 nm layer of amorphous SiN_x was applied to protect the film from oxidation during high-temperature treatment.

(#2) Epitaxial 1 μm thick films of $Al_{1-x}Sc_xN$ (x = 0, 0.09, 0.23, 0.32, 0.40) with *c*-axis-oriented columnar growth were deposited onto sapphire $Al_2O_3(0001)$ wafer substrates (diameter: 100 × 100 mm) by reactive pulsed-DC magnetron sputter epitaxy (MSE). For the depositions, an Evatec cluster sputter tool with a base pressure of ~5 × 10^{-6} Pa in a co-sputtering configuration of 99.9995% pure Al and 99.99% pure Sc targets (diameter: 100 mm) was used running a process described in previous studies [17,28]. The FWHMs of XRC scans for the AlScN 0002-reflection are in the range of 0.9 and 1.6° for $x = 0$ and $x = 0.40$, respectively.

(#3) Epitaxial thin films of $Al_{0.6}Sc_{0.4}N$ (thickness of 400 nm) with *c*-axis-oriented columnar growth on a template of Mo(110)/AlN(0001)/$Al_2O_3(0001)$. The films were deposited according to the procedure described in [29] and capped with a 100 nm thick layer of SiN_x for passivation during annealing experiments.

2.2. X-ray Diffraction Experiments

The temperature-dependent microstructural changes in all thin film samples were investigated via XRD experiments on as-deposited samples as well as during and after in situ experiments. The structural properties of the as-grown epitaxial films (#2) were analyzed on a PANalytical X'Pert Pro MRD diffractometer equipped with a multi-layer mirror and a 2 × Ge 220 monochromator, providing Cu-Kα_1 radiation (λ = 154.06 pm). XRD experiments on the sample sets #1 and #3 were conducted on a Rigaku SmartLab diffractometer operated under similar conditions. To characterize the as-deposited and annealed samples, phase analyses and mosaic tilt analyses were performed by recording $2\theta/\theta$-scans and rocking curve ω-scans of the AlScN 0002-reflection.

To study effects of high-temperature annealing in (#2) $Al_{1-x}Sc_xN/Al_2O_3$ films as a function of Sc concentration, we performed in situ XRD measurements in the temperature range of 30–1000 °C. The experiments were conducted at a base pressure of ~6 × 10^{-3} mbar achieved inside a graphite dome heating furnace (Anton Paar, Ostfildern-Scharnhausen, GermanyDHS 1100 Domed Hot Stage) placed in a PANalytical Empyrean diffractometer equipped with a PIXcel hybrid detector (Malvern Panbalytical, Kassel, Germany). $2\theta/\theta$ and $\omega/2\theta$ diffraction scans were recorded after 20 min hold time at selected temperatures for approximately 60 min of dwell time. The heating rate was set to 150 °C/min. Lattice parameters were estimated from scans of the symmetrical 0004-reflection and asymmetric 10–15-reflection with shallow (−) and steep (+) angles of incidence ω, similar to the technique described by Herres et al. [30]. For the analysis of sample set #1 and #3, simple $2\theta/\theta$-scans were conducted during in situ heating experiments under similar conditions using an AntonPaar (DHS 1100) hot stage placed on the Rigaku diffractometer and as described in [11].

In addition to the in situ examinations, we also conducted an ex situ annealing experiment to compare the high-temperature behavior of samples exposed to and protected from oxygen-contaminated environments during high-temperature annealing. Here, a $Al_{0.73}Sc_{0.27}N/Mo(110)/AlN(0001)/Si(001)$ sample was placed inside a self-built tube furnace keeping a quartz tube under vacuum ($3–4 \times 10^{-7}$ mbar). Initially, the quartz tube was annealed without the sample to 800 °C together with Ti getter material to reduce the amount of adsorbed oxygen on the tube walls. The actual annealing experiment was started at a base pressure of 1.2×10^{-6} mbar and the temperature was increased to 800 °C in steps of 150 °C and cooled down to room temperature in a single step.

3. Results

The Results section is structured in two parts. In this first part, in situ annealing experiments on the fiber textured $Al_{0.73}Sc_{0.27}N/Mo(110)/AlN(0001)/Si(001)$ (sample set #1) thin films are discussed with respect to intrinsic and extrinsic effects of observed anomalous thermal expansion. In the second part, the effect size of the thermal expansion behavior and the microstructural changes to epitaxial thin film samples of $Al_{1-x}Sc_xN/Al_2O_3(0001)$ are analyzed depending on Sc concentration ($0 < x < 0.40$) (sample set #2). Structural data recorded before and after annealing are discussed and data recorded during annealing will be compared. Eventually, we conclude on our hypotheses considering data from sample #3.

3.1. *Part A: Intrinsic and Extrinsic Effects of Anomalous Thermal Expansion in AlScN Thin Films*

The exceptional robustness of the wurtzite-type structure and its ferroelectric polarization switching have been recently demonstrated for 400 nm $Al_{0.73}Sc_{0.27}N/Mo(110)/AlN(0001)/Si(001)$ thin films during and after high-temperature treatment up to 1100 °C [11]. However, the experimental data suggested a minor degradation of the crystalline quality during the process, which is illustrated by highly comparable types of data collected on $Al_{0.73}Sc_{0.27}N$ thin films from the identical wafer as presented in Figure 1. The evolution with temperature of the $Al_{0.73}Sc_{0.27}N$ 0002-reflection profile centered at 2θ~36° shown in Figure 1a displays the decrease in the maximum diffracted intensity of the $Al_{0.73}Sc_{0.27}N$ component and the development of a hump as a shoulder at lower diffraction angles of the 0002-reflection. These observations indicate the reduction in the sizes of coherently scattering domains of the $Al_{0.73}Sc_{0.27}N$ phase and the formation of a top oxide layer of $(Al,Sc)(N,O)_x$, which is supported by energy-dispersive X-ray spectroscopy measurements on a scanning electron microscope before and after annealing (not shown). No further peaks indicating oxide formation were observed in the temperature-dependent $2\theta/\theta$-scans (30–100 °C) (not shown). Further, the temperature dependent 0002-reflection profiles in Figure 1a reveal the 0002-reflection of the AlN seed layer at a temperature of 1000 °C (see arrows) at higher diffraction angle with respect to the intensity maximum. This could be explained by differences in thermal expansion coefficients of AlN and AlScN, which are known for temperatures up to 400 °C [17]. However, when treating AlScN films at high temperatures, a transition in thermal expansion is observed. The comparisons between the thermal shifts in the 0002-reflection maxima for $Al_{0.73}Sc_{0.27}N$ and AlN thin films and the Mo(110)-reflection of the Mo layer underneath $Al_{0.73}Sc_{0.27}N$ are presented in Figure 1b. Here, the relative change in the lattice spacings $\Delta d[T]/d_0$ reveals strong non-linearity in the thermal lattice expansion of the $Al_{0.73}Sc_{0.27}N$ film. The curve for the $Al_{0.73}Sc_{0.27}N$ film (green curve in Figure 1b) shows a non-linearity in thermal expansion changing from a linear low-temperature branch to a high-temperature branch at a transition temperature of about T_{tr}~600 °C. The linear slope of the low-temperature branch is $(\Delta c[25\ °C–600\ °C]/c_0)/\Delta T$~$4.4 \times 10^{-6}/°C$ and increases by a factor of 3 to $(\Delta c[700\ °C–1000\ °C]/c_0)/\Delta T$~$13.7 \times 10^{-6}/°C$ in linear approximation to the high-temperature branch. Any influence of the underlayer is excluded by comparison to the relative change in the underlayer's Mo(110)-reflection which shows highly linear expansion. Further, an AlN film deposited under identical conditions was heated and shows a linear expansion of $(\Delta c[25\ °C–1100\ °C]/c_0)/\Delta T$~$3.5 \times 10^{-6}/°C$. Due to the rather rough linear approximation over the entire temperature interval, the value for AlN is somewhat smaller but still consistent with literature data of 4.2 [20–800 °C] – 4.65 [20–400 °C] × 10^{-6}/°C [17,31].

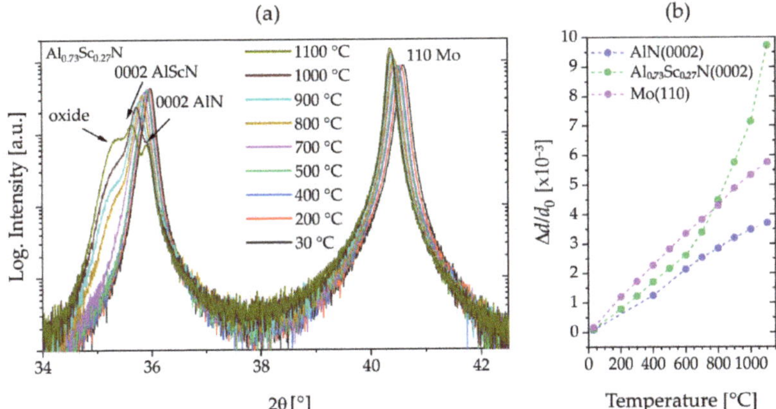

Figure 1. XRD in situ (with oxygen atmosphere) annealing experiments on $Al_{0.73}Sc_{0.27}N$ and AlN thin films grown on Mo(110)/AlN(0001)/Si(001) (sample set #1). (**a**) 2θ/θ-scan: Evolution of 0002-reflection profile of $Al_{0.73}Sc_{0.27}N$ and (**b**) relative changes in lattice spacings *d* with temperature calculated from AlN(0002) and $Al_{0.73}Sc_{0.27}N$(0002)/Mo(110)-reflections.

To our knowledge, this strongly non-linear transition has not been observed to date for AlScN thin films. In the case of many oxide materials, non-linear thermal expansion behavior can be discussed in the context of oxygen- or oxygen vacancy-related effects termed 'chemical expansion' [32]. These chemical expansion phenomena can be non-reversible or reversible in nature. For instance, non-reversibility is observed in the case of cation–oxygen networks forming in metallic glasses [33] or small structural transitions in $(Ba_{0.5}Sr_{0.5})TiO_3$ induced by a change in oxygen site occupancy [34]. Reversible phenomena are commonly observed in non-stoichiometric perovskite-based ion-conducting ceramics which show pronounced chemical expansion depending on the oxygen partial pressure and temperature [35,36].

Parallel to the strong oxidation of the AlScN film, the interaction with oxygen species and the intrinsic defect structure of the material could lead to the strong thermal expansion and irreversible changes. In this respect, the fiber textured columnar films could potentially provide a Sc- and oxygen-enriched grain boundary structure [37], which could promote pathways for atmospheric oxygen species into the material.

These preliminary experiments suggest the presence of residual oxygen contamination of the annealing atmosphere inside the graphitic dome placed in the diffractometer. Hence, the $Al_{0.73}Sc_{0.27}N$ film was capped with a 100 nm thick SiN_x layer to protect the film surface from oxidation-dependent effects, allowing us to investigate the purely intrinsic material contribution to the transition in thermal expansion behavior. With this experimental design, new in situ annealing experiments were conducted up to 1100 °C. The evolution of reflection profiles during the first and second annealing cycle and the corresponding $\Delta c(T)/c_0$ plots are shown in Figure 2. The reflection profiles depicted in Figure 2a exhibit neither an oxide hump, nor a strong decrease in the reflection intensity, which is a sign of improved structural stability due to avoiding surface oxidation. Instead, a negligible XRD reflection broadening is observed by the increase in background intensity at higher diffraction angles at 550 °C, which could be due to changes in the average crystallite size, accumulation of defects and local lattice strains. Upon further annealing, no further changes in the reflection profile (Figure 2b) are observed for a second temperature cycle, indicating completed activation of any intrinsic processes until the applied temperature of 1100 °C.

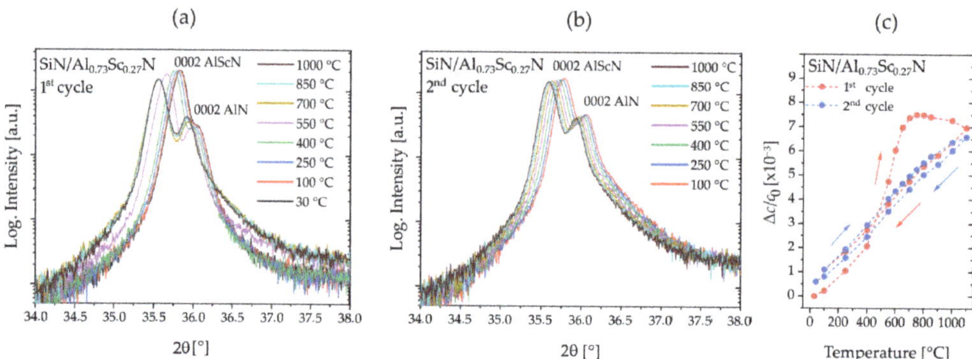

Figure 2. XRD in situ (with oxygen atmosphere) annealing experiments. Observation of anomalous intrinsic lattice expansion on oxygen-protected SiN passivated thin films of $Al_{0.73}Sc_{0.27}N(0001)/Mo(110)/AlN(0001)/Si(100)$ (sample set #1). (**a**) $2\theta/\theta$-scan: Evolution of the 0002-reflection profile during the first cycle. (**b**) $2\theta/\theta$-scan: Evolution of the 0002-reflection profile during the second cycle. (**c**) Relative change $\Delta c/c_0$ in lattice parameter with annealing temperature in two consecutive annealing cycles.

The corresponding $\Delta c(T)/c_0$ plots for the first and second full temperature cycle are displayed in Figure 2c. In the first cycle (red curve), the thermal expansion in the low-temperature regime is consistent with uncapped films with $(\Delta c[25\ ^\circ C–400\ ^\circ C]/c_0)/\Delta T{\sim}6.0 \times 10^{-6}/^\circ C$ (compare Figure 1b) and literature reference data ($6.38 \times 10^{-6}\ K^{-1}$) for $Al_{0.68}Sc_{0.32}N/Al_2O_3$ [17]. However, a new and purely intrinsic regime at intermediate temperatures is observed with $(\Delta c[450\ ^\circ C–650\ ^\circ C]/c_0)/\Delta T{\sim}22.2 \times 10^{-6}/^\circ C$ with much higher expansion and lower transition temperature $T_{tr}{\sim}450\ ^\circ C$. This value of thermal expansion is almost double compared to the uncapped film at high temperatures with $T_{tr} > 600\ ^\circ C$. In the high-temperature regime of >650 °C, the expansion slows down and reverses its sign $(\Delta c[750\ ^\circ C–1100\ ^\circ C])/c_0)/\Delta T{\sim}-1.5 \times 10^{-6}/^\circ C$. After cooling down to room temperature, linear thermal expansion over the entire temperature regime with $(\Delta c[1100\ ^\circ C–30\ ^\circ C]/c_0)/\Delta T{\sim} -6.0 \times 10^{-6}/^\circ C$ is observed and a small irreversible change in the c lattice parameter remains at $\Delta c/c_0{\sim}0.5 \times 10^{-3}$, which is much smaller than for the uncapped film. In a second temperature cycle (blue curve in Figure 2c), no anomalous thermal expansion is observed, consistent with the reflection profiles.

The comparison of both experiments suggests that oxidation effects or the interaction of material defects with oxygen species play a prominent role for high-temperature lattice expansion at >800 °C and the irreversible change in respective lattice parameters at room temperature.

This hypothesis is supported by the comparison of in situ XRD experiments with oxygen in the annealing atmosphere and ex situ XRD experiments without available oxygen in the atmosphere performed on uncapped films of $Al_{0.73}Sc_{0.27}N/Mo(110)/AlN(0001)/Si(001)$. In this study, the first sample was introduced into the in situ XRD analysis when performing two consecutive heating cycles. The second sample was placed into a quartz tube furnace which was evacuated to 10^{-7} mbar. Both samples were treated with identical temperature profiles. In the first heating cycles, a maximum temperature of 800 °C was applied, which is about the temperature at which the positive slope of the intrinsic expansion reverses, whereas a maximum of 1000 °C was used in the second in situ heating hysteresis. After the first ex situ cycle, a second temperature cycle was performed in situ to evaluate the 2θ shifts in the 0002-reflection at higher temperatures.

The reflection profiles and plots of the relative thermal expansion are summarized in Figure 3. Figure 3a shows the respective reflection profiles at selected stages during the experiment. As expected, no oxide hump is observed after the ex situ annealing (blue line), suggesting a purely intrinsic thermal expansion behavior. For the in situ annealed sample (red line), the oxide hump is observed. However, after performing the first cycles, both samples show an irreversible change in lattice parameter of about $\Delta c/c_0 \sim 1 \times 10^{-3}$ in the plots of the relative thermal expansion shown in Figure 3b. This could indicate that the observed oxidation does not have a major effect on the expansion at intermediate temperatures and that the intrinsic contribution is dominating. By stopping the first cycle at 800 °C, it is assumed that the intrinsic effects on the anomalously high positive thermal expansion have all been activated and no further reaction would be observed. Indeed, this holds true for both samples in the second cycle up to 800 °C (golden and turquoise lines). By passing the 800 °C mark, comparable thermal expansion is observed when residual oxygen is supplied by the annealing atmosphere which results in a strong oxidation and large irreversible lattice changes $\Delta c/c_0 \sim 3 \times 10^{-3}$ after annealing.

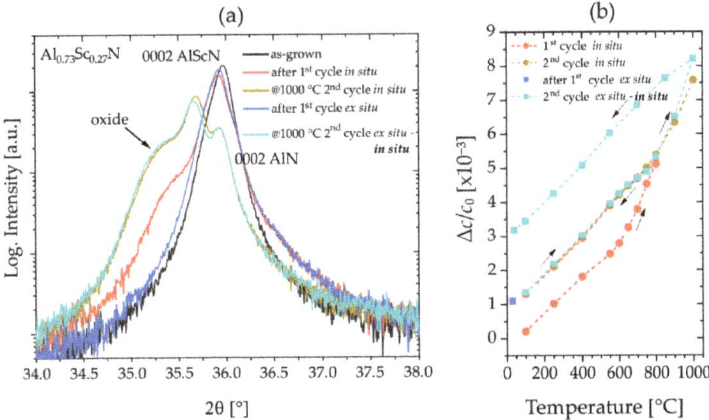

Figure 3. XRD annealing experiments on $Al_{0.73}Sc_{0.27}N$ thin films grown on Mo(110)/AlN(0001)/Si(001) (sample set #1). (**a**) 2θ/θ-scans: Evolution of 0002-reflection profile after ex situ (without oxygen atmosphere) and during in situ (with oxygen atmosphere) experiments. (**b**) Relative changes in lattice parameter c for $Al_{0.73}Sc_{0.27}N$ thin films annealed ex situ and in situ.

In summary, anomalous high thermal expansion and related irreversible lattice changes have been observed upon thermal activation by annealing of AlScN thin films. The experiments suggest an intrinsic material specific contribution activated at intermediate temperatures of >600–800 °C which is superimposed with extrinsic effects acting in parallel to oxidation of the films at temperatures exceeding 800 °C, if not protected by a surface covering layer.

3.2. Part B: Discussion of Scandium Concentration in $Al_{1-x}Sc_xN(0001)/Al_2O_3(0001)$ Thin Films

In this section, the discussion is turned towards epitaxial c-axis columnar grown thin films of 1 μm $Al_{1-x}Sc_xN$ on Al_2O_3 (sample set #2). The recorded XRD 2θ/θ-scans for 1 μm thick $Al_{1-x}Sc_xN$ ($0 < x < 0.40$) films before and after thermal annealing are shown in Figure 4a. All diffractograms of the as-grown sputtered $Al_{1-x}Sc_xN$ thin films demonstrate exclusive c-axis orientation as well as fixed in-plane orientation described by epitaxial relationships (0001)AlScN//(0001)Al_2O_3 and (10–10)lScN//(11–20)Al_2O_3, respectively [17]. After annealing in situ (with oxygen atmosphere) to 1000 °C, only the 000ℓ (ℓ = 2, 4, 6)-reflections are observed, indicating decent temperature stability of the wurtzite-type phase for all examined Sc concentrations x.

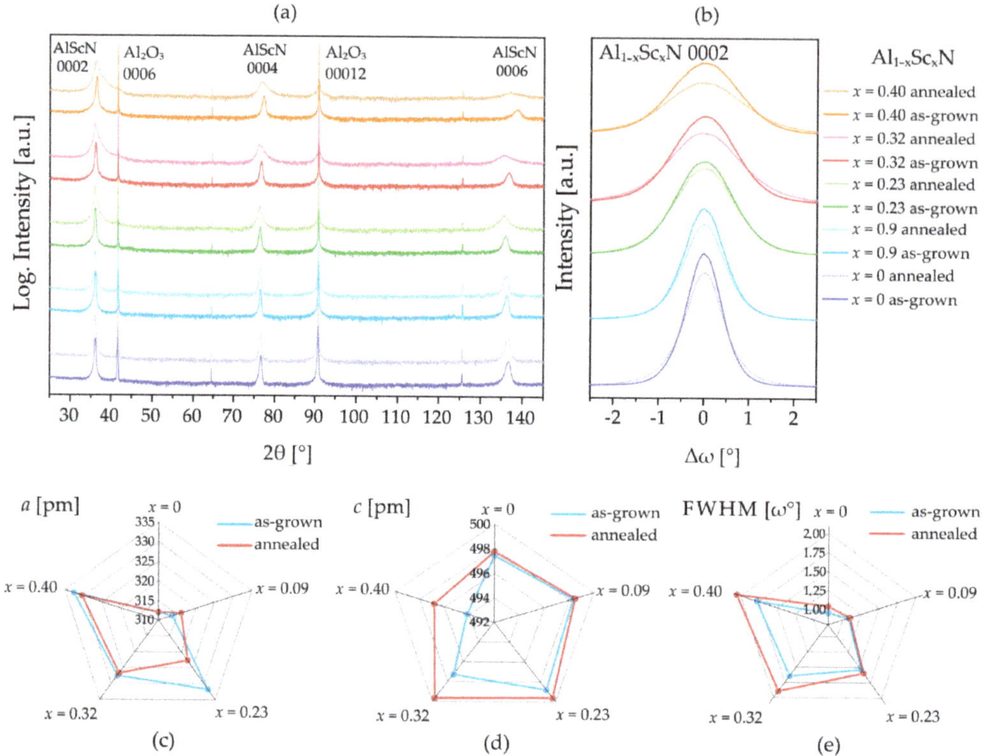

Figure 4. Analysis of $Al_{1-x}Sc_xN/Al_2O_3$ ($0 \leq x \leq 40.0$) before and after in situ thermal annealing (sample set #2). (**a**) XRD 2θ/θ diffractograms. (**b**) XRD rocking curve measurements of the 0002-reflections. Summary of changes in lattice parameters *a* (**c**) and *c* (**d**) calculated from symmetric and asymmetric measurements, ω-FWHM (**e**) calculated from the XRD rocking curves.

However, a considerable XRD reflection broadening of the 000ℓ-reflection profiles is observed after annealing for Sc concentrations $x \geq 0.23$. This is paired with a shift in the reflection position to lower 2θ values. The broadening of the 000ℓ-reflections features asymmetry, with the right tail of the Bragg reflection becoming extended relative to the left tail. No hump on the 0002-reflections is observed, which could indicate a better stability against oxidation for the epitaxial films with fewer grain boundaries. Indeed, rocking curve measurements of the 0002-reflection (Figure 4b) confirm this broadening and provide evidence for a structural degradation by the reduction in maximum diffracted intensity.

The XRD reflection broadening is mainly attributed to reduced sizes of coherently scattering domains, the accumulation of defects and local lattice strain. A shift in the diffraction angle to lower values of 2θ in consequence of annealing is related to the expansion of the *c* lattice parameter, as discussed in part A. However, the exact origin of the reflection broadening remains speculative without a structure model and Rietveld refinement. Instrumental broadening of the reflections can be neglected due to the high mosaicity of the AlScN films. A detailed analysis of the in-plane and out-of-plane lattice parameter changes using data from the symmetric 2θ/θ-scans and asymmetric ω/2θ 10–15(−) and 10–15(+)-reflection scans [30] is performed and the results are summarized in Figure 4c and Table 2. From the comparison, it is apparent that the high-temperature annealing induces irreversible changes to the lattice parameters of $Al_{1-x}Sc_xN$. These changes manifest in the reduction in the *a* parameter and the increase in the *c* parameter in combination with the degradation of the overall crystal quality (broadening of FWHM). A clear trend is visible in the magnitude of the effect which seems to scale with Sc concentration x indicating an increasing instability of high-Sc alloys at elevated temperatures. For instance, in the case of low-Sc content $Al_{0.91}Sc_{0.09}N$ films, the irreversible change in the c lattice parameter is $\Delta c/c_0$ 0.04% in contrast to high-Sc content $Al_{0.60}Sc_{0.40}N$ films showing $\Delta c/c_0$~0.5%, associated with a relative broadening of FWHM of 2.2% and 15.7%, respectively. We note that the film with x = 0.23 shows a very large unexpected change in the a parameter, but follows the general trend regarding the other parameters.

Table 2. Lattice parameters and mosaicity analyses of $Al_{1-x}Sc_xN$ ($0 \leq x \leq 0.4$) thin films by XRD before and after thermal annealing (sample set #2).

	a [pm]			c [pm]			ω-FWHM [°]		
Sample	**As-Grown**	**Annealed**	**$\Delta a/a_0$ [$\times 10^{-3}$]**	**As-Grown**	**Annealed**	**$\Delta c/c_0$ [$\times 10^{-3}$]**	**As-Grown**	**Annealed**	**$\Delta\omega/\omega_0$ [$\times 10^{-2}$]**
AlN	312.0	312.1	0.468	497.5	497.8	0.64	0.95	1.04	9.8
$Al_{0.91}Sc_{0.09}N$	313.6	316.0	7.64	498.3	498.5	0.39	1.07	1.10	2.2
$Al_{0.77}Sc_{0.23}N$	331.6	322.6	−27.13	498.8	499.6	1.66	1.53	1.59	3.7
$Al_{0.68}Sc_{0.32}N$	327.4	326.7	−2.38	497.3	499.7	4.80	1.64	1.89	14.9
$Al_{0.60}Sc_{0.40}N$	332.7	330.5	−6.61	494.2	496.8	5.33	1.78	2.06	15.7

The structural origin of the observed reflection broadening in high-Sc content films could also be related to the competition between the hexagonal wurtzite-type phase and the cubic rocksalt-type structure when approaching 46% Sc [23–25,38]. High-temperature annealing could result in local phase destabilization and formation of nanosized cubic domains in Sc-enriched regions, e.g., at defect sites or grain boundaries [37]. Such nanosized domains, as well as the migration of defects, e.g., dislocations, will lead to reflection broadening and asymmetry by diffuse scattering and the formation of low-intensity shoulders in diffraction patterns [39,40]. Typically, we would expect a phase transition to be reversible, but here the strong and non-reversible increase in lattice parameters seems to indicate irreversibility.

In further analysis of the in situ annealing experiments, the temperature-dependent changes in the reflection profiles are followed individually. In Figure 5, thin films of $Al_{0.77}Sc_{0.23}N$ and $Al_{0.60}Sc_{0.40}N$ are compared to demonstrate the difference between intermediate and high Sc contents. When comparing the reflection profiles, it is directly apparent that the Sc content influences the structural stability and the activation temperature of the observed degradation effects. The evolution of both the symmetric $2\theta/\theta$ 0002- and asymmetric $\omega/2\theta$ 10–15-reflection profiles shows a strong degradation of the initial crystalline quality. The loss of structural coherence is most pronounced in the low-intensity asymmetric reflections, which limits the precise calculation of in-plane parameters at higher temperatures and after annealing. Concerning the XRD reflection broadening, for Sc $x = 0.23$, the elevation of the background tails starts at a temperature of 850 °C, whereas for Sc $x = 0.40$ these features are already observed at 700 °C and with larger magnitude (Figure 5).

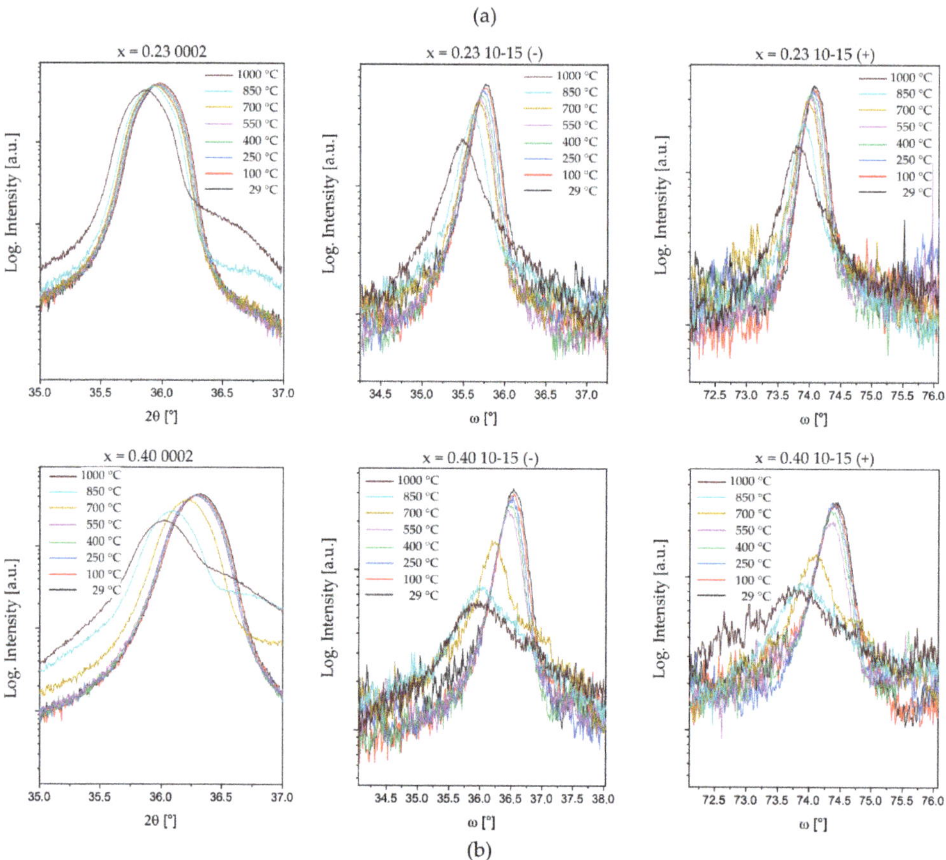

Figure 5. Temperature-dependent changes in symmetric 2θ/θ 0002- and asymmetric ω/2θ 10–15-reflection profiles for (**a**) $Al_{0.77}Sc_{0.23}N$ and (**b**) $Al_{0.60}Sc_{0.40}N$ compositions (sample set #2). 10–15-reflections are measured with shallow (−) and steep (+) angles of incidence ω, see [30] for more information on the method.

As before, the temperature-dependent changes in the reflection positions are used to calculate the relative thermal expansion of both *c* and *a* lattice parameters. The resulting thermal expansion is illustrated in Figure 6. Here, the relative changes in the *c* and *a* lattice parameters over the temperature range 25–1000 °C are presented for the investigated films in Figure 6a,b, respectively. In this analysis, the transition between the two expansion regimes is evidenced for all $Al_{1-x}Sc_xN$ compositions but not for AlN. Note that the films are not protected against available oxygen species, hence, intrinsic and extrinsic processes adding to the thermal expansion are superimposed. The transition from almost linear and low-value expansion at moderate temperatures into large values of expansion at high temperatures is observed for *a* and *c*. This expansion of the lattice is isotropic which is visible from the almost constant *c/a* ratio plotted in Figure 6c. The transition temperature T_{tr} seems to be related to the Sc content of the film, as well as the magnitude of the expansion. Roughly estimated values from the plots are T_{tr}~850 °C for $Al_{0.91}Sc_{0.09}N$ and T_{tr}~550 °C for $Al_{0.60}Sc_{0.40}N$ films as indicated by the vertical dotted lines in Figure 6b. In detail, the thermal expansion behavior in the low-temperature regime (<550 °C) is highly comparable for all Sc concentrations, in agreement with previous studies within this temperature range [17]. In strong contrast, the high-temperature regime is characterized by a manifold increase in the expansion, depending on the Sc concentration (up to ~8-fold for $Al_{0.60}Sc_{0.40}N$). After cooling back to room temperature, the lattice parameters show irreversible changes as discussed for Figure 4 and Table 2.

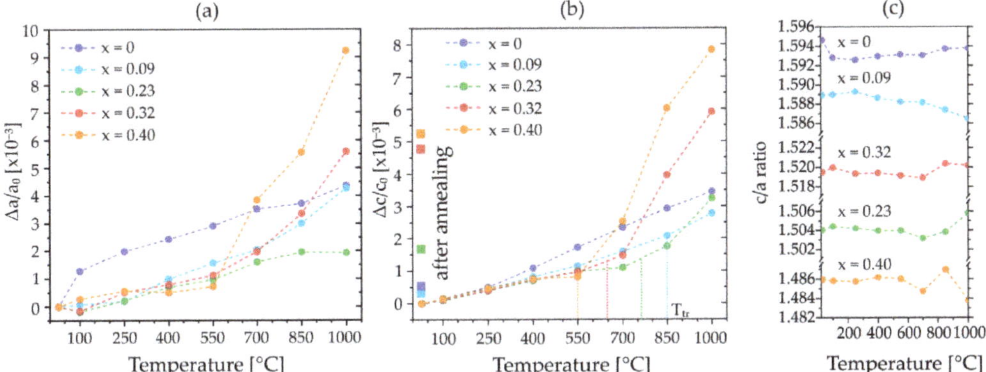

Figure 6. XRD in situ (with oxygen atmosphere) characterization of thermal evolution of the lattice parameters (sample set #2). (**a**) Relative change in the lattice parameter *a* and (**b**) lattice parameter *c* indicating a high-temperature regime of isotropic thermal expansion and estimated values of the transition temperature T_{tr}. Dashed lines serve as a guide to the eye only. Residual values of lattice expansion after annealing are shown as square data points as given in Table 2. (**c**) Calculated c/a ratios showing approximately isotropic thermal expansion for the investigated thin films across the full temperature range.

For direct comparison with fiber textured samples from set #1, XRD measurements during two temperature cycles were conducted for the $Al_{0.68}Sc_{0.32}N$ film, Figure 7. The displayed temperature cycle and reflection profiles are highly congruent to the data recorded on the fiber textured thin films with similar composition. An irreversible change of $\Delta c/c_0 \sim 3.5 \times 10^{-3}$ remains after annealing at 1000 °C (Figure 7a) and no further changes are observed in the second temperature hysteresis, congruent with the measured reflection profiles shown in Figure 7b,c.

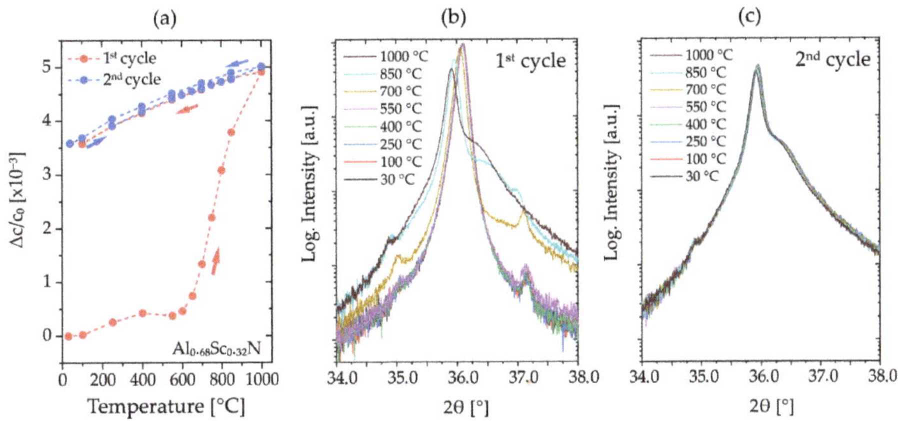

Figure 7. (**a**) XRD in situ (with oxygen atmosphere) temperature cycles in the example of $Al_{0.68}Sc_{0.32}N$ (sample set #2). (**b**) Evolution of 0002-reflection profile in the first annealing cycle. (**c**) 2θ/θ-scan: Evolution of 0002-reflection profile in the second annealing cycle.

4. Discussion

The described experiments reveal high-temperature effects related to structural degradation and irreversible anomalous non-linear thermal expansion behavior in AlScN thin films depending on Sc content and film microstructure. The non-linear thermal expansion is described by a transition from an initial expansion regime to a fast-expanding high-temperature regime which is divided into intrinsic and extrinsic contributions, dependent on the availability of oxygen. Available oxygen is believed to lead to the oxidation of the material dependent on the film microstructure (oxidation was only observed for fiber textured films with higher density of grain boundaries) but also enhances the expansion effect size at $T > 800$ °C and stabilizes the increased lattice parameters. The effect size of high-temperature expansion scales with the Sc content of the films, which could be related to a higher defect density leading to the increased destabilization and easier oxidation. A further indicator for the scaling defect density is the increasing magnitude of the peak broadening, which was discussed as an indicator for defect movement as well. Besides dislocations or grain boundaries, point defects such as nitrogen vacancies V_N are one major type of defect in AlN and AlScN thin films. Previous studies by Harris et al. have demonstrated that the wurtzite-type structure of AlN can incorporate up to 0.75 at% oxygen under thermal equilibrium conditions by substitution of N (O_N) associated with the formation of Al vacancies (V_{Al}) [41]. Further insights into the structure relation of AlN films and oxygen are provided by DFT modeling by Gasparotto et al. [42] showing that oxygen can have significant influence on the lattice parameters.

The discussed examples show potential interrelations of oxygen-induced defect structures and changes to the lattice parameters. As observed in this work on AlScN thin films, such phenomena could provide valid explanation for the discussed intrinsic and extrinsic contributions to the anomalous thermal expansion at high temperatures. In addition, Sc has strong affinity for oxygen [43–45], and the intrinsic oxygen contamination of AlScN thin films is supposed to scale with the Sc content as well. In this respect, Sc- and oxygen-enriched grain boundaries were already evidenced [37] and the photoluminescence emission of low-Sc films (x = 0.05) was already revealed to be dominated by oxygen defects [46]. Further, modeling of the point defects in rocksalt AlScN resulted in preferable defect complexes of substitutional and interstitial oxygen ($O_N + O_i$) [47]. The observed oxygen content-dependent destabilization of the lattice in Al-O-N is also consistent with the increasing degradation of the crystal quality with Sc concentration and the formation of oxide phases.

To support the above discussion, we designed a third sample of a SiN_x-capped AlScN film with high concentration of Sc $x = 0.4$ grown epitaxially on a sapphire substrate, i.e., $SiN_x/Al_{0.6}Sc_{0.4}N(0001)$/epi-Mo(110)/AlN(0001)/$Al_2O_3(0001)$. For such systems, previous work [29] demonstrated that the special epi-Mo(110) electrode provides a growth template for AlScN with one-dimensional single crystalline properties and low defect density. The following assumptions are made: First, the SiN barrier should protect the thin film from oxidation and interaction of oxygen species with the internal defect structure at high temperatures. Second, by choosing a high amount of Sc, any effect magnitude is expected to be large. Third, this AlScN thin film exhibits high crystalline quality with an FWHM~0.7° (from XRC) indicating highly oriented columnar grains originating from epitaxial growth and a low defect density. Hence, the effect of any intrinsic, defect-driven anomalous expansion should be strongly limited. Indeed, the recorded thermal expansion during the first temperature cycle shown in Figure 8 shows almost linear expansion featuring no anomalous behavior. This observation, although not expected so clearly, provides supporting evidence for the above hypothesis.

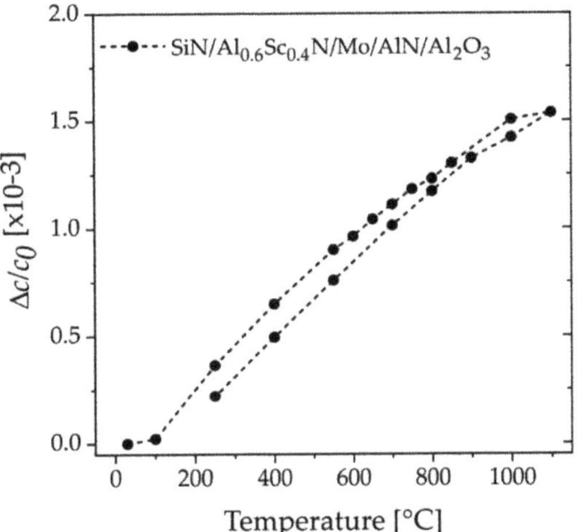

Figure 8. *XRD* in situ (with oxygen atmosphere) experiment. Relative change in lattice parameter *c* with temperature for epitaxial $SiN_x/Al_{0.6}Sc_{0.4}N(0001)$ on a template of epi-$Mo(110)/AlN(0001)/Al_2O_3(0001)$ (sample set #3).

In summary, the irreversible changes in the lattice parameters of $Al_{1-x}Sc_xN/Al_2O_3$ when exposed to atmospheric oxygen present in low-vacuum conditions are comparable to fiber textured systems. The estimated transition temperatures between the low-temperature expansion and high-temperature expansion regimes and the magnitude of the superimposed intrinsic and extrinsic contributions show a clear trend with the Sc content of the films. With increasing Sc content, the transition temperature is decreased and the effect sizes of thermal expansion and degradation of crystallinity are increased. The anomalous thermal expansion can only be activated by temperature once. Further, the microstructure and possibly the defect structure seem to impact the stability to oxidation of the films, which seems not to be interrelated with the extrinsic part of high-temperature thermal expansion. However, at very high temperatures >800 °C, hypothetically, oxygen species could diffuse into the material via defects and grain boundaries and interact with thermally activated defect sites to further drive the lattice expansion and act as obstacles to stabilize the expanded lattice. Scandium is known to destabilize the AlN lattice by its larger cation size and structural preference for octahedral coordination. Hence, increasing its concentration in AlN could induce higher defect densities, which are seemingly possible to activate at lower temperatures by the flattened energy landscape.

5. Conclusions

The thermal stability and temperature-induced effects of AlScN thin film samples with different microstructures based on the growth template were investigated by in situ XRD. The degradation of the crystalline quality and a remnant lattice expansion in the *c*-direction of up to 0.5% for epitaxial $Al_{0.60}Sc_{0.40}N(0001)/Al_2O_3(0001)$ were observed as a function of Sc concentration. There is first evidence that the remnant expansion is related to the activation of intrinsic defects and the films' oxygen affinity at elevated temperatures, which are accounted for as intrinsic and extrinsic sources of anomalous thermal expansion. The understanding of the exact details of these phenomena provides opportunity for further investigations on the exact type of defect structures using more advanced methods, such as positron annihilation spectroscopy [48–50]. The detailed understanding of the intrinsic defect structures and temperature activation effects is of high importance for the integration of sputtered AlScN layers into sandwich structures, e.g., for ferroelectric field-effect transistor memory capacitors or ferroelectric tunnel junctions. That is especially relevant with respect to downscaling of the layer thickness and high-temperature operation, where oxidation of the functional layer has to be prohibited and an anomalously high expansion of the AlScN crystal lattice puts increased stresses on any neighboring crystalline layers to avoid device failure via crack formation or delamination. In conclusion, our investigation emphasizes the requirement of a low density of material defects in AlScN thin films when operating at high temperatures and the benefit of integration with a protecting top layer, e.g., a temperature-resistant electrode such as Mo or NbN.

Author Contributions: Conceptualization, N.W. and A.Ž.; methodology, M.R.I. and L.K. (Lutz Kirste); validation of experiments, A.Ž., S.F., L.K. (Lutz Kirste) and L.K. (Lorenz Kienle); formal analysis, N.W., M.R.I. and L.K. (Lutz Kirste); investigation, N.W., M.R.I. and L.K. (Lutz Kirste); data curation, N.W., M.R.I. and L.K. (Lutz Kirste); writing—original draft preparation, N.W.; writing—review and editing, A.Ž., L.K. (Lutz Kirste), S.F. and L.K. (Lorenz Kienle); visualization, N.W.; supervision, A.Ž. and S.F.; project administration, F.L. and L.K. (Lorenz Kienle); funding acquisition, S.F., A.Ž. and L.K. (Lorenz Kienle). All authors have read and agreed to the published version of the manuscript.

Funding: This work was supported by the project 'ForMikro-SALSA' (grant no. 16ES1053) from the Federal Ministry of Education and Research (BMBF) and the DFG under the scheme of the Collaborative Research Center (CRC) 1261/A6.

Data Availability Statement: Original XRD data are available from the corresponding authors upon reasonable request.

Acknowledgments: The authors would like to thank Yuan Lu for his contribution to providing AlScN/Al_2O_3, Fredrik Eriksson for his contribution to performing in situ XRD annealing experiments on AlScN/Al_2O_3 samples and Maximilian Kessel for enthusiastic discussions on the presented topics of this paper. Financial support by the Fraunhofer Society is gratefully acknowledged.

Conflicts of Interest: The authors declare no conflict of interest.

References

1. Fichtner, S.; Wolff, N.; Lofink, F.; Kienle, L.; Wagner, B. AlScN: A III-V Semiconductor Based Ferroelectric. *J. Appl. Phys.* **2019**, *125*, 114103. [CrossRef]
2. Uehara, M.; Mizutani, R.; Yasuoka, S.; Shiraishi, T.; Shimizu, T.; Yamada, H.; Akiyama, M.; Funakubo, H. Demonstration of Ferroelectricity in ScGaN Thin Film Using Sputtering Method. *Appl. Phys. Lett.* **2021**, *119*, 172901. [CrossRef]
3. Wang, D.; Wang, P.; Wang, B.; Mi, Z. Fully Epitaxial Ferroelectric ScGaN Grown on GaN by Molecular Beam Epitaxy. *Appl. Phys. Lett.* **2021**, *119*, 111902. [CrossRef]
4. Ferri, K.; Bachu, S.; Zhu, W.; Imperatore, M.; Hayden, J.; Alem, N.; Giebink, N.; Trolier-McKinstry, S.; Maria, J.-P. Ferroelectrics Everywhere: Ferroelectricity in Magnesium Substituted Zinc Oxide Thin Films. *J. Appl. Phys.* **2021**, *130*, 044101. [CrossRef]
5. Hayden, J.; Hossain, M.D.; Xiong, Y.; Ferri, K.; Zhu, W.; Imperatore, M.V.; Giebink, N.; Trolier-McKinstry, S.; Dabo, I.; Maria, J.-P. Ferroelectricity in Boron-Substituted Aluminum Nitride Thin Films. *Phys. Rev. Mater.* **2021**, *5*, 044412. [CrossRef]
6. Liu, X.; Wang, D.; Kim, K.-H.; Katti, K.; Zheng, J.; Musavigharavi, P.; Miao, J.; Stach, E.A.; Olsson, R.H.; Jariwala, D. Post-CMOS Compatible Aluminum Scandium Nitride/2D Channel Ferroelectric Field-Effect-Transistor Memory. *Nano Lett.* **2021**, *21*, 3753–3761. [CrossRef]
7. Wang, P.; Wang, D.; Wang, B.; Mohanty, S.; Diez, S.; Wu, Y.; Sun, Y.; Ahmadi, E.; Mi, Z. N-Polar ScAlN and HEMTs Grown by Molecular Beam Epitaxy. *Appl. Phys. Lett.* **2021**, *119*, 082101. [CrossRef]
8. Hardy, M.T.; Downey, B.P.; Nepal, N.; Storm, D.F.; Katzer, D.S.; Meyer, D.J. (Invited) ScAlN: A Novel Barrier Material for High Power GaN-Based RF Transistors. *ECS Trans.* **2017**, *80*, 161. [CrossRef]
9. Manz, C.; Leone, S.; Kirste, L.; Ligl, J.; Frei, K.; Fuchs, T.; Prescher, M.; Waltereit, P.; Verheijen, M.A.; Graff, A.; et al. Improved AlScN/GaN Heterostructures Grown by Metal-Organic Chemical Vapor Deposition. *Semicond. Sci. Technol.* **2021**, *36*, 034003. [CrossRef]
10. Schönweger, G.; Petraru, A.; Islam, M.R.; Wolff, N.; Haas, B.; Hammud, A.; Koch, C.; Kienle, L.; Kohlstedt, H.; Fichtner, S. From Fully Strained to Relaxed: Epitaxial Ferroelectric Al1-xScxN for III-N Technology. *Adv. Funct. Mater.* **2022**, *32*, 2109632. [CrossRef]
11. Islam, M.R.; Wolff, N.; Yassine, M.; Schönweger, G.; Christian, B.; Kohlstedt, H.; Ambacher, O.; Lofink, F.; Kienle, L.; Fichtner, S. On the Exceptional Temperature Stability of Ferroelectric Al1-xScxN Thin Films. *Appl. Phys. Lett.* **2021**, *118*, 232905. [CrossRef]
12. He, X.; Wang, T.; Li, X.; Das, S.; Liu, X.; Chen, Z. Enhanced Piezoelectricity and Excellent Thermal Stability in Sm3+-Doped $BiFeO_3$-$PbTiO_3$ Ceramics. *ACS Appl. Electron. Mater.* **2022**, *4*, 807–813. [CrossRef]
13. Zhang, S.; Yu, F. Piezoelectric Materials for High Temperature Sensors. *J. Am. Ceram. Soc.* **2011**, *94*, 3153–3170. [CrossRef]
14. Zhu, W.; Hayden, J.; He, F.; Yang, J.-I.; Tipsawat, P.; Hossain, M.D.; Maria, J.-P.; Trolier-McKinstry, S. Strongly Temperature Dependent Ferroelectric Switching in AlN, Al1-xScxN, and Al1-XBxN Thin Films. *Appl. Phys. Lett.* **2021**, *119*, 062901. [CrossRef]

15. Mizutani, R.; Yasuoka, S.; Shiraishi, T.; Shimizu, T.; Uehara, M.; Yamada, H.; Akiyama, M.; Sakata, O.; Funakubo, H. Thickness Scaling of $(Al_{0.8}Sc_{0.2})N$ Films with Remanent Polarization beyond 100 MC Cm^{-2} around 10 Nm in Thickness. *Appl. Phys. Express* **2021**, *14*, 105501. [CrossRef]
16. Kurz, N.; Lu, Y.; Kirste, L.; Reusch, M.; Žukauskaitė, A.; Lebedev, V.; Ambacher, O. Temperature Dependence of the Pyroelectric Coefficient of AlScN Thin Films. *Phys. Status Solidi (A)* **2018**, *215*, 1700831. [CrossRef]
17. Lu, Y.; Reusch, M.; Kurz, N.; Ding, A.; Christoph, T.; Prescher, M.; Kirste, L.; Ambacher, O.; Žukauskaitė, A. Elastic Modulus and Coefficient of Thermal Expansion of Piezoelectric Al1−xScxN (up to x = 0.41) Thin Films. *APL Mater.* **2018**, *6*, 076105. [CrossRef]
18. Ding, A.; Reusch, M.; Lu, Y.; Kurz, N.; Lozar, R.; Christoph, T.; Driad, R.; Ambacher, O.; Žukauskaitė, A. Investigation of Temperature Characteristics and Substrate Influence on AlScN-Based SAW Resonators. In Proceedings of the 2018 IEEE International Ultrasonics Symposium (IUS), IEEE International, Kobe, Japan, 22–25 October 2018; pp. 1–9. [CrossRef]
19. Kurz, N.; Ding, A.; Urban, D.F.; Lu, Y.; Kirste, L.; Feil, N.M.; Žukauskaitė, A.; Ambacher, O. Experimental Determination of the Electro-Acoustic Properties of Thin Film AlScN Using Surface Acoustic Wave Resonators. *J. Appl. Phys.* **2019**, *126*, 075106. [CrossRef]
20. Österlund, E.; Ross, G.; Caro, M.A.; Paulasto-Kröckel, M.; Hollmann, A.; Klaus, M.; Meixner, M.; Genzel, C.; Koppinen, P.; Pensala, T.; et al. Stability and Residual Stresses of Sputtered Wurtzite AlScN Thin Films. *Phys. Rev. Mater.* **2021**, *5*, 035001. [CrossRef]
21. Urban, D.F.; Ambacher, O.; Elsässer, C. First-Principles Calculation of Electroacoustic Properties of Wurtzite (Al,Sc)N. *Phys. Rev. B* **2021**, *103*, 115204. [CrossRef]
22. Tasnádi, F.; Alling, B.; Höglund, C.; Wingqvist, G.; Birch, J.; Hultman, L.; Abrikosov, I.A. Origin of the Anomalous Piezoelectric Response in Wurtzite ScxAl1−xN Alloys. *Phys. Rev. Lett.* **2010**, *104*, 137601. [CrossRef] [PubMed]
23. Zywitzki, O.; Modes, T.; Barth, S.; Bartzsch, H.; Frach, P. Effect of Scandium Content on Structure and Piezoelectric Properties of AlScN Films Deposited by Reactive Pulse Magnetron Sputtering. *Surf. Coat. Technol.* **2017**, *309*, 417–422. [CrossRef]
24. Akiyama, M.; Kamohara, T.; Kano, K.; Teshigahara, A.; Takeuchi, Y.; Kawahara, N. Enhancement of Piezoelectric Response in Scandium Aluminum Nitride Alloy Thin Films Prepared by Dual Reactive Cosputtering. *Adv. Mater.* **2009**, *21*, 593–596. [CrossRef] [PubMed]
25. Höglund, C.; Bareño, J.; Birch, J.; Alling, B.; Czigány, Z.; Hultman, L. Cubic Sc1−xAlxN Solid Solution Thin Films Deposited by Reactive Magnetron Sputter Epitaxy onto ScN(111). *J. Appl. Phys.* **2009**, *105*, 113517. [CrossRef]
26. Fichtner, S.; Wolff, N.; Krishnamurthy, G.; Petraru, A.; Bohse, S.; Lofink, F.; Chemnitz, S.; Kohlstedt, H.; Kienle, L.; Wagner, B. Identifying and Overcoming the Interface Originating C-Axis Instability in Highly Sc Enhanced AlN for Piezoelectric Micro-Electromechanical Systems. *J. Appl. Phys.* **2017**, *122*, 035301. [CrossRef]
27. Fichtner, S.; Reimer, T.; Chemnitz, S.; Lofink, F.; Wagner, B. Stress Controlled Pulsed Direct Current Co-Sputtered Al1−xScxN as Piezoelectric Phase for Micromechanical Sensor Applications. *APL Mater.* **2015**, *3*, 116102. [CrossRef]
28. Baeumler, M.; Lu, Y.; Kurz, N.; Kirste, L.; Prescher, M.; Christoph, T.; Wagner, J.; Žukauskaitė, A.; Ambacher, O. Optical Constants and Band Gap of Wurtzite Al1−xScxN/Al2O3 Prepared by Magnetron Sputter Epitaxy for Scandium Concentrations up to x = 0.41. *J. Appl. Phys.* **2019**, *126*, 045715. [CrossRef]
29. Wolff, N.; Fichtner, S.; Haas, B.; Islam, M.R.; Niekiel, F.; Kessel, M.; Ambacher, O.; Koch, C.; Wagner, B.; Lofink, F.; et al. Atomic Scale Confirmation of Ferroelectric Polarization Inversion in Wurtzite-Type AlScN. *J. Appl. Phys.* **2021**, *129*, 034103. [CrossRef]
30. Herres, N.; Kirste, L.; Obloh, H.; Köhler, K.; Wagner, J.; Koidl, P. X-ray Determination of the Composition of Partially Strained Group-III Nitride Layers Using the Extended Bond Method. *Mater. Sci. Eng. B* **2002**, *91–92*, 425–432. [CrossRef]
31. Yim, W.M.; Paff, R.J. Thermal Expansion of AlN, Sapphire, and Silicon. *J. Appl. Phys.* **1974**, *45*, 1456–1457. [CrossRef]
32. Tyunina, M.; Pacherova, O.; Kocourek, T.; Dejneka, A. Anisotropic Chemical Expansion Due to Oxygen Vacancies in Perovskite Films. *Sci. Rep.* **2021**, *11*, 15247. [CrossRef] [PubMed]
33. Onodera, Y.; Kohara, S.; Masai, H.; Koreeda, A.; Okamura, S.; Ohkubo, T. Formation of Metallic Cation-Oxygen Network for Anomalous Thermal Expansion Coefficients in Binary Phosphate Glass. *Nat. Commun.* **2017**, *8*, 15449. [CrossRef] [PubMed]
34. Bauer, S.; Rodrigues, A.; Baumbach, T. Real Time in Situ X-ray Diffraction Study of the Crystalline Structure Modification of $(Ba_{0.5}Sr_{0.5})TiO_3$ during the Post-Annealing. *Sci. Rep.* **2018**, *8*, 11969. [CrossRef] [PubMed]
35. Perry, N.H.; Kim, J.J.; Bishop, S.R.; Tuller, H.L. Strongly Coupled Thermal and Chemical Expansion in the Perovskite Oxide System $Sr(Ti,Fe)O_{3-\alpha}$. *J. Mater. Chem. A* **2015**, *3*, 3602–3611. [CrossRef]
36. Swallow, J.G.; Lee, J.K.; Defferriere, T.; Hughes, G.M.; Raja, S.N.; Tuller, H.L.; Warner, J.H.; Van Vliet, K.J. Atomic Resolution Imaging of Nanoscale Chemical Expansion in PrxCe1-xO2−δ during In Situ Heating. *ACS Nano* **2018**, *12*, 1359–1372. [CrossRef] [PubMed]
37. Sandu, C.S.; Parsapour, F.; Mertin, S.; Pashchenko, V.; Matloub, R.; LaGrange, T.; Heinz, B.; Muralt, P. Abnormal Grain Growth in AlScN Thin Films Induced by Complexion Formation at Crystallite Interfaces. *Phys. Status Solidi (A)* **2019**, *216*, 1800569. [CrossRef]
38. Höglund, C.; Birch, J.; Alling, B.; Bareño, J.; Czigány, Z.; Persson, P.O.Å.; Wingqvist, G.; Žukauskaitė, A.; Hultman, L. Wurtzite Structure Sc1−xAlxN Solid Solution Films Grown by Reactive Magnetron Sputter Epitaxy: Structural Characterization and First-Principles Calculations. *J. Appl. Phys.* **2010**, *107*, 123515. [CrossRef]
39. Wiskel, J.B.; Lu, J.; Omotoso, O.; Ivey, D.G.; Henein, H. Characterization of Precipitates in a Microalloyed Steel Using Quantitative X-ray Diffraction. *Metals* **2016**, *6*, 90. [CrossRef]

40. Seymour, T.; Frankel, P.; Balogh, L.; Ungár, T.; Thompson, S.P.; Jädernäs, D.; Romero, J.; Hallstadius, L.; Daymond, M.R.; Ribárik, G.; et al. Evolution of Dislocation Structure in Neutron Irradiated Zircaloy-2 Studied by Synchrotron X-ray Diffraction Peak Profile Analysis. *Acta Mater.* **2017**, *126*, 102–113. [CrossRef]
41. Harris, J.H.; Youngman, R.A.; Teller, R.G. On the Nature of the Oxygen-Related Defect in Aluminum Nitride. *J. Mater. Res.* **1990**, *5*, 1763–1773. [CrossRef]
42. Gasparotto, P.; Fischer, M.; Scopece, D.; Liedke, M.O.; Butterling, M.; Wagner, A.; Yildirim, O.; Trant, M.; Passerone, D.; Hug, H.J.; et al. Mapping the Structure of Oxygen-Doped Wurtzite Aluminum Nitride Coatings from Ab Initio Random Structure Search and Experiments. *ACS Appl. Mater. Interfaces* **2021**, *13*, 5762–5771. [CrossRef] [PubMed]
43. Mah, A.D.; Grain, C.F.; Smith, D.F.; Singleton, E.L.; Peters, F.A.; Alley, J.K.; Tyrrell, M.E.; Cole, W.A.; Campbell, W.J.; King, E.G.; et al. *Heats and Free Energies of Formation of Gallium Sesquioxide and Scandium Sesquioxide*; U.S. Department of the Interior, Bureau of Mines: Washington, DC, USA, 1962.
44. Moram, M.A.; Barber, Z.H.; Humphreys, C.J. The Effect of Oxygen Incorporation in Sputtered Scandium Nitride Films. *Thin Solid Film.* **2008**, *516*, 8569–8572. [CrossRef]
45. Casamento, J.; Xing, H.G.; Jena, D. Oxygen Incorporation in the Molecular Beam Epitaxy Growth of ScxGa1−xN and ScxAl1−xN. *Phys. Status Solidi (B)* **2020**, *257*, 1900612. [CrossRef]
46. Wang, P.; Wang, B.; Laleyan, D.A.; Pandey, A.; Wu, Y.; Sun, Y.; Liu, X.; Deng, Z.; Kioupakis, E.; Mi, Z. Oxygen Defect Dominated Photoluminescence Emission of ScxAl1−xN Grown by Molecular Beam Epitaxy. *Appl. Phys. Lett.* **2021**, *118*, 032102. [CrossRef]
47. Kumar, R.; Nayak, S.; Garbrecht, M.; Bhatia, V.; Indiradevi Kamalasanan Pillai, A.; Gupta, M.; Shivaprasad, S.M.; Saha, B. Clustering of Oxygen Point Defects in Transition Metal Nitrides. *J. Appl. Phys.* **2021**, *129*, 055305. [CrossRef]
48. Chichibu, S.F.; Miyake, H.; Ishikawa, Y.; Tashiro, M.; Ohtomo, T.; Furusawa, K.; Hazu, K.; Hiramatsu, K.; Uedono, A. Impacts of Si-Doping and Resultant Cation Vacancy Formation on the Luminescence Dynamics for the near-Band-Edge Emission of Al0.6Ga0.4N Films Grown on AlN Templates by Metalorganic Vapor Phase Epitaxy. *J. Appl. Phys.* **2013**, *113*, 213506. [CrossRef]
49. Eldrup, M.; Singh, B.N. Studies of Defects and Defect Agglomerates by Positron Annihilation Spectroscopy. *J. Nucl. Mater.* **1997**, *251*, 132–138. [CrossRef]
50. Youngman, R.A.; Harris, J.H. Luminescence Studies of Oxygen-Related Defects In Aluminum Nitride. *J. Am. Ceram. Soc.* **1990**, *73*, 3238–3246. [CrossRef]

micromachines

MDPI

Article

High-Temperature Ferroelectric Behavior of $Al_{0.7}Sc_{0.3}N$

Daniel Drury [1,2,3,*], Keisuke Yazawa [1,2], Andriy Zakutayev [2], Brendan Hanrahan [3] and Geoff Brennecka [1,*]

1 Colorado School of Mines, 1500 Illinois Ave., Golden, CO 80401, USA; keisuke.yazawa@nrel.gov
2 National Renewable Energy Laboratory, 15013 Denver West Parkway, Golden, CO 80401, USA; andriy.zakutayev@nrel.gov
3 U.S. Army Combat Capabilities Development Command—Army Research Laboratory, Adelphi, MD 20783, USA; brendan.m.hanrahan.civ@army.mil
* Correspondence: ddrury@mines.edu (D.D.); gbrennec@mines.edu (G.B.)

Abstract: Currently, there is a lack of nonvolatile memory (NVM) technology that can operate continuously at temperatures > 200 °C. While ferroelectric NVM has previously demonstrated long polarization retention and $>10^{13}$ read/write cycles at room temperature, the largest hurdle comes at higher temperatures for conventional perovskite ferroelectrics. Here, we demonstrate how AlScN can enable high-temperature (>200 °C) nonvolatile memory. The c-axis textured thin films were prepared via reactive radiofrequency magnetron sputtering onto a highly textured Pt (111) surface. Photolithographically defined Pt top electrodes completed the capacitor stack, which was tested in a high temperature vacuum probe station up to 400 °C. Polarization–electric field hysteresis loops between 23 and 400 °C reveal minimal changes in the remanent polarization values, while the coercive field decreased from 4.3 MV/cm to 2.6 MV/cm. Even at 400 °C, the polarization retention exhibited negligible loss for up to 1000 s, demonstrating promise for potential nonvolatile memory capable of high−temperature operation. Fatigue behavior also showed a moderate dependence on operating temperature, but the mechanisms of degradation require additional study.

Keywords: AlScN; ferroelectric; high temperature; nonvolatile memory; retention; fatigue; wurtzite; film; sputter deposition

Citation: Drury, D.; Yazawa, K.; Zakutayev, A.; Hanrahan, B.; Brennecka, G. High-Temperature Ferroelectric Behavior of $Al_{0.7}Sc_{0.3}N$. *Micromachines* **2022**, *13*, 887. https://doi.org/10.3390/mi13060887

Academic Editor: Faisal Mohd-Yasin

Received: 12 May 2022
Accepted: 30 May 2022
Published: 31 May 2022

1. Introduction

Nonvolatile memories, which do not require a permanent/frequent voltage supplied to maintain the bit state, are important for storing information indefinitely and reliably. The concept of high−operating−temperature nonvolatile memory (HOT−NVM) has been an elusive capability. Current computational and data storage limitations in harsh environments either require on−board cooling or locating sensors/computations away from the heat source. These harsh environments for electronics are a focus of the NASA HOTTech project and can be found within jet turbines, within deep−well drilling, and on the surface of Venus [1].

Nonvolatile random−access memory (RAM) based on FLASH, magnetic, phase−change, and resistive mechanisms degrade quickly even at moderate temperatures (<200 °C) [2–5]. Microelectromechanical (MEM) and nanogap resistance switching (NGS) offer promise for elevated temperatures NVM, but there are downsides with moving parts and operating in various atmospheres [6,7]. Ferroelectric technology based on perovskite or fluorite structures (e.g., $Pb(Zr,Ti)O_3$ or $(Hf,Zr,Si)O_2$) is currently limited to temperatures < 200 °C [8,9]. This is a result of a destabilized polar structure, chemical instabilities, and increased domain wall mobility in $Pb(Zr,Ti)O_3$ [10,11], whereas increased pyroelectric contributions and depolarization fields plague $(Hf,Zr)O_2$ compositions when increasing temperature [12,13]. Here, we demonstrate promising initial results when using ferroelectric $Al_{0.7}Sc_{0.3}N$ to meet the HOT−NVM demands.

Since its discovery in 2019 as the first wurtzite nitride ferroelectric, $Al_{1-x}Sc_xN$ (AlScN) has received much attention from both the ferroelectric and piezoelectric communities [14]. The high remanent polarization (P_r) associated with the ideal wurtzite structure is an attractive property for increasing either device density or extracted charge density. Previous studies on AlN−based ferroelectrics reported the high coercive field (E_c) necessary to switch between the two polarization states. While this high E_c is problematic for integration with low−voltage devices, it can be valuable for retention characteristics as temperatures increase [8]. The E_c of AlScN can be tuned through various modifications such as composition [14–16], stress [17], epitaxial alignment [18], and temperature [19–22]. In addition to thickness scaling, these trends are important for integrating AlScN into a high−temperature chip to reduce the operating voltages.

To assess polarization stability for a nonvolatile memory, Islam et al. poled the sample at room temperature (RT), baked the package at 1100 °C, and remeasured the P_r once cooled back to RT [23]. While promising for an NVM since the polarization state is maintained after exposure to extreme temperatures, these ex situ results did not address the P_r or switching behavior at elevated temperatures. Liu et al. focused on the retention state at RT for an AlScN ferroelectric−gated field effect transistor (FET) device by monitoring changes in the resistance state over time [24]. Furthermore, an NVM must endure switching cycles without compromising the P_r or E_c values, which can impact deciphering between either polarization state, and the device must avoid electrical shorts. Two recent reports have noted that AlScN has low endurance, a measure of repeated bit flips, at $<10^5$ cycles before dielectric failure, which may be limited by the slim margin between coercive and breakdown fields (E_{bd}) for the samples, but a true mechanistic understanding of fatigue in AlScN is still lacking [25,26]. Thus, these reports do not represent a fundamental limitation of the ferroelectric switching endurance. In this article, we report device relevant behavior (i.e., leakage, retention, and fatigue) between 23 and 400 °C to introduce AlScN as a candidate material for HOT−NVM.

2. Materials and Methods

The 225 nm thick ferroelectric AlScN film was deposited by radiofrequency (rf) reactive magnetron sputtering onto a highly textured Pt (111) bottom electrode. The base pressure of the chamber reached 2×10^{-6} Torr prior to flowing the deposition gases. The pressure was maintained at 2 mTorr while flowing 5 sccm N_2 and 15 sccm Ar. Furthermore, 300 W of rf power was supplied to the 5 cm diameter $Al_{0.7}Sc_{0.3}$ alloy target. The substrate was heated to 400 °C and positioned 16.5 cm from the target. Details of sputter deposition processes using in situ optical emission spectrometry were previously described [27]. The target was conditioned with a 1 h pre−sputter to remove surface oxidation and stabilize the reactive target mode. Following the nitride layer deposition, lithographically defined 50 μm diameter platinum top electrodes were sputter−deposited to form a metal−ferroelectric−metal (MFM) structure for electrical testing.

The resultant AlScN film was structurally characterized by X−ray diffraction (XRD) using Cu Kα radiation. A Panalytical X'Pert MRD diffractometer was used to scan the $\theta-2\theta$ and ω space. The sample was tested in a high−temperature vacuum probe station to mitigate possible oxidation of the nitride. Electrical properties of the MFM capacitors were measured using a ferroelectric analyzer (Radiant Precision Premier II).

The MFM capacitors were evaluated with DC current−voltage (IV) sweeps to assess the resistivity at various temperatures; this is important for evaluating polarization values that may be inflated by charge leakage. Polarization−electric field (P–E) hysteresis loops were collected between probe station temperatures of 23−400 °C using an excitation frequency of 10 kHz. This frequency was chosen to reduce the leakage contribution to the P_r values, and 10 kHz is a relatively high frequency when compared to previous reports on ferroelectric AlScN, which range between 333 Hz and 100 kHz [19,21]. Retention testing was used to measure the stability of a polarization state over time to inspect the volatility of that state. The test involved a positive or negative pre−pulse to the bottom electrode to pole the sample in either a metal or a nitrogen polar orientation, respectively. This was followed by a dwell for a length of time, and then the MFM capacitor was pulsed twice with the same voltage profile (Figure 1). The measured $2P_r$ elucidates whether the capacitor underwent ferroelectric polarization switching between pulse 1 and pulse 2 or not. In this study, four different pulse sequences were necessary to characterize the same state and opposite state retention characteristics for each polarization state—nitrogen (N) and metal (M) polar. A 'same state' (SS) test indicates that the bias of the read and write pulses is the same, while an 'opposite state' (OS) test switches pulse polarity. The difference between a switched and unswitched polarization response is $2P_r$. If the $2P_r$ shrinks over time, then the bit state will eventually be indecipherable; therefore, a stable polarization value is essential for NVM memory applications. Ten virgin capacitors with an active (top electrode) diameter of 50 µm were assessed at each temperature. P–E loops (10 kHz) were collected prior to each 1000 cycle (1 kHz triangular waveform applied for 1 s) fatigue to inspect the ferroelectric properties throughout the process.

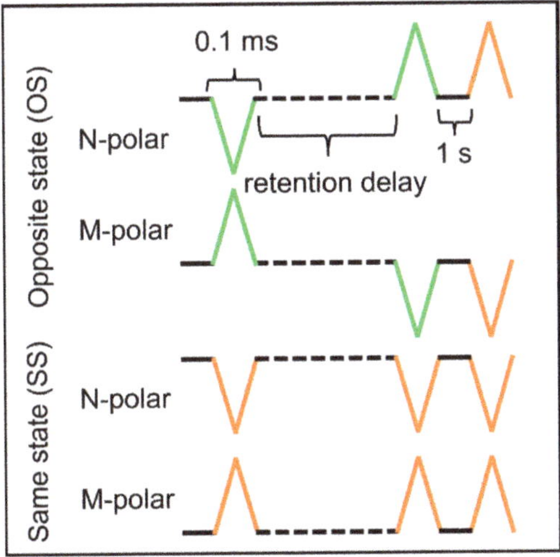

Figure 1. Retention test protocols for OS and SS in either a nitrogen or metal polar surface state. Green and orange traces represent switching and non−switching pulses, respectively.

3. Results and Discussion

3.1. Structural and Leakage Characteristics

In Figure 2a, the XRD scan reveals a wurtzite phase of AlScN with c−axis texture. The (0002) reflection appeared at 36.056° in 2θ with a full width at half maximum (FWHM) value of 0.158°. The ω−rocking scan of the (0002) position resulted in an FWHM value of 2.694°, which is comparable to prior reports (Figure 2b) [16,22]. Lack of additional wurtzite peaks associated with other orientations indicates that the film had a purely c−axis texture. Since the c−axis is the polar axis of the wurtzite structure, it is important to maximize the degree of c−axis out of plane texture when investigating the ferroelectric properties of an MFM capacitor (inset of Figure 2a).

The P–E loops measured at 10 kHz between 23 and 400 °C reveal a strong dependence of the E_c on temperature (Figure 3a). The square−like hysteresis loops were a result of a relatively low permittivity, defined energy barriers between polarization states, and minimal loss mechanisms. Levels of polarization saturation similar to previous reports were reached at each temperature [14,16]. At 400 °C, the derivative dP/dE became negative after saturation, indicating decreased capacitor resistance. This observation is supported by the current density–electric field (J–E) sweeps in Figure 3b. When subjected to a 1 $MV \cdot cm^{-1}$ DC field, the current density rose from 1×10^{-7} $A \cdot cm^{-2}$ at 23 °C to 3×10^{-4} $A \cdot cm^{-2}$ at

400 °C. Furthermore, J increased exponentially with E and maintained this relationship at all temperatures. The increasing leakage with temperature is important for compensation when possibly measuring changes in polarization (e.g., during retention testing).

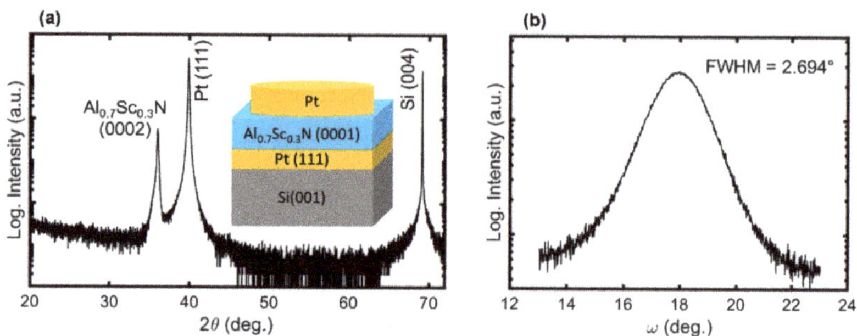

Figure 2. (**a**) The 2θ–θ XRD scan of the investigated sample revealing a pure wurtzite phase with c−axis texture. The inset depicts the stacking sequence of the metal–ferroelectric–metal capacitor. (**b**) ω rocking scan of the $Al_{0.7}Sc_{0.3}N$ (0002) peak.

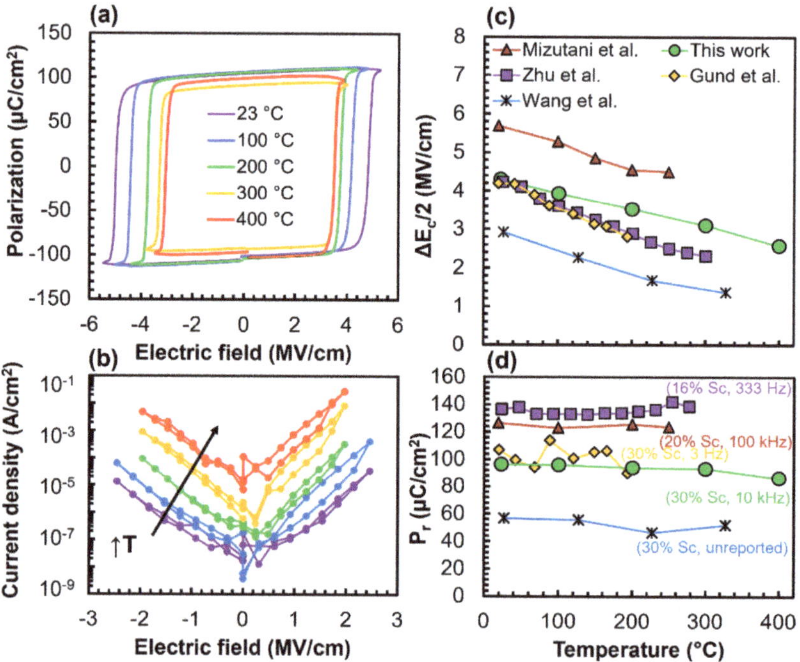

Figure 3. (**a**) P−E loops between 23 and 400 °C collected with 10 kHz bipolar voltage waveform. (**b**) Temperature−dependent leakage current density (J) over bias between 23 and 400 °C. Comparison of (**c**) $\Delta E_c/2$ and (**d**) P_r results from this work with Mizutani et al. [21], Zhu et al. [20], Gund et al. [22], and Wang et al. [20]. The results in (**a**,**b**) are from this work, while (**c**,**d**) are a comparison of our results with previous reports.

The $\Delta E_c/2$ decreased linearly as temperature increased, describing a reduced activation barrier due to thermal energy contribution. An E_c temperature coefficient of 4.5 $kV \cdot cm^{-1} \cdot K^{-1}$ describes our results in Figure 3c, which showed a decrease from 4.3 to

2.6 MV·cm^{-1} between 23 and 400 °C. In the limited number of reports that analyzed the E_c and P_r dependence on temperature for $Al_{1-x}Sc_xN$, there were a range of compositions studied with various frequencies. Our data (30% Sc, 10 kHz) showed a slightly lower E_c temperature coefficient in comparison to the literature. The overall E_c results (apart from the reduced E_c values reported by Wang et al.) are consistent with the literature in terms of a lower E_c due to increasing Sc content or decreasing excitation frequency. This may explain both the coincidental values between Gund et al. (30% Sc, 3 kHz) and Zhu et al. (16% Sc, 333 Hz) and the increased E_c for Mizutani et al. (20% Sc, 100 kHz) [19–22]. Our work shows that the P_r values for $Al_{0.7}Sc_{0.3}N$ remained relatively stable up to 400 °C (Figure 3d). The temperature−independent P_r behavior of an $Al_{1-x}Sc_xN$ composition was also noted by the mentioned reports. In general, the P_r decrease with increasing x was seen in a majority of the reports [16].

3.2. Polarization Retention at Elevated Temperatures

Retention studies were carried out relative to the surface polar state, nitrogen or metal, termed SS_N, SS_M, OS_N, and OS_M. Polarization changes were investigated with durations of pulses up to 1000 s to reveal if any changes occurred. A reduction in measured polarization would indicate that an internal process occurred, reducing the net polarization via mechanisms such as back−switching, domain pinning, or phase degradation. However, our results revealed no P_r loss, indicating a kinetically stable polarization direction and a negligible change in the polarization−pinning defect environment (Figure 4). This contrasts perovskite and fluorite−based ferroelectric MFM capacitors, which showed a clear reduction in P_r when subjected to a bake between the write and read voltage pulses [9,12,28]. For both SS tests, there was a marginal level of polarization (<6 μC·cm^{-2}), which may have been the effect of changes in the leakage contribution between the first and second read pulses. Since the OS and SS tests for the metal and nitrogen poled devices showed a steady polarization retention between 1 and 1000 s, the estimated time until the $2P_r$ was reduced enough for the bit state to become indecipherable requires further retention testing before a reasonable estimate can be established. Our results are also in agreement with the RT retention tests performed by Fichtner et al., who reported steady retention up to 10^5 s [14]. Overall, the ferroelectric AlScN retained all the polarization that was written into the MFM capacitor, even while operating at 400 °C. This characteristic is a promising factor for enabling high−temperature nonvolatile memory.

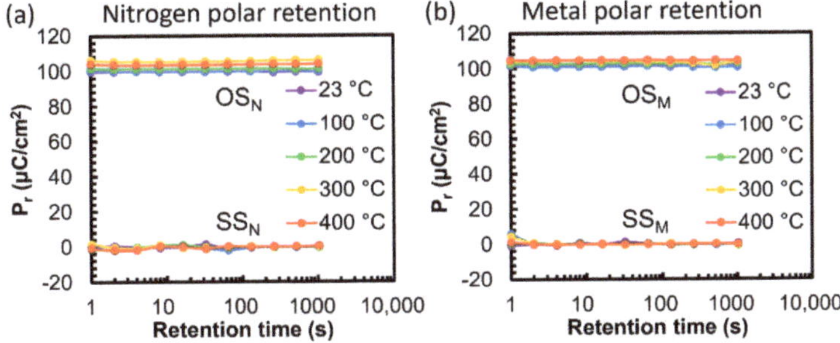

Figure 4. Polarization retention after poling in either (**a**) nitrogen or (**b**) metal surface state and dwelling up to 1000 s at the indicated temperature.

3.3. Device Fatigue at Elevated Temperatures

To ascertain the switching endurance of the sample, 10 MFM capacitors from the same sample were tested at each temperature. The maximum field applied during the fatigue cycles and P–E loop measurements was scaled to 105% of the E_c. Figure 5a provides a summary of the fatigue results, which were also temperature−dependent. A P–E loop measurement followed every 1000 cycles to track the changes in the polarization throughout the fatigue subjection. One standard deviation is denoted by the error bars in Figure 5a. Device failure was defined at the point when the device shorted. The highest average endurance occurred at 200 °C ($2.7 \times 10^5 \pm 1.9 \times 10^5$ cycles) with a maximum of 6.0×10^5 cycles. While this is an increased fatigue endurance relative to previous reports on AlScN, it still indicates that effort is needed to increase the margin between E_c and E_{bd} and to investigate and mitigate the underlying causes of fatigue in these materials. Because E_c decreased with increased temperature faster than E_{bd} decreased, the increased margin between the two was presumably an important contributor to the increase in fatigue life from RT to 200 °C. The degraded performance at temperatures > 200 °C, however, was associated with larger leakage currents in the MFM capacitor that occurred during each fatigue cycle.

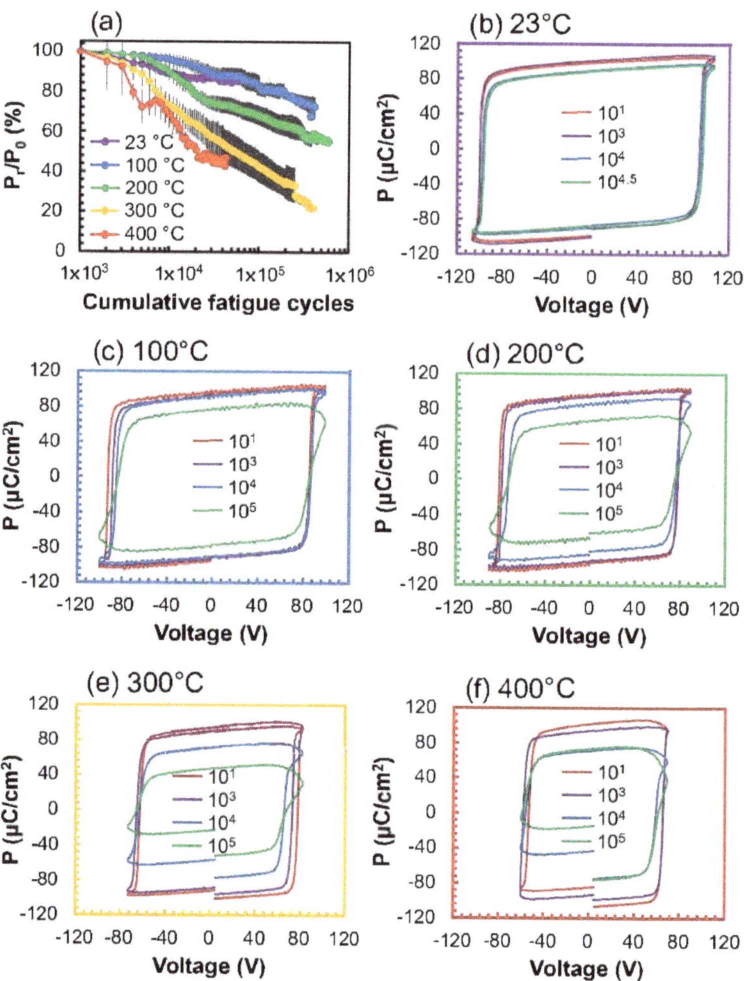

Figure 5. Ferroelectric fatigue exhibits a strong dependence on the measurement temperature. (**a**) Normalized P_r dependence on cycles for temperatures between 23 and 400 °C with error bars of a standard deviation. Representative P–E loops throughout the fatigue cycling for (**b**) 23 °C, (**c**) 100 °C, (**d**) 200 °C, (**e**) 300 °C, and (**f**) 400 °C. The colors of the plot borders for (**b**–**f**) correspond to the temperature of the measurement in (**a**), and the different traces in (**b**–**f**) are consistent for ease of comparison.

The least change in polarization versus cycling happened at 100 °C. Figure 5b–f are representative snapshots of the P–E loop measured after the given number of cycles. Especially for the higher−temperature panels, the initially square−like hysteresis loops deteriorated into a combination of lower P_r and inflated polarization values due to an uncompensated leakage contribution. Additional study into the cycle−dependent leakage characteristics may point toward a specific leakage mechanism that increases in effect with cycling. Furthermore, a systematic investigation on the AlScN switching fatigue behavior for various electrode types will be critical for applications that require more resilient endurance.

4. Conclusions

In summary, a thin film, c−axis textured AlScN was electrically tested in a metal–ferroelectric–metal capacitor for polarization hysteresis, retention, and fatigue up to 400 °C. The leakage current density increased by four orders of magnitude between RT and 400 °C, which revealed the importance of compensating for the leakage contribution to polarization, especially as temperature increased. While a strong E_c dependence on temperature ($4.5\ kV \cdot cm^{-1} \cdot K^{-1}$) led to a 60% reduction at 400 °C compared to RT, the P_r remained stable over all temperatures. As an important characteristic for NVM, in situ polarization retention testing up to 1000 s revealed negligible (<2%) polarization loss for either the metal or the nitrogen polar state at all temperatures tested up to 400 °C. On average, the highest endurance before complete failure occurred at 200 °C with 2.7×10^5 switching cycles; however, the reduction in P_r versus cycles was lowest at 100 °C. Our results showing >10^5 switching cycles are encouraging for using ferroelectric AlScN in HOT−NVM applications where <10^3 switching cycles would be required. Future work on the fatigue mechanisms at different temperatures and expanding the retention duration time are vital avenues for understanding the intrinsic limitations of AlScN.

Author Contributions: Conceptualization, D.D., B.H. and G.B.; methodology, D.D.; validation, D.D., B.H. and G.B.; formal analysis, D.D.; investigation, D.D.; resources, K.Y., A.Z. and B.H.; data curation, D.D.; writing—original draft preparation, D.D., B.H. and G.B.; writing—review and editing, D.D., B.H. and G.B.; supervision, K.Y., A.Z., B.H. and G.B.; project administration; funding acquisition, A.Z., B.H. and G.B. All authors read and agreed to the published version of the manuscript.

Funding: This research was sponsored by the Army Research Laboratory and was accomplished under Cooperative Agreement Number W911NF-21-2-0210. The views and conclusions contained in this document are those of the authors and should not be interpreted as representing the official policies, either expressed or implied, of the Army Research Laboratory or the US Government. The US Government is authorized to reproduce and distribute reprints for Government purposes notwithstanding any copyright notation herein. This work was coauthored by the Colorado School of Mines and the National Renewable Energy Laboratory, operated by the Alliance for Sustainable Energy, LLC, for the US Department of Energy (DOE) under Contract No. DE-AC36-08GO28308. Funding was provided by the DARPA Tunable Ferroelectric Nitrides (TUFEN) program (DARPA-PA-19-04-03) as a part of Development and Exploration of Ferroelectric Nitride Semiconductors (DEFENSE) project (diffraction, microscopy, and electrical characterization), and by Office of Science (SC), Office of Basic Energy Sciences (BES) as part of the Early Career Award "Kinetic Synthesis of Metastable Nitrides" (material synthesis and in situ monitoring).

Data Availability Statement: The data that support the findings of this study are available from the corresponding author upon reasonable request.

Acknowledgments: The authors also express their appreciation to Wanlin Zhu and Susan Trolier-McKinstry of the Pennsylvania State University for providing $Pt/TiO_2/SiO_2/Si$ substrates. We also thank Jeff Alleman for assistance with setting up the sputter system at NREL.

Conflicts of Interest: The authors declare no conflict of interest.

References

1. Neudeck, P.G.; Meredith, R.D.; Chen, L.; Spry, D.J.; Nakley, L.M.; Hunter, G.W. Prolonged Silicon Carbide Integrated Circuit Operation in Venus Surface Atmospheric Conditions. *AIP Adv.* **2016**, *6*, 125119. [CrossRef]
2. Lee, K.; Kang, S.H. Design Consideration of Magnetic Tunnel Junctions for Reliable High-Temperature Operation of STT-MRAM. *IEEE Trans. Magn.* **2010**, *46*, 1537–1540. [CrossRef]
3. Wong, H.-S.P.; Raoux, S.; Kim, S.; Liang, J.; Reifenberg, J.P.; Rajendran, B.; Asheghi, M.; Goodson, K.E. Phase Change Memory. *Proc. IEEE* **2010**, *98*, 2201–2227. [CrossRef]
4. Ye, C.; Wu, J.; He, G.; Zhang, J.; Deng, T.; He, P.; Wang, H. Physical Mechanism and Performance Factors of Metal Oxide Based Resistive Switching Memory: A Review. *J. Mater. Sci. Technol.* **2016**, *32*, 1–11. [CrossRef]
5. Taito, Y.; Nakano, M.; Okimoto, H.; Okada, D.; Ito, T.; Kono, T.; Noguchi, K.; Hidaka, H.; Yamauchi, T. 7.3 A 28 nm Embedded SG-MONOS Flash Macro for Automotive Achieving 200 MHz Read Operation and 2.0MB/S Write Throughput at Ti, of 170 °C. In Proceedings of the 2015 IEEE International Solid-State Circuits Conference-(ISSCC) Digest of Technical Papers, San Francisco, CA, USA, 22–26 February 2015; pp. 1–3.
6. Suga, H.; Suzuki, H.; Shinomura, Y.; Kashiwabara, S.; Tsukagoshi, K.; Shimizu, T.; Naitoh, Y. Highly Stable, Extremely High-Temperature, Nonvolatile Memory Based on Resistance Switching in Polycrystalline Pt Nanogaps. *Sci. Rep.* **2016**, *6*, 34961. [CrossRef] [PubMed]
7. Singh, P.; Arya, D.S.; Jain, U. MEM-FLASH Non-Volatile Memory Device for High-Temperature Multibit Data Storage. *Appl. Phys. Lett.* **2019**, *115*, 43501. [CrossRef]
8. Stolichnov, I.; Tagantsev, A.K.; Colla, E.; Setter, N.; Cross, J.S. Physical Model of Retention and Temperature-Dependent Polarization Reversal in Ferroelectric Films. *J. Appl. Phys.* **2005**, *98*, 84106. [CrossRef]
9. Mueller, S.; Muller, J.; Schroeder, U.; Mikolajick, T. Reliability Characteristics of Ferroelectric Si:HfO_2 Thin Films for Memory Applications. *IEEE Trans. Device Mater. Relib.* **2013**, *13*, 93–97. [CrossRef]
10. Kounga, A.B.; Granzow, T.; Aulbach, E.; Hinterstein, M.; Rödel, J. High-Temperature Poling of Ferroelectrics. *J. Appl. Phys.* **2008**, *104*, 24116. [CrossRef]
11. Kamel, T.M.; Kools, F.X.N.M.; de With, G. Poling of Soft Piezoceramic PZT. *J. Eur. Ceram. Soc.* **2007**, *27*, 2471–2479. [CrossRef]
12. Ali, T.; Kuhnel, K.; Czernohorsky, M.; Mart, C.; Rudolph, M.; Patzold, B.; Lehninger, D.; Olivo, R.; Lederer, M.; Muller, F.; et al. A Study on the Temperature-Dependent Operation of Fluorite-Structure-Based Ferroelectric HfO_2 Memory FeFET: Pyroelectricity and Reliability. *IEEE Trans. Electron. Devices* **2020**, *67*, 2981–2987. [CrossRef]
13. Mohan, J.; Hernandez-Arriaga, H.; Jung, Y.C.; Onaya, T.; Nam, C.-Y.; Tsai, E.H.R.; Kim, S.J.; Kim, J. Ferroelectric Polarization Retention with Scaling of $Hf_{0.5}Zr_{0.5}O_2$ on Silicon. *Appl. Phys. Lett.* **2021**, *118*, 102903. [CrossRef]
14. Fichtner, S.; Wolff, N.; Lofink, F.; Kienle, L.; Wagner, B. AlScN: A III-V Semiconductor Based Ferroelectric. *J. Appl. Phys.* **2019**, *125*, 114103. [CrossRef]
15. Hayden, J.; Hossain, M.D.; Xiong, Y.; Ferri, K.; Zhu, W.; Imperatore, M.V.; Giebink, N.; Trolier-McKinstry, S.; Dabo, I.; Maria, J.-P. Ferroelectricity in Boron-Substituted Aluminum Nitride Thin Films. *Phys. Rev. Mater.* **2021**, *5*, 44412. [CrossRef]
16. Yasuoka, S.; Shimizu, T.; Tateyama, A.; Uehara, M.; Yamada, H.; Akiyama, M.; Hiranaga, Y.; Cho, Y.; Funakubo, H. Effects of Deposition Conditions on the Ferroelectric Properties of $(Al_{1-x}Sc_x)N$ Thin Films. *J. Appl. Phys.* **2020**, *128*, 114103. [CrossRef]
17. Schönweger, G.; Petraru, A.; Islam, M.R.; Wolff, N.; Haas, B.; Hammud, A.; Koch, C.; Kienle, L.; Kohlstedt, H.; Fichtner, S. From Fully Strained to Relaxed: Epitaxial Ferroelectric $Al_{1-x}Sc_x$ N for III-N Technology. *Adv. Funct. Mater.* **2022**, *32*, 2109632. [CrossRef]
18. Yazawa, K.; Drury, D.; Zakutayev, A.; Brennecka, G.L. Reduced Coercive Field in Epitaxial Thin Film of Ferroelectric Wurtzite $Al_{0.7}Sc_{0.3}N$. *Appl. Phys. Lett.* **2021**, *118*, 162903. [CrossRef]
19. Zhu, W.; Hayden, J.; He, F.; Yang, J.-I.; Tipsawat, P.; Hossain, M.D.; Maria, J.-P.; Trolier-McKinstry, S. Strongly Temperature Dependent Ferroelectric Switching in AlN, $Al_{1-x}Sc_xN$, and $Al_{1-x}B_xN$ Thin Films. *Appl. Phys. Lett.* **2021**, *119*, 062901. [CrossRef]
20. Wang, J.; Park, M.; Ansari, A. High-Temperature Acoustic and Electric Characterization of Ferroelectric $Al_{0.7}Sc_{0.3}N$ Films. *J. Microelectromech. Syst.* **2022**, *31*, 234–240. [CrossRef]
21. Mizutani, R.; Yasuoka, S.; Shiraishi, T.; Shimizu, T.; Uehara, M.; Yamada, H.; Akiyama, M.; Sakata, O.; Funakubo, H. Thickness Scaling of $(Al_{0.8}Sc_{0.2})N$ Films with Remanent Polarization beyond 100 MC Cm^{-2} around 10 Nm in Thickness. *Appl. Phys. Express* **2021**, *14*, 105501. [CrossRef]
22. Gund, V.; Davaji, B.; Lee, H.; Asadi, M.J.; Casamento, J.; Xing, H.G.; Jena, D.; Lal, A. Temperature-Dependent Lowering of Coercive Field in 300 Nm Sputtered Ferroelectric $Al_{0.70}Sc_{0.30}N$. In Proceedings of the 2021 IEEE International Symposium on Applications of Ferroelectrics (ISAF), Sydney, Australia, 16 May 2021; pp. 1–3.
23. Islam, M.d.R.; Wolff, N.; Yassine, M.; Schönweger, G.; Christian, B.; Kohlstedt, H.; Ambacher, O.; Lofink, F.; Kienle, L.; Fichtner, S. On the Exceptional Temperature Stability of Ferroelectric $Al_{1-x}Sc_xN$ Thin Films. *Appl. Phys. Lett.* **2021**, *118*, 232905. [CrossRef]
24. Liu, X.; Wang, D.; Kim, K.-H.; Katti, K.; Zheng, J.; Musavigharavi, P.; Miao, J.; Stach, E.A.; Olsson, R.H.; Jariwala, D. Post-CMOS Compatible Aluminum Scandium Nitride/2D Channel Ferroelectric Field-Effect-Transistor Memory. *Nano Lett.* **2021**, *21*, 3753–3761. [CrossRef] [PubMed]

25. Mikolajick, T.; Slesazeck, S.; Mulaosmanovic, H.; Park, M.H.; Fichtner, S.; Lomenzo, P.D.; Hoffmann, M.; Schroeder, U. Next Generation Ferroelectric Materials for Semiconductor Process Integration and Their Applications. *J. Appl. Phys.* **2021**, *129*, 100901. [CrossRef]
26. Tsai, S.-L.; Hoshii, T.; Wakabayashi, H.; Tsutsui, K.; Chung, T.-K.; Chang, E.Y.; Kakushima, K. Room-Temperature Deposition of a Poling-Free Ferroelectric AlScN Film by Reactive Sputtering. *Appl. Phys. Lett.* **2021**, *118*, 082902. [CrossRef]
27. Drury, D.; Yazawa, K.; Mis, A.; Talley, K.; Zakutayev, A.; Brennecka, G.L. Understanding Reproducibility of Sputter-Deposited Metastable Ferroelectric Wurtzite $Al_{0.6}Sc_{0.4}N$ Films Using In Situ Optical Emission Spectrometry. *Phys. Status Solidi RRL* **2021**, *15*, 2100043. [CrossRef]
28. Rodriguez, J.A.; Remack, K.; Boku, K.; Udayakumar, K.R.; Aggarwal, S.; Summerfelt, S.R.; Celii, F.G.; Martin, S.; Hall, L.; Taylor, K.; et al. Reliability Properties of Low-Voltage Ferroelectric Capacitors and Memory Arrays. *IEEE Trans. Device Mater. Relib.* **2004**, *4*, 436–449. [CrossRef]

micromachines

Article

Laser Ultrasound Investigations of AlScN(0001) and AlScN(11-20) Thin Films Prepared by Magnetron Sputter Epitaxy on Sapphire Substrates

Elena A. Mayer [1], Olga Rogall [1], Anli Ding [2], Akash Nair [2], Agnė Žukauskaitė [2,3,*], Pavel D. Pupyrev [1], Alexey M. Lomonosov [4] and Andreas P. Mayer [1,*]

1 B + W Department, Offenburg University of Applied Sciences, 77652 Offenburg, Germany
2 Fraunhofer Institute for Applied Solid State Physics IAF, 79108 Freiburg, Germany
3 Fraunhofer Institute for Organic Electronics, Electron Beam and Plasma Technology FEP, 01277 Dresden, Germany
4 School of Science, Nanjing University of Science and Technology, Nanjing 210094, China
* Correspondence: agne.zukauskaite@fep.fraunhofer.de (A.Ž.); andreas.mayer@hs-offenburg.de (A.P.M.)

Abstract: The laser ultrasound (LU) technique has been used to determine dispersion curves for surface acoustic waves (SAW) propagating in AlScN/Al_2O_3 systems. Polar and non-polar $Al_{0.77}Sc_{0.23}N$ thin films were prepared by magnetron sputter epitaxy on Al_2O_3 substrates and coated with a metal layer. SAW dispersion curves have been measured for various propagation directions on the surface. This is easily achieved in LU measurements since no additional surface structures need to be fabricated, which would be required if elastic properties are determined with the help of SAW resonators. Variation of the propagation direction allows for efficient use of the system's anisotropy when extracting information on elastic properties. This helps to overcome the complexity caused by a large number of elastic constants in the film material. An analysis of the sensitivity of the SAW phase velocities (with respect to the elastic moduli and their dependence on SAW propagation direction) reveals that the non-polar AlScN films are particularly well suited for the extraction of elastic film properties. Good agreement is found between experiment and theoretical predictions, validating LU as a non-destructive and fast technique for the determination of elastic constants of piezoelectric thin films.

Keywords: laser ultrasound; AlScN; surface acoustic waves; magnetron sputter epitaxy; elastic properties; thin films

Citation: Mayer, E.A.; Rogall, O.; Ding, A.; Nair, A.; Žukauskaitė, A.; Pupyrev, P.D.; Lomonosov, A.M.; Mayer, A.P. Laser Ultrasound Investigations of AlScN(0001) and AlScN(11-20) Thin Films Prepared by Magnetron Sputter Epitaxy on Sapphire Substrates. *Micromachines* **2022**, *13*, 1698. https://doi.org/10.3390/mi13101698

Academic Editor: Alexandra Joshi-Imre

Received: 12 August 2022
Accepted: 29 September 2022
Published: 9 October 2022

1. Introduction

Among the various fields of application of $Al_{1-x}Sc_xN$ films, analog signal processing devices for mobile communication play a preeminent role. The working principle of such devices, such as frequency filters and delay lines, is based on bulk or surface acoustic waves (BAW and SAW, respectively). The main reason for the growing interest in AlScN films constitutes their favorable piezoelectric and mechanical properties [1–3]. Piezoelectric properties of this ternary nitride can be enhanced by increasing the scandium concentration x, which has consequences for the electromechanical coupling and also for the elastic properties and hence the velocities of acoustic waves [4–6]. For the design of signal processing devices containing AlScN films, knowledge of the elastic constants is of paramount importance.

For the determination of elastic constants of AlScN films, a fast, non-destructive, and easy-to-handle method is called for. Optimally, such a method could also be employed for quality control of AlScN films. A particular challenge in the case of epitaxial AlScN films is the large number of material constants that have to be determined. The main goal of this study is to demonstrate that laser ultrasound (LU) is an attractive technique for this purpose.

In the past, elastic constants of AlScN films have been determined with Brillouin scattering [7–9] and via electrical excitation of acoustic vibrations, especially with the help of SAW resonators (called resonator method in the following) [10,11] or delay lines [12] (for a review see for example [10]). However, preparing SAW or other types of resonators requires advanced fabrication capabilities to obtain the full dispersion curves from multiple test structures (phase velocity as a function of frequency).

The piezoelectric constants are yet out of reach for our laser ultrasound set-up. This problem is shared with Brillouin scattering, another non-destructive optical method, which (as with LU) does not require any specific modifications of the sample, unlike the resonator method. However, the LU approach needs much shorter measurement times, especially when being partially automated. A full SAW dispersion curve is obtained in one measurement cycle, taking about 30 min. The propagation direction on the surface can be varied without extra effort. This allows to efficiently exploit the anisotropy of the substrate to determine film properties from SAW dispersion curves, as was done here to investigate the highly anisotropic AlScN-on-sapphire systems. We show here that valuable information can be gained with comparatively robust and inexpensive instrumentation. For additional benchmarking of our experimental approach, a comparison to the results of ab initio calculations, using the density functional theory can be performed [5,13–15]. Here, we compare SAW dispersion curves obtained by LU to those calculated with the most recently published theoretical data on material constants of AlScN [5].

In the past, the determinations of elastic properties of thin films by laser ultrasound were mostly confined to isotropic films. (The only exception known to us are diamond films [16] and a pre-study of AlScN films [17]).

In our previous study, we showed that LU can be used to characterize textured AlScN(0001) films on silicon Si(001) substrates [17]. In these samples, the acoustic pulses were generated by laser pulses at the interface between the transparent AlScN film and the opaque silicon substrate. The fact that sapphire substrates are also transparent to the laser light constitutes an additional challenge and requires a metal layer to absorb the laser pulse. More recently, in-plane oriented AlScN(0001) films were grown on sapphire Al_2O_3(0001) substrates by magnetron sputter epitaxy [18] and then used for SAW resonator fabrication. The motivation was two-fold: sapphire substrate enables higher phase velocity and higher frequency of SAW and at the same time the higher material quality leads to improved electromechanical coupling and quality factor Q [19]. In order to further increase electromechanical coupling k_{eff}^2, epitaxial a-plane (non-polar) AlScN(11-20) SAW structures were successfully achieved on r-plane Al_2O_3(1-102) [20,21]. By aligning the SAW propagation direction with the piezoelectric constant d_{33}, an additional 85% improvement in k_{eff}^2 was demonstrated. In these earlier studies, a substrate off-cut angle was shown to strongly influence the crystalline quality of AlScN. As the ability to deposit high-quality non-polar AlScN is a rather recent discovery [22], no measurements of elastic constants have been reported so far. In this work we propose using LU to estimate the elastic properties of $Al_{0.77}Sc_{0.23}N$(11-20)/Al_2O_3(1-102), especially focusing on high anisotropy in this material system.

The paper is organized in the following way. In Section 2, the geometries of the systems investigated in this work are defined and their fabrication process and characterization are described. Moreover, details of the LU technique, as applied to the investigation of AlScN films, are given.

In the first part of Section 3, the results of our investigations of the elastic properties of the sapphire substrates are presented. This knowledge is a prerequisite for the determination of film properties by SAW-based LU. Values for the elastic moduli of sapphire were compiled from the literature. They were used as input for simulations of the SAW slowness curves of c-plane and r-plane sapphire. Simulated SAW velocities are compared with corresponding LU measurement results.

Next, experimental data are presented for dispersion curves of SAW propagating in $Al_{0.77}Sc_{0.23}N$ films. Both AlScN(0001)/Al_2O_3(0001) and AlScN(11-20)/Al_2O_3(1-102) were investigated. These first LU measurement results for AlScN films on sapphire are compared with simulated dispersion curves using the ab initio elastic and piezoelectric constants of [5]. Very good agreement is found between theory and experiment.

In order to assess the significance of this agreement for the individual material constants of $Al_{0.77}Sc_{0.23}N$, a sensitivity analysis is presented for the SAW dispersion curves in the frequency range accessible for our LU setup for both geometries studied.

2. Materials and Methods

2.1. Substrate and Film Geometries

As already mentioned, for the LU investigations of elastic properties of AlScN films reported here, two different layered structures were considered, i.e., AlScN(0001)/Al_2O_3(0001) and AlScN(11-20)/Al_2O_3(1-102). In Figure 1, the epitaxial relationship between the film and substrate material is shown schematically for the two systems. The Cartesian coordinate systems introduced after rotation with the x_3-axis normal to the surface correspond to the Euler angles $(\lambda, \mu, \theta) = (0°, 0°, \theta)$ for c-plane sapphire and $(\lambda, \mu, \theta) = (60°, 57.6°, \theta)$ for r-plane sapphire, respectively. The Euler angle $\mu = 57.6°$ for perfect r-plane orientation results from a ratio of lattice constants c/a = 2.73 for sapphire. For optimal growth conditions of the non-polar AlScN film, substrate off-cut angle χ is needed, which is related to the second Euler angle via $\chi = \mu - 57.6°$.

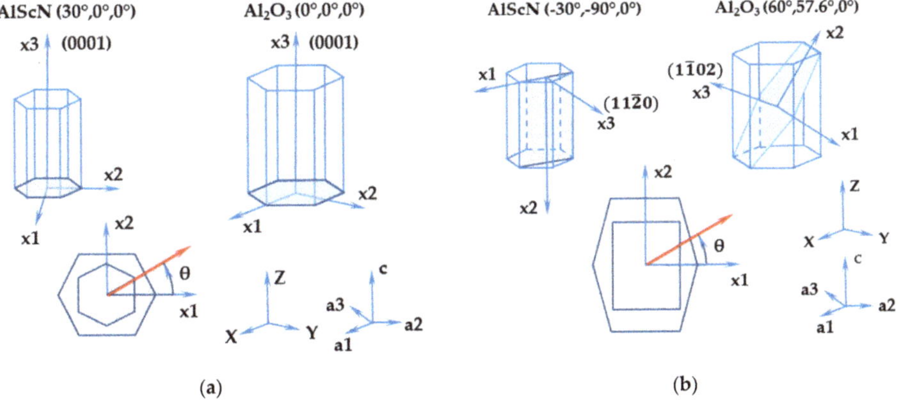

Figure 1. Crystal cuts of Al_2O_3 substrate and AlScN film, defined in the hexagonal crystal system (a1,a2,a3,c) and with the help of the two first Euler angles λ and μ (rotation with respect to Cartesian coordinate system XYZ). The third Euler angle θ defines the direction of SAW propagation (red arrows) on the surface planes (x1x2). (**a**) The c-plane (0001) cuts of both film and substrate; (**b**) a-plane AlScN(11-20) film on the r-plane Al_2O_3(1-102) substrate.

We note that the angle θ defines the wavevector direction of SAW propagating on the corresponding surface.

2.2. Fabrication of AlScN Films

AlScN(0001)/Al_2O_3(0001) and AlScN(11-20)/Al_2O_3(1-102) thin films were grown by the magnetron sputter epitaxy method [18,20,22]. In the case of non-polar AlScN, 3° substrate off-cut was found to be the best for high crystalline quality [20], a detailed growth optimization study including a proposed growth model for non-polar III-nitrides and different off-cut angles is published elsewhere [22]. All films were grown on ∅ = 100 mm substrates in an Evatec sputter cluster tool (base pressure ~5×10^{-6} Pa), using reactive pulsed-DC magnetron co-sputtering. Substrate rotation ensured the composition and thickness uniformity of the films. The scandium concentration x = 0.23 was achieved by setting the P(Al, 99.9995% pure) = 684 W and P(Sc, 99.99% pure) = 316 W. This specific Sc concentration was chosen as it allowed us to deposit AlScN thin films with very high crystalline quality and low density of abnormally oriented grains, leading to reliable evaluation of material properties by LU. Prior to deposition, the sapphire substrates were cleaned in-situ using Ar inductively coupled plasma (ICP) etching and the targets were pre-sputtered in Ar behind a closed shutter. More details about the growth conditions can be found in [18,20,22], all parameters except for the N_2 gas flow were kept the same, as summarized in Table 1.

Table 1. Growth parameters of $Al_{0.77}Sc_{0.23}N$.

Parameter	Value
N2 gas flow (sccm)	20 [1], 30 [2]
Target-to-substrate distance (mm)	65
Chuck temperature (°C)	450
Base pressure (Pa)	$<9 \times 10^{-6}$
$P_{Al} + P_{Sc}$ (W)	1000

[1] for AlScN(0001)/Al_2O_3(0001). [2] for AlScN(11-20)/Al_2O_3(1-102)

The scandium concentration (+/−2% error) was estimated using energy dispersive x-ray (scanning electron microscope Zeiss Auriga Crossbeam FIB-SEM with EDX spectroscopy from Bruker Quantax) on AlScN(0001)/Si(001) films deposited under the same conditions. This was done in order to avoid the overlap of Al emission peaks from the sapphire substrate and AlScN film [18]. No variation of composition was observed between the edge and the center of the wafers. The average film thickness (+/−3% error) was determined by spectroscopic ellipsometry (J.A. Woollam M-2000X), using a model optimized for AlScN from [23]. Thickness uniformity analysis indicated <3% variation in thickness across the wafer. The surface roughness of the films and the thickness of the metal layers were evaluated by atomic force microscopy (AFM, Bruker Dimension Icon) in tapping mode (not shown). The crystalline quality and in-plane orientation of AlScN films were confirmed by X-ray diffraction (XRD) [18,20] (not shown). Based on XRD pole figures analysis, the epitaxial relationship for c-plane AlScN could be defined as $[10\text{-}10]_{AlScN}//[11\text{-}20]_{sapphire}$ and $(0001)_{AlScN}//(0001)_{sapphire}$ and for a-plane AlScN $[0001]_{AlScN}//[1\text{-}101]_{sapphire}$ and $[1\text{-}100]_{AlScN}//[11\text{-}20]_{sapphire}$, respectively.

Details on the samples investigated by LU in this work are given in Table 2.

Table 2. Samples investigated in this work.

Material	Sample 1	Sample 2	Sample 3
Substrate	Al_2O_3(0001)	Al_2O_3(1-102)	Al_2O_3(1-102)
Piezoelectric film, thickness (nm)	$Al_{0.77}Sc_{0.23}N$(0001) 1035	- -	$Al_{0.77}Sc_{0.23}N$(11-20) 860
Metal film, thickness (nm)	Mo 50	Mo 50	Mo 50

2.3. Application of the Laser Ultrasound Approach

A schematic drawing of the LU set-up is shown in Figure 2. Acoustic pulses, traveling along the sample surface, are excited by laser pulses via the thermoelastic effect. This excitation mechanism is explained in detail in [24,25]. The optical pulses are generated by a passively Q-switched Nd:YAG laser (pulse duration: 1 ns; wavelength: 1064 nm, frequency-doubled to 532 nm). They are focused on a straight line on the surface with a cylindrical lens to minimize diffraction of the excited SAW and to achieve a well-defined wavevector direction.

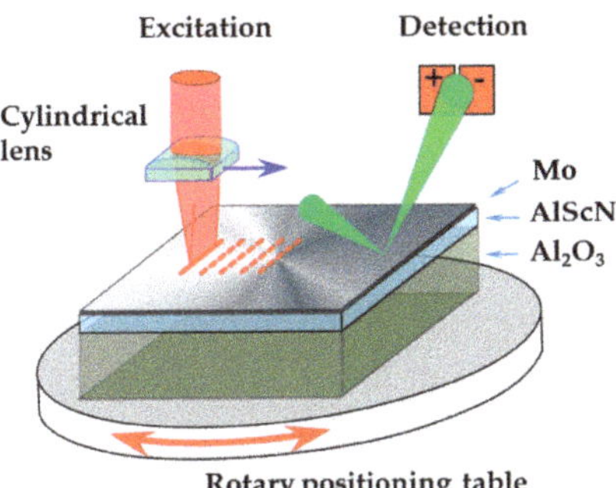

Figure 2. Experimental setup.

With the help of a continuous-wave laser, the acoustic surface pulses are detected with the probe-beam deflection method at a fixed observation point at the surface for various positions of the movable line source. Their shapes are recorded along with the distances of the observation points from the source. The presence of a film on the substrate surface causes the pulse shape to change with distance from the source. The quantity measured by probe-beam deflection at each observation point is the local surface slope as a function of time [25]. If the film and substrate are both transparent at the frequency of laser light used for the excitation of ultrasound pulses, the surface has to be coated with a thin metal film to ensure absorption of the laser pulses and enable thermoelastic excitation of SAW pulses. The effect of this additional layer has to be accounted for in the data interpretation and analysis.

By Fourier decomposition of the pulse shapes at the different distances between the line source and observation point, the phase velocities of each individual Fourier component are determined. In this way, a full dispersion curve of SAW with wavevectors vertical to the line source is obtained in one measurement cycle. Using a translation stage with a controlled stepper motor, the measurements are partly automated. The broad-band character of the LU method and its semi-automated operation renders it a fast tool for the characterization of near-surface elastic properties of materials. The achievable frequency range is partly material-dependent with an upper limit of 400 to 600 MHz for the materials investigated here with our set-up. With the help of a rotary positioning table for the sample, SAW dispersion curves can be measured for any wavevector direction on the surface.

Since even the third harmonic of the carrier frequency of the Nd:YAG pulse laser is below the absorption edges of sapphire and AlN, a thin absorbing layer is needed to enable thermoelastic excitation of acoustic waves. Test measurements have been carried out to find a suitable surface metallization Main criteria were signal quality, availability, and especially the influence of the metal coating on the SAW dispersion, which should be kept as small as possible. After careful consideration, molybdenum and titanium were identified as suitable metal coatings, but only data with Mo are shown here (Table 2).

The thickness of the AlScN film, together with the frequency range of the average SAW phase velocity, and to some extent, the mode character, determine the minimal penetration depth of the SAW in the substrate. For a rough estimate, we assume that the displacement field associated with a SAW has an appreciable magnitude up to distances of a wavelength from the surface. Our investigations refer to samples with AlScN films with a thickness of $\leq$1000 nm and reached frequencies up to 600 MHz. This suggests that even at the upper edge of the frequency range accessible for our LU experiments, the SAWs penetrate deeply into the substrate and consequently, their dispersion is strongly influenced by the density and elastic constants of the substrate. Therefore, these quantities have to be known to a high precision in order to be able to extract the elastic properties of the film from the SAW dispersion curves. The determination of elastic constants of the sapphire using LU is described in Section 3.1.1 and the off-cut angle assessment in Section 3.1.2, respectively.

For data processing and evaluation and the interpretation of the measurement results, simulations of dispersion curves and displacement fields have been carried out with a computer program based on a semi-analytic Greens function approach (see e.g., [26]). In these simulations, a time (t) and position (x) dependent traction vector

$$\boldsymbol{T} = \hat{\boldsymbol{k}}\, T_0 \exp[i(\omega t - \boldsymbol{k}\cdot\boldsymbol{x})] \tag{1}$$

is applied to the surface, where $\boldsymbol{k}$ is a wavevector in the surface plane, $\hat{\boldsymbol{k}}$ the unit vector pointing into the wavevector direction, $\omega/(2\pi)$ a frequency, and T_0 an arbitrary traction amplitude, respectively.

3. Results

3.1. Substrates

3.1.1. Elastic Constants of Sapphire Substrates

As mentioned in Section 2, for extraction of the elastic properties of AlScN films with thicknesses smaller than 1 μm from SAW dispersion curves in a frequency range below 600 MHz, the elastic moduli and the density of the substrate have to be known with high precision. In Table 3, data from the literature are compiled for Al_2O_3. In the case of c_{12}, the values of the different sources are at variance by more than 3%, and the value of c_{14} by more than 7%, in addition to the problem of finding the correct sign for this constant, which was resolved in [27,28].

Table 3. Elastic constants $c_{\mu\nu}$ for sapphire, GPa.

	Tefft et al. 1966 [29] *	Gieske et al. 1968 [30]	Goto et al. 1989 [31]	Landolt-Börnstein [32]	Auld 1990 [33]	Gladden 2004 [27]
c_{11}	497.4	497.6	497.3	496	494	497.5
c_{12}	164.0	162.6	162.8	159	158	162.7
c_{13}	112.3	117.2	116	114	114	115.5
c_{14}	−23.6	−22.9	−21.9	−23	−23	22.5
c_{33}	499.4	501.9	500.9	499	496	503.3
c_{44}	147.4	147.2	146.8	146	145	147.4

* Calculated from compliances. In all calculations we took the same density ρ = 3982 kg/m^3 and dielectric permittivities ε_{11} = 9.34 and ε_{33} = 11.54 from [33].

LU measurements were carried out on two samples with sapphire substrates in two different orientations, Al_2O_3(0001) (Sample 1 in Table 2) and Al_2O_3(1-102) (Sample 2 in Table 2). The surfaces of both samples were coated with a Mo layer of ~50 nm thickness. Sample 1 had an additional AlScN film of 1 μm thickness between the substrate and the Mo layer. Figure 3b shows the dependence of surface acoustic pulses in the Al_2O_3(1-102) sample, coated with a Mo layer only (Sample 2 in Table 2), detected at a fixed distance from the line source, on the wavevector direction (surface slope at observation point as a function of the arrival time and angle θ). In Figure 4, SAW dispersion curves obtained from the LU measurements on Sample 2 in Table 2 are shown for five different wavevector directions on the surface. Because of the small thickness of the metal layer in comparison to the SAW wavelength, the dispersion curves are essentially straight lines. The SAW phase velocity for the uncoated sapphire surface is obtained by extrapolating these straight lines to zero frequency. In this way, the dependence of the SAW phase velocity on the wavevector direction and hence the slowness curve of SAWs on this surface of sapphire can be determined experimentally.

Figure 3d shows the results of simulations of the local surface displacement u_3 for Al_2O_3(1-102) as a response to a surface traction (1). Note that the quantity measured by probe-beam deflection in our LU setup is the local surface slope, i.e., the directional derivative of u_3 along the SAW wavevector. In a homogeneous medium with planar surface, this response depends on the frequency $\omega/(2\pi)$ and wavelength $2\pi/|\boldsymbol{k}|$ of the excitation (surface traction in (1)) via the ratio $\omega/|\boldsymbol{k}|$ only (phase velocity in Figure 3c,d). The SAW phase velocities correspond to the maxima of $|u_3|$ (bright curves emerging in Figure 3c,d). Comparison with Figure 3b confirms the expected inverse behavior of the SAW phase velocity and the delay time of the SAW pulses.

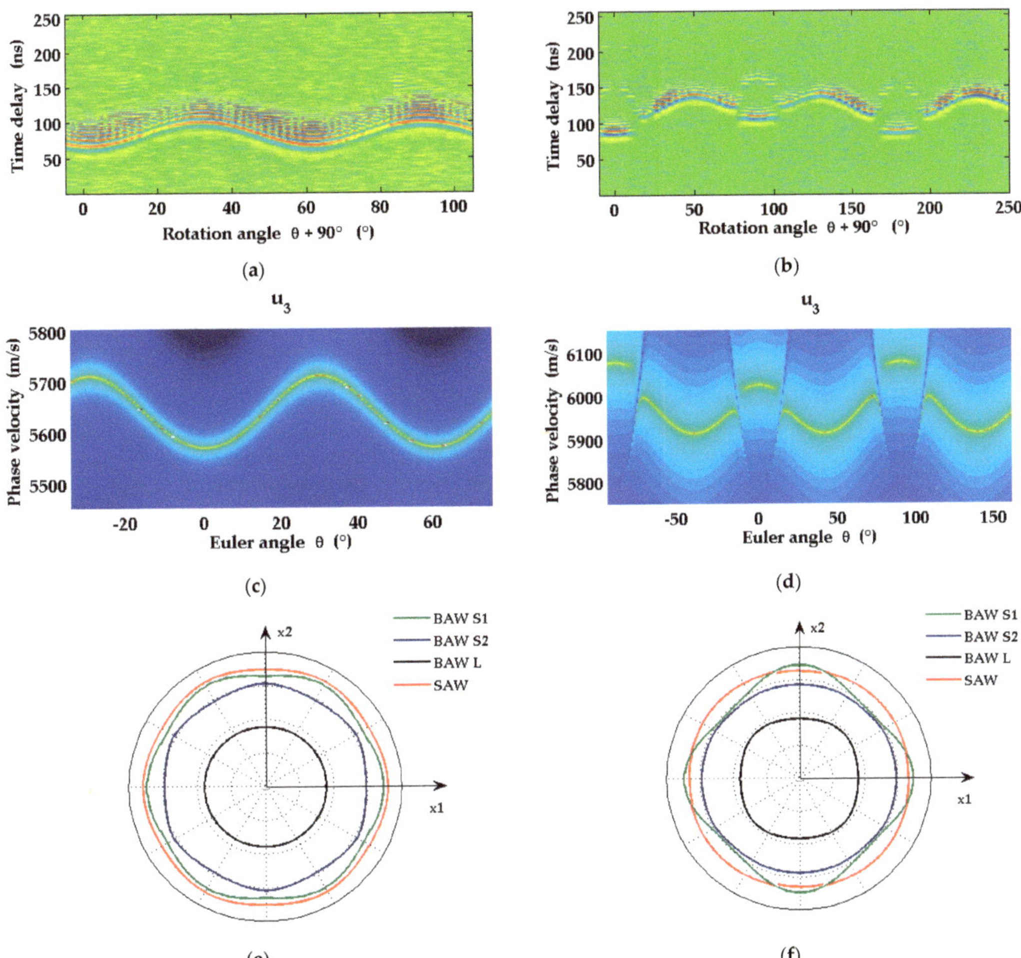

Figure 3. Dependencies on SAW wavevector direction for the sapphire cuts: c-plane Al_2O_3(0001) (**a**,**c**,**e**) with Euler angles (0°, 0°, θ) and r-plane Al_2O_3(1-102) (**b**,**d**,**f**) with Euler angles (60°, 57.6°, θ). Measured signal on the surface of sample 1 in Table 2 (**a**) and of sample 2 in Table 2 (**b**). Measurements taken at various angles using a rotary positioning table; (**c**,**d**) show the calculated SAW phase velocities on the surface of pure sapphire crystal as functions of Euler angle θ; (**e**,**f**) present the intersections of the slowness surface of bulk acoustic waves with the surface plane (quasi-longitudinal sheet (BAW L), two quasi-shear sheets (BAW S1 and BAW S2)) and the slowness curve for Rayleigh waves (SAW) on pure sapphire as functions of the Euler angle θ.

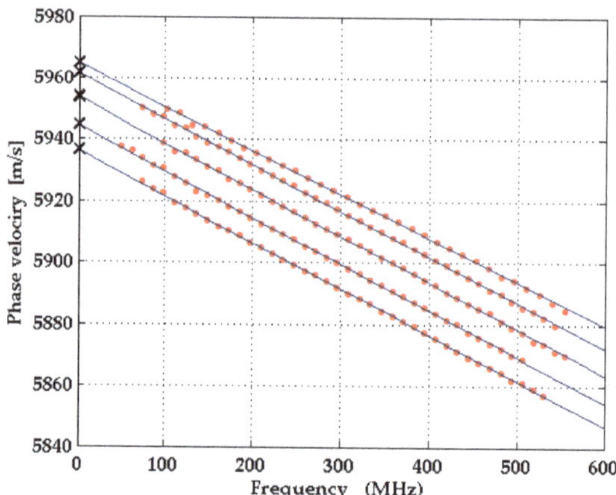

Figure 4. Experimental dispersion curves for the directions $\theta = 15°, 18°, 21°, 24°, 27°$ on r-plane sapphire with a 50 nm molybdenum film on top (Sample 2 in Table 2). Extrapolation of fitted straight lines to zero frequency gives the values of phase velocity (black crosses) on pure sapphire.

The weakness of the detected signal in certain intervals of angle θ, for example in the neighborhood of $\theta = 12°$ and $\theta = 78°$ in Figure 3d, can be explained by the smallness of the out-of-plane displacement amplitude $|u_3|$ in these angular intervals. The unusual features in the dependence of the SAW phase velocity on the wavevector direction in the case of the r-plane surface are related to a phenomenon visible in Figure 3f. Here, the slowness curve of SAW propagating on the r-plane surface of sapphire is shown together with the intersection curves of the surface and the three sheets of the slowness surface of bulk acoustic waves (quasi-longitudinal, fast quasi-shear, and slow quasi-shear). In the angular regions corresponding to wavevector directions with faint signals (two at the same wavevector direction) and comparatively strong variation of the SAW phase velocity, the SAW slowness curve and the intersection curve of the sheet of slow quasi-shear bulk waves approach, and a repulsion of these two modes can be seen. Here, a transition occurs from the usual situation of the SAW velocity being the lowest phase velocity of acoustic modes for a given wavevector direction parallel to the surface, to an interval of wavevector directions where the phase velocity of SAW is larger than that of the slow quasi-shear bulk waves. This effect does not occur in the case of $Al_2O_3(0001)$ (Figure 3c,e).

The experimental results for the SAW phase velocities in Al_2O_3(1-102) were then compared with the results of calculations performed with the semi-analytic Greens function method. Input data for these calculations are the sets of elastic moduli listed in Table 3. For the density of sapphire, the value $\rho = 3982$ kg/m^3 [31] was used, and for the dielectric constants we used the values $\varepsilon_{11} = 9.34$ and $\varepsilon_{33} = 11.54$ [33] in all our calculations. The off-cut angle of 3° was accounted for. It was found that the set of elastic constants reported by Gladden et al. [27] and also the constants provided by Gieske and Barsch [30], the latter after correction of the sign of c_{14}, fit best to the LU data and lead to very good agreement between calculated and measured phase velocities. A comparison of calculated phase velocities with three different input sets and measured values as functions of wavevector direction is provided in Figure 5.

This finding is confirmed by measurements and calculations of SAW dispersion curves for 1000 nm thick $Al_{0.77}Sc_{0.23}N(0001)/Al_2O_3(0001)$ coated with ~50 nm Mo (Sample 1 in Table 2). Two different wavevector directions were considered. The dispersion due to the presence of the AlScN film leads to a broadening of the detected SAW pulse shapes (Figure 3a), and the dispersion curves in Figure 6 are no longer straight lines. In the calculations, the elastic, piezoelectric, and dielectric constants of [5] were used. The material constants of the molybdenum layer, $c_{11} = 440$ GPa, $c_{44} = 126$ GPa, and density $\rho = 10{,}280$ kg/m^3, were taken from [34]. A comparison in Figure 6 shows very good agreement between experimental dispersion curves and those calculated with the elastic constants of sapphire taken from [30] with the corrected sign of c_{14}.

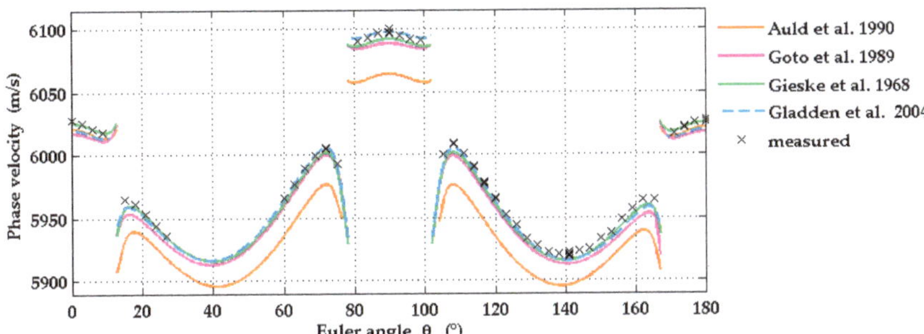

Figure 5. Theoretical dependencies of SAW phase velocities on wavevector direction for r-plane sapphire (solid and dashed lines) calculated using different sets of elastic constants taken from the literature (see Table 3), and compared with experimental data (black crosses).

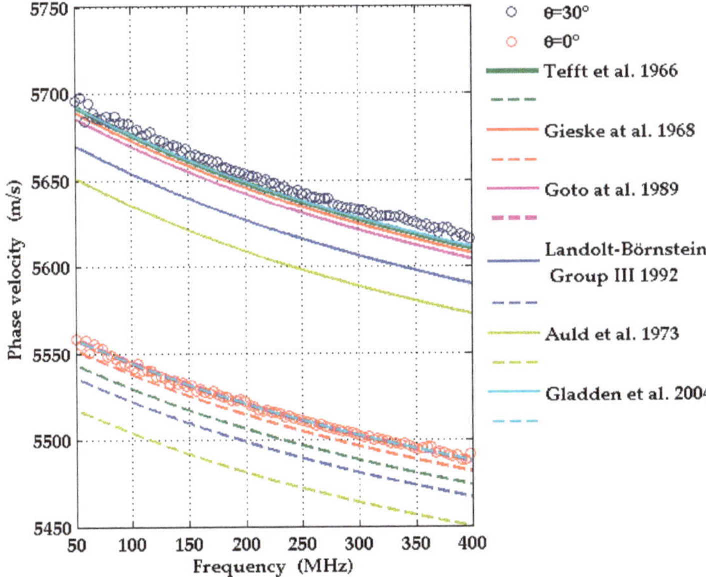

Figure 6. Experimentally obtained dispersion curves for the directions $\theta = 0°$ (red circles) and 30° (black circles) on the structure consisting of c-plane sapphire with a 1 µm thick $Al_{0.77}Sc_{0.23}N$ film and a 50 nm molybdenum film on top, compared with corresponding simulated dispersion curves (solid and dashed lines for $\theta = 30°$ and for $\theta = 0°$, respectively). The latter were obtained with material constants of sapphire from different authors (see Table 3).

3.1.2. Off-Cut Angle in Al_2O_3(1-102) Substrates

In order to find out to what accuracy the LU method is able to determine the off-cut angle χ of the r-plane sapphire substrate, the SAW phase velocity was calculated as a function of wavevector direction for the values $\chi = -3°, 0°, +3°$. A comparison of the calculated data with the measured SAW phase velocities in Figure 7 proves that an offset angle of $\pm 3°$ is clearly detectable. It also reveals in which regions of wavevector directions the phase velocity is particularly sensitive to the off-cut. The wavevector directions around $\theta = 90°$ are particularly suitable for the detection of off-cut angles larger or equal to 3°, whereas directions with θ near 0° seem to be more suitable for values of the off-cut angle smaller than −3°. However, the detected signals of the SAW pulses were comparatively weak for θ near 0°.

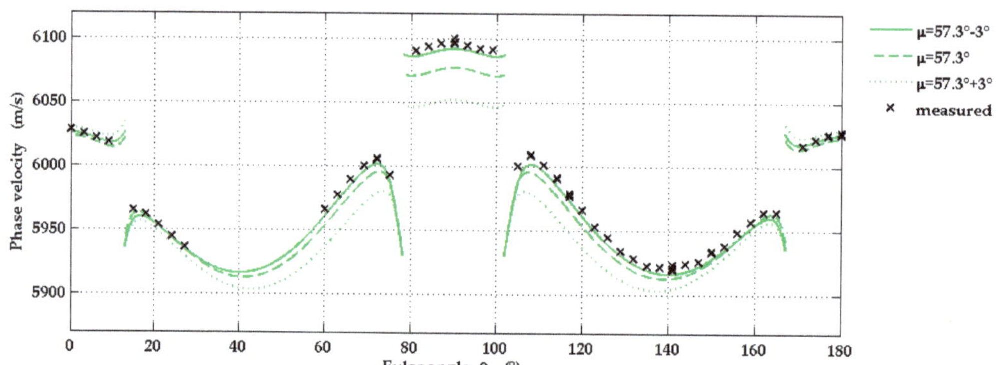

Figure 7. Experimental values of the SAW phase velocities for various propagation directions on r-plane sapphire (black crosses) and corresponding calculated curves, taking into account off-cut angles $\chi = \pm 3°$. Here the elastic constants from [30] were used for calculations.

Figure 8 shows calculated SAW phase velocities for the correct off-cut angle $\chi = -3°$ and four additional values of χ. The experimental data for the phase velocities shown in this figure are downshifted by 3 m/s as an attempt to correct a systematic error in our measurements due to a misalignment between the pump beam and the moving direction of the translation stage mentioned in Section 2.3.

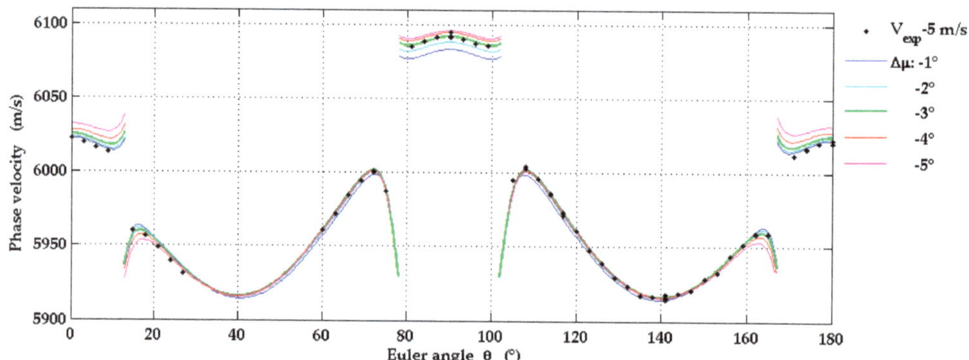

Figure 8. Influence of the off-cut angle χ on the angular dependence of the SAW phase velocity on Al_2O_3(1-102). Here, the experimental results are downshifted by 3 m/s.

The comparison of the data in Figure 8 suggests that the off-cut angle of the r-plane geometry can be determined by LU with an accuracy of almost 1° with an optimized alignment of the optical and mechanical components. Obviously the accuracy reachable in the determination of the off-cut angle by LU strongly depends on the precision to which the elastic constants and the density of sapphire are known. A 0.1 % change of the mass density translates into a phase velocity shift of approximately 3 m/s.

We note here that the SAW slowness curve on the r′-plane in Al_2O_3 (Euler angles (0°, 57.6°, θ)) differs substantially from that on the r-plane (Euler angles (60°, 57.6°, θ)), because the elastic constant c_{14} has a non-negligible magnitude in comparison to the other elastic constants (Table 3). In the case of hexagonal symmetry, the two planes would be equivalent [35]. On the r′-plane, the SAW slowness curve does not cross an intersection curve of the bulk wave slowness surface with the crystal surface. Therefore, r-plane and r′-plane samples can easily and quickly be distinguished by an LU measurement.

The correct sign of c_{14} has been clarified by ab initio calculations and new experiments ([27,28] and references therein). A change of sign of c_{14} causes an interchange of the elastic properties of r-plane and r′-plane.

3.2. Film on Substrate: $Al_{0.77}Sc_{0.23}N/Al_2O_3$

In order to acquire information about the elastic constants of AlScN films on sapphire substrates, LU measurements have been carried out for the two systems $Al_{0.77}Sc_{0.23}N$(0001)/ Al_2O_3(0001) and $Al_{0.77}Sc_{0.23}N$(11-20)/Al_2O_3(1-102) (in the following called "c-plane samples" and "a-plane samples" for short, respectively, Samples 1 and 3 in Table 2).

SAW dispersion curves were simulated with the semi-analytic Greens function method. Input data for the substrate material in these calculations were the set of elastic constants from [30] with the corrected sign of c_{14} along with the mass density of sapphire from [31] and dielectric constants from [33] (see Section 3.1.1).

The material constants chosen for the isotropic metal layer molybdenum are given in Section 3.1.1.

For the elastic and piezoelectric constants of the $Al_{0.77}Sc_{0.23}N$ films, we used the data obtained in ab initio calculations, based on the density functional theory, by Urban et al. [5]. In [5], an interpolation formula is given for the mass density and for each individual elastic constant $c_{\mu\nu}(x)$ and piezoelectric constant $e_{i\mu}(x)$ of $Al_{1-x}Sc_xN$ containing terms of up to second order in the scandium concentration x. The coefficients in these formulas were obtained by a fit to the corresponding material constants from ab initio calculations for a set of scandium concentrations x. As a second variant, Urban et al. suggest a rescaling of their interpolation formulas such that in the limit x to zero, the material constants take the values calculated for pure AlN. The SAW dispersion curves simulated with these two variants of material constants for the $Al_{0.77}Sc_{0.23}N$ films are indistinguishable on the scales of the graphs in Figures 6, 9 and 10.

For both independent dielectric constants ε_{11}, ε_{33} the value 17.56 was used in the calculations. It follows from an interpolation formula of experimental data for ε_{33}, provided in [36] (see also [10]), applied to the scandium concentration x = 0.23.

3.2.1. C-Plane Samples

The symmetry of the c-plane geometry implies that [0°, 30°] is an irreducible interval of angles θ for the SAW wavevector directions. LU measurements were taken for nine different wavevector directions corresponding to angles θ in this interval. For each direction, the SAW pulse shapes were recorded at 32 different distances from the source. From these data, the SAW dispersion curves were determined in a frequency range from 50 up to 400 MHz, shown in Figure 9. They exhibit a small, but non-negligible curvature.

Simulated dispersion curves for the nine wavevector directions are also presented in Figure 9 for comparison.

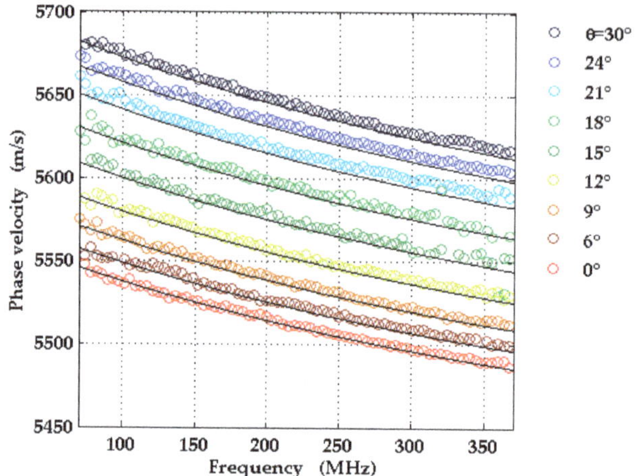

Figure 9. Experimental and theoretical dispersion curves for the different directions θ for $Ti/Al_{0.77}Sc_{0.23}N(0001)/Al_2O_3(0001)$. In the calculations, the elastic constants from [30] for sapphire and material constants from [5] for AlScN were used.

3.2.2. A-Plane Samples

Because of the lower symmetry of this configuration, an irreducible interval of SAW wavevector directions on this surface is the range of angles θ between 0° and 90°. In Figure 10, experimental dispersion curves are shown for wavevector directions with angles θ from 15° to 70° in steps of 5° and, in addition, for θ = 90°. For angles in the vicinity of 10° or 80°, the detected signals are very weak. The reason is presumably that here the surface displacements in the direction normal to the surface are much smaller than for other wavevector directions and the same frequency, since the SAW penetrates deeply into the substrate. Here, two very weak signals emerge because of mode repulsion (see Section 3.1.1) instead of one strong signal, leading to inaccuracies in the data processing.

The dispersion curves show a modest amount of curvature which is favorable for the extraction of elastic constants of the AlScN film. The experimental dispersion curves are in good agreement with the results of calculations, carried out with input data from [5].

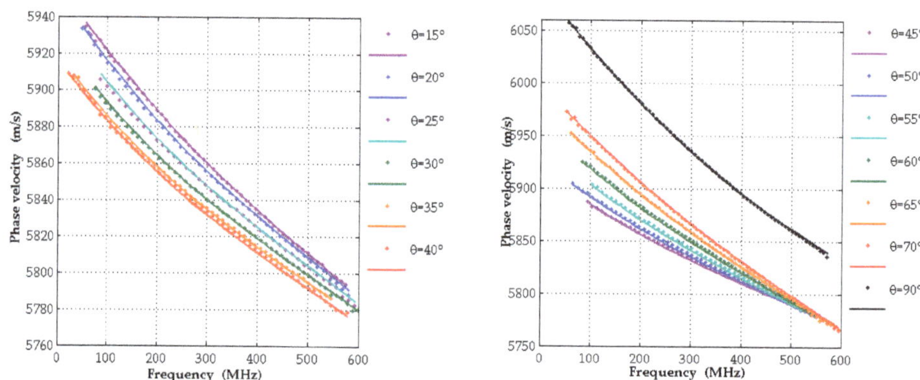

Figure 10. Experimental (dots) and theoretical (lines) dispersion curves for various SAW propagation directions for $Mo/Al_{0.77}Sc_{0.23}N(11\text{-}20)/Al_2O_3(1\text{-}102)$. For the calculations the constants from [5,30] and were used.

3.3. Sensitivity Analysis

In order to assess to what extent the agreement between the dispersion curves measured with the LU method and those simulated with the calculated material constants of [5] can be taken as a confirmation of the latter, the sensitivity of the dispersion curves with respect to changes of each single elastic constant of the AlScN film has been analyzed (Figure 11). With the same material data used for the simulation of the dispersion curves in Figures 9 and 10, we calculated the relative change $\Delta v/v$ of phase velocity v with a 1% increase for each of the five independent elastic constants $c_{\mu\nu}$ (see Table 3), while leaving the remaining material constants unchanged. Figure 11a,b show results of this calculation for the c-plane geometry and the a-plane geometry, respectively. $\Delta v/v$ is plotted as a function of wavevector direction at the fixed frequency 400 MHz. This frequency value has been chosen since it corresponds to the upper edge of the frequency interval of the measured dispersion curves for the c-plane samples and is located in the upper third of the frequency band for the a-plane geometry. With increasing frequency, the fraction of the SAW displacement field localized in the AlScN film is expected to rise as well. When comparing the relative velocity changes in Figure 11a,b, one has to account for the slightly different thicknesses of AlScN layer in the two types of samples (1 μm for the c-plane, 860 nm for a-plane samples).

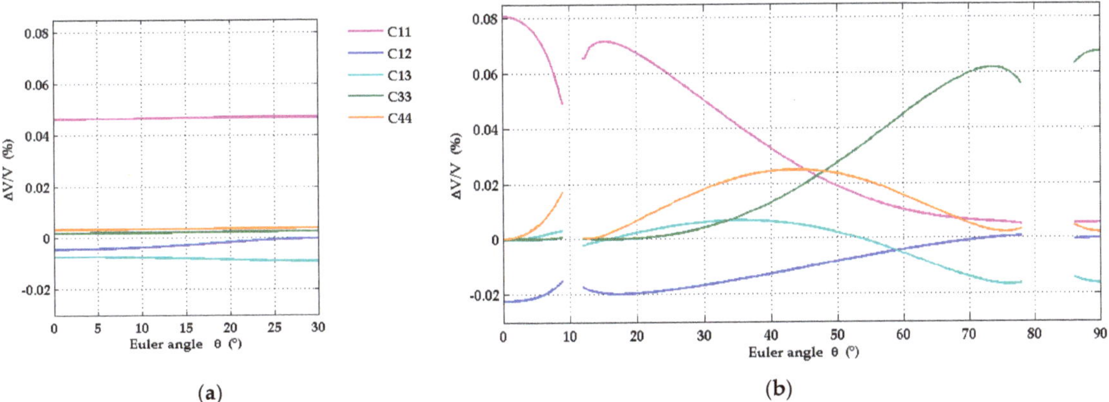

Figure 11. Calculated relative change of phase velocity with 1% increase of elastic constants for $Mo/Al_{0.77}Sc_{0.23}/Al_2O_3$ structures with 1 μm thick $Al_{0.77}Sc_{0.23}N$ film and additional Mo coating with 50 nm thickness, as a function of SAW propagation direction on the surface. (**a**) case of AlScN(0001)/Al_2O_3(0001) and (**b**) AlScN(11-20)/Al_2O_3(1-102).

In the case of the c-plane samples, the sensitivities of the SAW phase velocity with respect to changes in the elastic constants of the AlScN film are comparatively small, except for the constant c_{11}. The isotropy of the hexagonal film in the surface plane and the moderate deviations of the SAW slowness curve from a circle on c-plane sapphire (Figure 3e) are reflected in a very weak dependence of the sensitivities with respect to the wavevector direction (Figure 11a), with the exception of the sensitivity to changes of c_{12}, which vanishes at the wavevector direction with Euler angle $\theta = 30°$.

In the case of the a-plane geometry, the sensitivities exhibit a remarkable dependence on the wavevector direction. This is due to the strong anisotropy of the film in the surface plane, and it is also associated with the strong variation of the SAW mode pattern in the neighborhood of the Euler angles θ, where the SAW slowness curve of r-plane sapphire crosses the intersection curve of the slowness surface of acoustic bulk waves. Moreover, the sensitivities are on average clearly larger than those on the c-plane samples.

A feature of particular interest is the relative size of the sensitivities for the directions with Euler angles θ in the vicinity of 0° on the one hand and in the neighborhood of 90° on the other. In the first range of wavevector directions, the sensitivities with respect to c_{11} and c_{12} dominate, while the sensitivities with respect to the other elastic moduli are largely negligible. In the second range, the dispersion curve is mainly sensitive to c_{33} and c_{13} and the other elastic constants play a largely negligible role. Knowledge of this behavior should be very helpful for fitting strategies to extract the elastic constants from measured dispersion curves. At wavevector directions with Euler angles θ around 45°, the relative velocity variations with relative changes of c_{11}, c_{33}, and c_{44} are of comparable size. In general, one may notice that for each elastic constant there are ranges of wavevector directions where this constant has a non-negligible influence on the SAW dispersion curves. Figure 12 shows how the sensitivities vary as functions of frequency for the fixed wavevector directions with Euler angles θ = 0°, 45° and 90°. The sensitivities with respect to almost all elastic constants increase at higher frequencies because of the increasing localization of the SAW displacement field in the AlScN film. In the case of θ = 0°, the sensitivities with respect to c_{11} and c_{12} dominate over the whole frequency range from zero to 600 MHz, and likewise the sensitivities with respect to c_{33} and c_{13} in the case of θ = 90°.

The sensitivities of the dispersion curves for SAW in AlScN films on sapphire, discussed above, may be compared with those for a c-plane $Al_{0.68}Sc_{0.32}N$ film on a Si(001) substrate, presented in [17] for two different wavevector directions. (Note that the data in Figure 3 of [17] refer to a relative change of 10% of the elastic and piezoelectric constants. The thickness of the AlScN(0001)/Si(001) was ~1 µm.) Remarkably, the sensitivities for the c-plane film on silicon are of comparable size to those of the a-plane sample. However, in both wavevector directions in the c-plane film on the silicon substrate, the influence of the elastic constant c_{11} of AlScN dominates.

Figure 3b in [17] clearly shows that the sensitivities of the SAW phase velocities with respect to the piezoelectric constants (from D. Urban et al. in Reference [5]) are by more than one order of magnitude smaller than the ones with respect to the elastic constants.

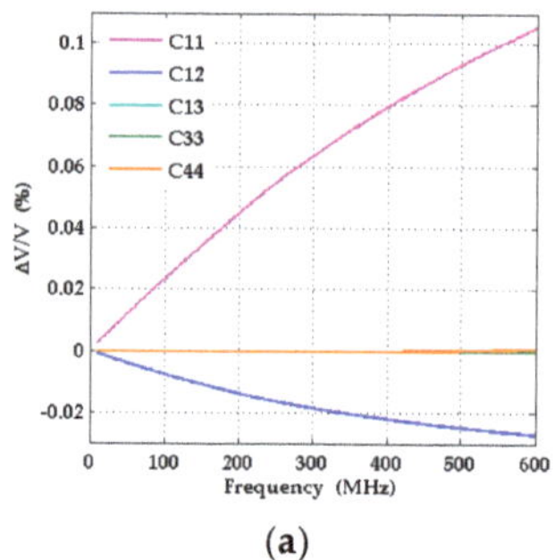

(a)

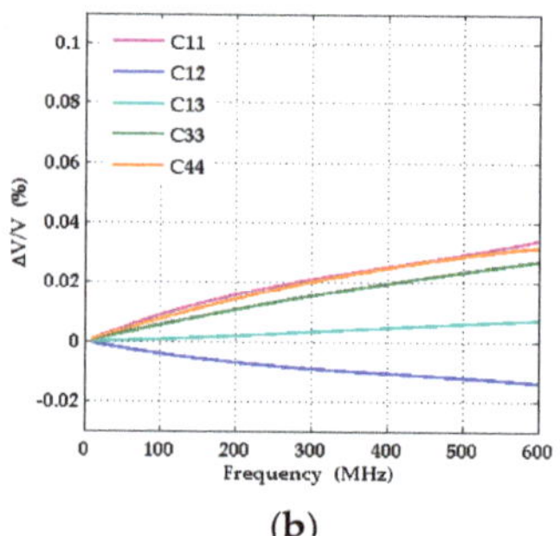

(b)

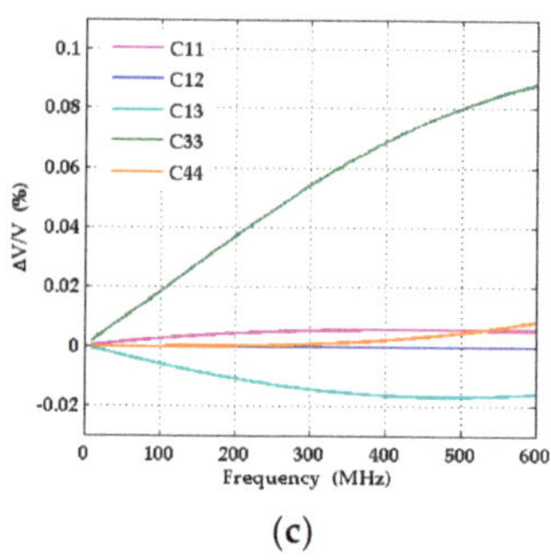

(c)

Figure 12. Relative change of phase velocity with 1% increase of elastic constants in the structure AlScN(11-20)/Al_2O_3(1-102) with 1000 nm thick $Al_{0.77}Sc_{0.23}N$ film and 50 nm molybdenum film as function of frequency for the cases (**a**) θ = 0°; (**b**) θ = 45°; (**c**) θ = 90°.

4. Discussion and Conclusions

The main results of the investigations reported in this contribution may be summarized as follows:

- SAW dispersion curves were measured by laser ultrasound for various wavevector directions in c-plane and a-plane $Al_{0.77}Sc_{0.23}N$ films on sapphire substrates. They are in very good agreement with the corresponding theoretical dispersion curves computed with the elastic moduli and piezo-electric constants obtained in ab initio calculations by Urban et al. [5]. The theoretical elastic constants for $Al_{1-x}Sc_xN$ in [5] are given in the form of interpolation formulas quadratic in the parameter x. For the Sc concentrations, x = 0.14 and x = 0.32, the authors of [5] compare their calculated elastic constants with corresponding data determined by Kurz et al. [10] with the help of SAW resonators. For both concentrations, the agreement is very good (2% deviation on average, less than 5% in the worst case). This confirms the high-quality of the theoretical data in [5] and may also be regarded as an additional, indirect confirmation of the measured SAW dispersion curves presented here.

- The off-cut angle of 3° in the r-plane sapphire substrate, needed to provide optimal growth conditions for a-plane AlScN, can easily be detected with the laser ultrasound setup described above.
- Sensitivities of the SAW dispersion curves with respect to variations of each of the independent elastic constants of the AlScN film have been calculated as functions of the SAW wavevector direction and frequency. Due to the strong anisotropy of the a-plane geometry in the surface plane, the sensitivities show strong variations with wavevector direction. In certain intervals of wavevector directions, only a few elastic constants predominantly influence the dispersion curve, while the sensitivity with respect to the others is much smaller. In other wavevector directions, certain elastic constants change their role with respect to their influence on the SAW dispersion curves. This finding can be made use of in the development of fit strategies for an accurate determination of all elastic constants of AlScN films.

The size of the sensitivities at around 400 MHz leads us to the following conclusion. Assuming that a relative change of velocity by 6 m/s is resolvable in the LU experiment, a change of 2% in the elastic constants c_{11} of c_{33} of the AlScN should be detectable. With an efficient use of the anisotropy of both the substrate and the film and with sufficiently small attainable SAW wavelengths on the scale of the film thickness (i.e., a sufficiently large frequency range), a determination of the elastic constants of AlScN films should be achievable. Even in the long-wavelength limit, when the dispersion curves are essentially straight lines, certain combinations of material constants can be extracted.

If the minimal achievable SAW wavelength is larger or of the order of the film thickness, such that the SAW displacement field penetrates deeply into the substrate, the elastic properties of the substrate as well as its orientation (e.g., off-cut angles) have to be known to a high precision. On the other hand, if the frequency range can be considerably extended to higher values, higher-order guided acoustic modes can be used in addition to the lowest SAW to gain information about the elastic moduli of the film [11].

The sensitivity analysis in [17] confirms that the piezoelectric constants have a very small influence on the SAW dispersion curves if compared to the elastic constants. However, depending on the required accuracy in the determination of the elastic moduli, auxiliary measurements with an alternative technique, such as the resonator method [10,11] will be needed.

The measured SAW dispersion curves for c-plane and a-plane $Al_{0.77}Sc_{0.23}N$ compare very favorably with dispersion curves calculated with the data for the elastic and piezoelectric constants in [5] for all SAW wavevector directions. These data were obtained in ab initio calculations treating the atomic positions as static, which means that phononic thermal contributions are disregarded. Very rough estimates on the basis of data for the temperature dependence of Young's modulus of AlN ceramics [37] and predictions of the temperature dependence of the elastic moduli of AlN with highly simplifying assumptions [38] suggest that the thermal contributions at room temperature to the elastic moduli are smaller than 1% of their total value. The sensitivities of the SAW dispersion curves, discussed above, imply that such small variations cannot be resolved with our current laser ultrasound setup.

In conclusion, the results for AlScN films on sapphire substrates, presented in this work, confirm that laser ultrasound can be applied as a viable tool for the determination of elastic properties of anisotropic, including piezoelectric films on anisotropic substrates. This requires measurements of SAW dispersion curves for various propagation directions, which can be pre-selected by simulations and a detailed sensitivity analysis. In comparison to isotropic films, anisotropy poses an additional challenge because of an increased number of independent elastic constants. At the same time, the anisotropy of the film and of the substrate offers the possibility of gaining additional information on different elastic constants from certain different SAW propagation directions.

Author Contributions: Conceptualization, E.A.M., A.D., A.P.M., A.Ž.; methodology, A.M.L., E.A.M., P.D.P.; software, A.M.L., E.A.M., A.P.M., P.D.P.; validation, E.A.M.; formal analysis, E.A.M., O.R.; investigation, E.A.M., O.R.; resources, A.D., A.N., A.Ž.; data curation, E.A.M., O.R.; writing, A.P.M., E.A.M., A.D., A.N., A.Ž.; visualization, E.A.M., A.M.L. All authors have read and agreed to the published version of the manuscript.

Funding: This research was funded by Fraunhofer Society, project no. 005-601187. A.N. was funded by the COMET Centre ASSIC Austrian Smart Systems Integration Research Center, which is funded by BMK, BMDW, and the Austrian provinces of Carinthia and Styria, within the framework of COMET—Competence Centers for Excellent Technologies. The COMET program is run by FFG.

Institutional Review Board Statement: Not applicable.

Informed Consent Statement: Not applicable.

Data Availability Statement: Data are available from the corresponding authors upon reasonable request.

Acknowledgments: The authors thank Lutz Kirste and the Structural analysis team at Fraunhofer IAF for their help with the X-ray diffraction experiments and Niclas Feil and Oliver Ambacher for fruitful discussions.

Conflicts of Interest: The authors declare no conflict of interest. The funders had no role in the design of the study; in the collection, analyses, or interpretation of data; in the writing of the manuscript, or in the decision to publish the results.

References

1. Akiyama, M.; Kamohara, T.; Kano, K.; Teshigahara, A.; Takeuchi, Y.; Kawahara, N. Enhancement of piezoelectric response in scandium aluminum nitride alloy thin films prepared by dual reactive cosputtering. *Adv. Mater.* **2009**, *21*, 593–596. [CrossRef] [PubMed]
2. Wingqvist, G.; Tasnádi, F.; Žukauskaitė, A.; Birch, J.; Arwin, H.; Hultman, L. Increased electromechanical coupling in w-$Sc_xAl_{1-x}N$. *Appl. Phys. Lett.* **2010**, *97*, 112902. [CrossRef]
3. Yanagitani, T.; Arakawa, K.; Kano, K.; Teshigahara, A.; Akiyama, M. Giant shear mode electromechanical coupling coefficient k15 in c-axis tilted ScAlN films. In Proceedings of the 2010 IEEE International Ultrasonics Symposium, San Diego, CA, USA, 11–14 October 2010; pp. 2095–2098.
4. Tasnádi, F.; Alling, B.; Höglund, C.; Wingqvist, G.; Birch, J.; Hultman, L.; Abrikosov, I.A. Origin of the anomalous piezoelectric response in wurtzite $Sc_xAl_{1-x}N$ alloys. *Phys. Rev. Lett.* **2010**, *104*, 137601. [CrossRef] [PubMed]
5. Urban, D.F.; Ambacher, O.; Elsässer, C. First-principles calculation of electroacoustic properties of wurtzite (Al, Sc)N. *Phys. Rev. B* **2021**, *103*, 115204. [CrossRef]
6. Olsson, R.H.; Tang, Z.; D'Agati, M. Doping of aluminum nitride and the impact on thin film piezoelectric and ferroelectric device performance. In Proceedings of the 2020 IEEE Custom Integrated Circuits Conference (CICC), Boston, MI, USA, 22–25 March 2020; pp. 1–6. [CrossRef]
7. Ichihashi, H.; Yanagitani, T.; Suzuki, M.; Takayanagi, S.; Matsukawa, M. Effect of Sc concentration on shear wave velocities in ScAlN films measured by micro-Brillouin scattering technique. In Proceedings of the 2014 IEEE International Ultrasonics Symposium, Chicago, IL, USA, 3–6 September 2014; pp. 2521–2524.
8. Ichihashi, H.; Yanagitani, T.; Suzuki, M.; Takayanagi, S.; Kawabe, M.; Tomita, S.; Matsukawa, M. Acoustic-wave velocities and refractive indices in an m-plane GaN single-crystal plate and c-axis-oriented ScAlN films measured by Brillouin scattering techniques. *IEEE Trans. Ultrason. Ferroelectr. Freq. Control* **2016**, *63*, 717–725. [CrossRef]
9. Carlotti, G.; Sadhu, J.; Dumont, F. Dependence of the different elastic constants of ScAlN films on Sc content: A Brillouin scattering study with polarization analysis. In Proceedings of the 2017 IEEE International Ultrasonics Symposium, Washington, DC, USA, 6–9 September 2017; pp. 1–5.
10. Kurz, N.; Ding, A.; Urban, D.F.; Lu, Y.; Kirste, L.; Feil, N.M.; Žukauskaitė, A.; Ambacher, O. Experimental determination of the electroacoustic properties of thin film AlScN using surface acoustic wave resonators. *J. Appl. Phys.* **2019**, *126*, 075106. [CrossRef]
11. Feil, N.M.; Mayer, E.; Nair, A.; Christian, B.; Ding, A.; Sun, S.; Mihalic, S.; Kessel, M.; Žukauskaitė, A.; Ambacher, O. Properties of higher-order surface acoustic wave modes in $Al_{1-x}Sc_xN$/sapphire structures. *J. Appl. Phys.* **2021**, *130*, 164501. [CrossRef]
12. Soluch, W.; Brzozowski, E.; Lysakowska, M.; Sadura, J. Determination of mass density, dielectric, elastic, and piezoelectric constants of bulk GaN crystal. *IEEE Trans. Ultrason. Ferroelectr. Freq. Control* **2011**, *58*, 2469–2474. [CrossRef]

13. Zhang, S.; Fu, W.Y.; Holec, D.; Humphreys, C.J.; Moram, M.A. Elastic constants and critical thicknesses of ScGaN and ScAlN. *J. Appl. Phys.* **2013**, *114*, 243516. [CrossRef]
14. Caro, M.A.; Zhang, S.; Ylilammi, M.; Riekkinen, T.; Moram, M.A.; Lopez-Acevedo, O.; Molarius, J.; Laurila, T. Piezoelectric coefficients and spontaneous polarization of ScAlN. *J. Phys. Condens. Matter* **2015**, *27*, 245901, Erratum in *J. Phys. Condens. Matter* **2015**, *27*, 279602. [CrossRef]
15. Momida, H.; Teshigahara, A.; Oguchi, T. Strong enhancement of piezoelectric constants in $Sc_xAl_{1-x}N$: First-principles calculations. *AIP Adv.* **2016**, *6*, 065006. [CrossRef]
16. Shen, Z.H.; Lomonosov, A.M.; Hess, P.; Fischer, M.; Gsell, S.; Schreck, M. Multimode photoacoustic method for the evaluation of mechanical properties of heteroepitaxial diamond layers. *J. Appl. Phys.* **2010**, *108*, 083524. [CrossRef]
17. Rogall, O.; Feil, N.M.; Ding, A.; Mayer, E.; Pupyrev, P.D.; Lomonosov, A.M.; Žukauskaitė, A.; Ambacher, O.; Mayer, A.P. Determining elastic constants of AlScN films on silicon substrates by laser ultrasonics. In Proceedings of the 2020 IEEE International Ultrasonics Symposium, Las Vegas, NV, USA, 6–11 September 2020; pp. 1–4.
18. Lu, Y.; Reusch, M.; Kurz, N.; Ding, A.; Christoph, T.; Prescher, M.; Kirste, L.; Ambacher, O.; Žukauskaitė, A. Elastic modulus and coefficient of thermal expansion of piezoelectric $Al_{1-x}Sc_xN$ (up to x = 0.41) thin films. *APL Mater.* **2018**, *6*, 076105. [CrossRef]
19. Ding, A.; Reusch, M.; Lu, Y.; Kurz, N.; Lozar, R.; Christoph, T.; Driad, R.; Ambacher, O.; Žukauskaitė, A. Investigation of temperature characteristics and substrate influence on AlScN-based SAW resonators. In Proceedings of the 2018 IEEE International Ultrasonics Symposium, Kobe, Japan, 22–25 October 2018; pp. 1–9.
20. Ding, A.; Kirste, L.; Lu, Y.; Driad, R.; Kurz, N.; Lebedev, V.; Christoph, T.; Feil, N.M.; Lozar, R.; Metzger, T.; et al. Enhanced electromechanical coupling in SAW resonators based on sputtered non-polar $Al_{0.77}Sc_{0.23}N$ (11-20) thin films. *Appl. Phys. Lett.* **2020**, *116*, 101903. [CrossRef]
21. Ding, A.; Driad, R.; Lu, Y.; Feil, N.M.; Kirste, L.; Christoph, T.; Ambacher, O.; Žukauskaitė, A. Non-polar a-plane AlScN (1120) thin film based SAW resonators with significantly improved electromechanical coupling. In Proceedings of the 2020 IEEE International Ultrasonics Symposium, Las Vegas, NV, USA, 6–11 September 2020; pp. 1–4.
22. Nair, A.; Kessel, M.; Kirste, L.; Žukauskaitė, A. Growth optimization of non-polar $Al_{0.7}Sc_{0.3}N(11\text{-}20)/Al_2O_3$ ($Al_2O_3(1\text{-}102)$) thin films using reactive magnetron sputter epitaxy. *Phys. Status Solidi* **2022**. [CrossRef]
23. Baeumler, M.; Lu, Y.; Kurz, N.; Kirste, L.; Prescher, M.; Christoph, T.; Wagner, J.; Žukauskaitė, A.; Ambacher, O. Optical constants and band gap of wurtzite $Al_{1-x}Sc_xN/Al_2O_3$ prepared by magnetron sputter epitaxy for scandium concentrations up to x = 0.41. *J. Appl. Phys.* **2019**, *126*, 045715. [CrossRef]
24. Gusev, V.E.; Karabutov, A.A. *Laser Optoacoustics*; American Institute of Physics: New York, NY, USA, 1993.
25. Lomonosov, A.; Mayer, A.P.; Hess, P. Laser-based surface acoustic waves in materials science. In *Experimental Methods in the Physical Sciences*; Levy, M., Bass, H.E., Stern, R., Eds.; Academic Press: San Diego, CA, USA, 2001; Volume 39, pp. 65–134.
26. Peach, R. On the existence of surface acoustic waves on piezoelectric substrates. *IEEE Trans. Ultrason. Ferroelectr. Freq. Control* **2001**, *48*, 1308–1320. [CrossRef]
27. Gladden, J.R.; So, J.H.; Maynard, J.D.; Saxe, P.W.; Le Page, Y. Reconciliation of ab initio theory and experimental elastic properties of Al_2O_3. *Appl. Phys. Lett.* **2004**, *85*, 392–394. [CrossRef]
28. Hovis, D.B.; Reddy, A.; Heuer, A.H. X-ray elastic constants for α-Al_2O_3. *Appl. Phys. Lett.* **2006**, *88*, 131910. [CrossRef]
29. Tefft, W.E. Elastic constants of synthetic single crystal corundum. *J. Res. Natl. Bur. Stand.-A Phys. Chem. (Natl. Bur. Stand.)* **1966**, *70A*, 277–280. [CrossRef]
30. Gieske, J.H.; Barsch, G.R. Pressure dependence of the elastic constants of single crystalline aluminum oxide. *Phys. Status Solidi* **1968**, *29*, 121–131. [CrossRef]
31. Goto, T.; Anderson, O.L.; Ohno, I.; Yamamoto, S. Elastic constants of corundum up to 1825 K. *J. Geophys. Res.* **1989**, *94*, 7588–7602. [CrossRef]
32. Every, A.G.; McCurdy, A.K. Second and higher order elastic constants. In *Landolt-Börnstein Group III Condensed Matter (Second and Higher Order Elastic Constants)*; Nelson, D.F., Ed.; Springer: Berlin, Germany, 1992; Volume 29A.
33. Auld, B.A. *Acoustic Fields and Waves in Solids*, 2nd ed.; Krieger Publishing: Malabar, FL, USA, 1990; Volume 1.
34. Wikipedia. Available online: https://en.wikipedia.org/wiki/Molybdenum (accessed on 7 June 2022).
35. Grundmann, M.; Lorenz, M. Azimuthal anisotropy of rhombohedral (corundum phase) heterostructures. *Phys. Status Solidi B* **2021**, *258*, 2100104. [CrossRef]
36. Ambacher, O.; Christian, B.; Feil, N.; Urban, D.F.; Elsässer, C.; Prescher, M.; Kirste, L. Wurtzite ScAlN, InAlN, and GaAlN crystals, a comparison of structural, elastic, dielectric, and piezoelectric properties. *J. Appl. Phys.* **2021**, *130*, 045102. [CrossRef]
37. Bruls, R.J.; Hintzen, H.T.; de With, G.; Metselaar, R. The temperature dependence of the Young's modulus of $MgSiN_2$, AlN and Si_3N_4. *J. Eur. Ceram. Soc.* **2001**, *21*, 263–268. [CrossRef]
38. Reeber, R.R.; Wang, K. High temperature elastic constant prediction of some group III nitrides. *MRS Internet J. Nitride Semicond. Res.* **2001**, *6*, 3. [CrossRef]

micromachines

Article

Vertical and Lateral Etch Survey of Ferroelectric AlN/$Al_{1-x}Sc_xN$ in Aqueous KOH Solutions

Zichen Tang [1], Giovanni Esteves [2], Jeffrey Zheng [3] and Roy H. Olsson III [1,*]

1 Electrical and Systems Engineering Department, University of Pennsylvania, Philadelphia, PA 19104, USA; zichent@seas.upenn.edu
2 Microsystems Engineering, Science and Applications (MESA), Sandia National Laboratories, Albuquerque, NM 87123, USA; gesteve@sandia.gov
3 Material Science and Engineering Department, University of Pennsylvania, Philadelphia, PA 19104, USA; jxzheng@seas.upenn.edu
* Correspondence: rolsson@seas.upenn.edu

Abstract: Due to their favorable electromechanical properties, such as high sound velocity, low dielectric permittivity and high electromechanical coupling, Aluminum Nitride (AlN) and Aluminum Scandium Nitride ($Al_{1-x}Sc_xN$) thin films have achieved widespread application in radio frequency (RF) acoustic devices. The resistance to etching at high scandium alloying, however, has inhibited the realization of devices able to exploit the highest electromechanical coupling coefficients. In this work, we investigated the vertical and lateral etch rates of sputtered AlN and $Al_{1-x}Sc_xN$ with Sc concentration x ranging from 0 to 0.42 in aqueous potassium hydroxide (KOH). Etch rates and the sidewall angles were reported at different temperatures and KOH concentrations. We found that the trends of the etch rate were unanimous: while the vertical etch rate decreases with increasing Sc alloying, the lateral etch rate exhibits a V-shaped transition with a minimum etch rate at x = 0.125. By performing an etch on an 800 nm thick $Al_{0.875}Sc_{0.125}N$ film with 10 wt% KOH at 65 °C for 20 min, a vertical sidewall was formed by exploiting the ratio of the $\{10\bar{1}\bar{1}\}$ planes and $\{1\bar{1}00\}$ planes etch rates. This method does not require preliminary processing and is potentially beneficial for the fabrication of lamb wave resonators (LWRs) or other microelectromechanical systems (MEMS) structures, laser mirrors and Ultraviolet Light-Emitting Diodes (UV-LEDs). It was demonstrated that the sidewall angle tracks the trajectory that follows the $\{\bar{1}2\bar{1}2\}$ of the hexagonal crystal structure when different c/a ratios were considered for elevated Sc alloying levels, which may be used as a convenient tool for structure/composition analysis.

Keywords: aluminum scandium nitride (AlScN); aluminum nitride (AlN); wet etch; potassium hydroxide (KOH); ferroelectric

Citation: Tang, Z.; Esteves, G.; Zheng, J.; Olsson, R.H., III. Vertical and Lateral Etch Survey of Ferroelectric AlN/$Al_{1-x}Sc_xN$ in Aqueous KOH Solutions. *Micromachines* **2022**, *13*, 1066. https://doi.org/10.3390/mi13071066

Academic Editor: Agnė Žukauskaitė

Received: 28 May 2022
Accepted: 27 June 2022
Published: 2 July 2022

Publisher's Note: MDPI stays neutral with regard to jurisdictional claims in published maps and institutional affiliations.

1. Introduction

Acoustic filters are key components in the evolution of radio frequency communication systems. Quartz crystals, lithium tantalate ($LiTiO_3$) and lithium niobate ($LiNbO_3$) surface acoustic wave (SAW) filters made the first two generations (Global System for Mobile communication, or GSM and Code-Division Multiple Access, or CDMA) of mobile networks possible [1,2], followed by AlN Bulk Acoustic Wave (BAW) resonators for the 3G (WCDMA) and 4G-LTE networks [3]. As the road map to 5G and beyond unfolded, filter requirements have increased and call for filters that exhibit lower insertion loss, higher temperature stability, steeper skirts and wider bandwidth. To achieve these metrics, a new material with enhanced piezoelectric coefficients is needed, since it can lead to more efficient coupling that directly transfers to increased filter bandwidth. In 2001, Takeuchi identified, by first principal calculation, that wurtzite AlScN had the potential [4] to achieve higher piezoelectric coefficients than AlN. Akiyama et al. later showed this through measurement of co-sputtered AlScN films [5–7] that demonstrated a peak d_{33} of 27.6 pC/N [6].

Therefore, AlScN has become a competitive candidate for different filter designs targeting 5G NR mmWave (K_a band, 26 GHz) [8–10] or higher. As a result, extensive research has been conducted on its material properties, growth, characterization and device fabrication.

Etching is a key step in the fabrication of AlScN devices. Like other III–V nitride alloys, existing etching techniques on AlN/$Al_{1-x}Sc_xN$ can be grouped into two categories: dry etching and wet etching. For dry etching, ion-milling that uses Ar exclusively and Inductively Coupled Plasma Reactive Ion Etching (ICP-RIE), which utilizes $BCl_3/Cl_2/Ar$ mixtures, are common methods [11]. The latter is the more routinely used technique for dry etching polycrystalline AlN, and etch rates can reach up to 420 nm/min [12] for an ICP power of 800 W. Nevertheless, this etch rate drops dramatically with increasing scandium concentration. For an $Al_{0.85}Sc_{0.15}N$ film, the etch rate declined to 64% of that of AlN [13]; for $Al_{0.73}Sc_{0.27}N$, 42% under the same etch condition [14]; and for $Al_{0.64}Sc_{0.36}N$, 10% of that for AlN [15]. As for single-crystalline $Al_{1-x}Sc_xN$, the reduction in etch rate occurred much faster: at x = 0.02, the etch rate already reduced to 15% of AlN, and at x = 0.15, it was 12.7% [16]. Not only has the existence of scandium retarded the etch rate, but its non-volatile etching by-products also re-deposit during the etch process, resulting in a roughened and tapered side wall less than 76° if Ion Beam Etching (IBE) is not used [17,18]. The poor selectivity requires very thick, hard masks during processing and makes it challenging to stop the AlScN etch on underlying metal electrode materials. Both the slow etch rate and low selectivity can limit the maximum AlScN film thickness realizable in a MEMS process. Table 1 summarizes selected studies on the dry etch rates of AlN/$Al_{1-x}Sc_xN$:

Table 1. Published dry etch rates of AlN/$Al_{1-x}Sc_xN$ in ICP-RIE.

Material	Crystallinity	Etchant Flow (sccm)	ICP/RF Power (W)	Pressure (m Torr)	Etch Rate ($nm \cdot s^{-1}$)	Rate to AlN	Ref.
AlN	Polycrystal	$BCl_3/Cl_2/Ar$ (10/14/6)	800/NA	5	420	1.00	[12]
$Al_{0.85}Sc_{0.15}N$		$BCl_3/Cl_2/He$ (30/90/100)	550/150	NA	160	0.64	[13]
$Al_{0.73}Sc_{0.27}N$		$SiCl_4$ (NA)	150/225	15	10	0.42	[14]
$Al_{0.64}Sc_{0.36}N$		$BCl_3/Cl_2/Ar$ (30/90/70)	400/120	NA	30	0.10	[15]
AlN $Al_{0.98}Sc_{0.02}N$ $Al_{0.84}Sc_{0.16}N$	Single Crystal	$BCl_3/Cl_2/Ar$ (10/20/10)	200/50	5	86.0 13.3 11.0	1.00 0.15 0.127	[16]

Compared to dry etching, wet etching can be rather advantageous owing to its less expensive tooling. Common etchants include tetramethylammonium hydroxide (TMAH) [19–21], phosphoric acid (H_3PO_4) [22,23] or phosphoric acid-based solution [21] (PWS, containing 80% H_3PO_4, 16% H_2O and 4% HNO_3), potassium hydroxide (KOH) [22,24,25] and AZ400K Developer (contain 10 wt% of KOH) [26]. Among them, we found KOH to be rather attractive due to its availability and non-toxicity with a relatively high etch rate and high etch selectivity to silicon nitride (SiN_x). The current research focusing on AlN etching in KOH is abundant, yet very few details were disclosed regarding the etching of $Al_{1-x}Sc_xN$, let alone a complete survey of its etch rate as a function of Sc alloying. For the data available, the etch rate varies substantially based on the temperature, crystallinity and etchant concentration. Table 2 summarizes previous studies on the etch rates of AlN/$Al_{1-x}Sc_xN$ in aqueous KOH.

Table 2. Published vertical etch rates of AlN/$Al_{1-x}Sc_xN$ in aqueous KOH solutions.

Material	Crystallinity	Etchant	Temp (°C)	Etch Rate ($nm{\cdot}s^{-1}$)	Activation Energy ($kcal{\cdot}mol^{-1}$)	Rate at 45 °C ($nm{\cdot}s^{-1}$)	Ref.
AlN	Single Crystal	AZ400K	60	1.1	15.13	0.8	[26]
	Single Crystal (with high defect density)	AZ400K	60	25.1	15.24	10.3	
	Polycrystal	AZ400K	32	185.0	15.65	561.40	[26]
		AZ400K	40	92.20	2.0 ± 0.5	107.93	[27]
		45 wt% KOH	60	5.83	NA	1.89	[28]
		1 wt% KOH	70	8.33	NA	1.34	[29]
$Al_{0.80}Sc_{0.20}N$	Polycrystal	25 wt% KOH	RT	0.66	15.85	3.59	[25]
		25 wt% KOH	40	2.38			
$Al_{0.64}Sc_{0.36}N$		25 wt% KOH	80	33.33	NA	2.77	[23]
$Al_{0.63}Sc_{0.37}N$		20 wt% KOH	20	0.42	NA	3.56	[30]

Note: For comparison, the etch rates were converted to 45 °C with activation energy E_a given by the author, or assuming E_a = 15.85 kcal/mol if data were unavailable.

The lack of variable control and the limited data make it difficult to draw solid conclusions on the factors affecting the etch rate, though from the AlN/$Al_{1-x}Sc_xN$ dry etch results and the KOH wet etch of its III–V alloy kin $Al_{1-x}In_xN$ [27]/$Al_{1-x}Ga_xN$ [29], one would assume that the etch rate decreases with the increasing Sc concentration. Moreover, even though prior etching studies reported the anisotropic nature of the etch, with preferential etching of the c-plane $\{000\bar{1}\}$ in N-polar AlN/$Al_{1-x}Sc_xN$ [28,29] while exposing the $\{10\bar{1}\bar{1}\}$ planes [31] from whose boundary the sidewall that follows is the $\{\bar{1}2\bar{1}2\}$ of the hexagonal crystal structure, lateral etch rates have not been reported. To date, there have been but a few papers [32,33] discussing KOH etching of AlN/$Al_{1-x}Sc_xN$ in detail, and only AlN and $Al_{0.80}Sc_{0.20}N$ were studied. Therefore, we report a thorough survey on the vertical and lateral etch rates of sputtered AlN/$Al_{1-x}Sc_xN$ in aqueous KOH solutions vs. scandium concentrations, where the KOH concentrations and solution temperatures were strictly regulated. We found that the vertical etch rate declines steadily with increasing scandium concentration, whereas the lateral etch rate experiences a V-shaped transition with a minimum value of 0.043 ± 0.002 nm/s at x = 0.125. By etching the $Al_{0.875}Sc_{0.125}N$ film in 10 wt% KOH at 45 °C for 10 min, a nearly 90° sidewall was produced by exposing the $\{1\bar{1}00\}$ planes. This technique is capable of generating a vertical sidewall without pre-treatment and could be beneficial for the fabrication of numerous kinds of microelectromechanical systems (MEMS) device such as lamb wave resonators (LWRs), laser mirrors and Ultraviolet Light-Emitting Diodes (UV-LEDs).

The findings are presented in four sections: (1) Introduction, containing the problem selected, literature review, novelty and section description; (2) Experiment, containing the deposition methods, film characterization, etch mask patterning, wet etching process steps and data interpretation methodology; (3) Results and Analysis, containing the illustration of results and detailed analysis; (4) Conclusion, summarizes the paper and the novelty of the research.

2. Experiment

2.1. Film Deposition

2.1.1. Growth

Based on the crystallinity of AlN/$Al_{1-x}Sc_xN$, several methods can be used for film growth. Single-crystal AlN can be grown via physical vapor transport (PVT) on substrates up to 60 mm in diameter [34–36]. Single-crystal $Al_{1-x}Sc_xN$ can be grown by molecular beam epitaxy (MBE) [9] on 100 mm wafers [37] with scandium concentration x from 0.06 to 0.36 [38]. The growth of Sc alloyed AlN via metal–organic chemical vapor deposition

(MOCVD) had suffered from the lack of Sc precursors [39]; however, research in this field is catching up quickly, and the ability to grow 36% Sc alloyed $Al_{1-x}Sc_xN$ films on 100 mm wafers has been demonstrated [40,41]. High-quality polycrystalline $Al_{1-x}Sc_xN$ films can be deposited with physical vapor deposition (PVD) methods such as magnetron sputtering. This method features a high deposition rate, low growth temperature, and the capability of up-scaling in substrate size [42–46], and has been adopted by a variety of tool manufacturers for industrial mass production [47].

$Al_{1-x}Sc_xN$ depositions were performed in an Evatec CLUSTERLINE® 200 II Physical Vapor Deposition System (Evatec AG, Trübbach, Switzerland) at a substrate temperature of 350 °C with 150 kHz pulsed DC bias with an 88% duty cycle. No RF substrate bias was applied and the AlScN materials were deposited directly onto silicon substrates. No pre-cleans were applied to the substrates. Two 100 mm metal targets were used, Al (99.999%) and Sc (99.99%), with a target-to-substrate distance of 88.5 mm. Before deposition, the chamber was pumped to a base pressure lower than 1.0×10^{-7} mbar. A 15 nm AlN film was deposited first onto a 100 mm Si (100) wafer as the seed layer by sputtering Al in a pure nitrogen environment with a target power of 1000 W and N_2 flow of 20 sccm. Subsequently, a 35 nm linearly graded $Al_{1-x}Sc_xN$ layer was deposited by gradually increasing the Sc target power while maintaining the Al target power constant. Finally, a 750 nm bulk $Al_{1-x}Sc_xN$ layer was deposited by fixing both the Al and Sc target powers. The 20 sccm N_2 flow was maintained during the process and no Ar was used throughout. The chamber pressure remained close to 8.0×10^{-4} mbar during the deposition. A total of 15 films were deposited with Sc concentration ranging from 0% to 42%. Table 3 summarizes the correlation between Sc target power and Sc concentration based on our previous research [48].

Table 3. Scandium concentration vs. Sc target power.

Sc Alloying in Film (%)	**0**	**5**	**10**	**12.5**	**15**	**20**	**25**	**28**	**30**	**32**	**34**	**36**	**38**	**40**	**42**
Sc Target Power (W)	0	40	80	130	185	300	400	450	510	555	610	655	685	710	770

2.1.2. Surface Metrology

The adoption of the AlN seed layer, the $Al_{1-x}Sc_xN$ gradient layer, and pure nitrogen sputtering environment greatly reduced the occurrence of abnormally oriented grains (AOGs). AOGs are a series of wurtzite $Al_{1-x}Sc_xN$ crystals that do not have their c-axis perpendicular to the substrate [49]. They could erupt from the crystalline interface if grown under unfavorable conditions, especially for films with higher Sc concentration [50]. If not suppressed, they may occupy the entire film surface [49,51,52], severely degrading device performance [53,54] and locally slowing the etch rate. To examine the film quality, atomic force microscopy (AFM) scans were conducted using a Bruker Icon AFM, and most of the films measured showed a roughness of <2 nm. Figure 1a shows the surface of an $Al_{0.64}Sc_{0.36}N$ film within a 20×20 μm² field. The root mean square (RMS) surface roughness R_q is 0.840 nm and R_a is 0.641 nm. The film quality is also supported by the rocking curve measurements, which were performed by a Rigaku Smart Lab X-RAY Diffractometer (XRD) with a high resolution Parallel Beam (PB) Ge (220) × 2 monochromator (Rigaku Corporation, Tokyo, Japan). The Omega scan data are centered at 18.13 ° with a full width at half maximum (FWHM) of 1.80 ° for a 500 nm film deposited on Si (100) using the same process, indicating that the film is highly *c*-axis textured. All films had a similar quality based on the AFM and XRD measurement.

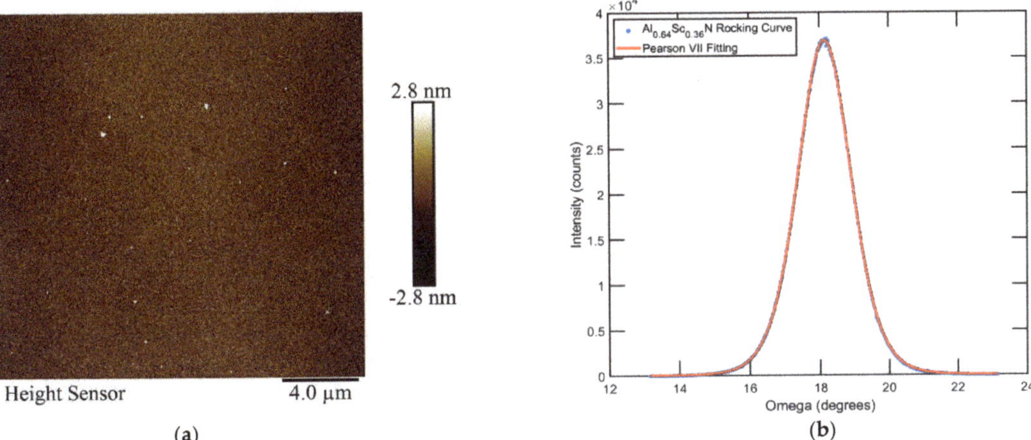

Figure 1. (**a**) AFM of the $Al_{0.64}Sc_{0.36}N$ Film. (**b**) Rocking curve measurement (Omega Scan) of an $Al_{0.64}Sc_{0.36}N$ film.

2.2. Film Patterning

To selectively etch $AlN/Al_{1-x}Sc_xN$, the film must be patterned so that only locations of interest would be exposed in the etchant. Silicon nitride (SiN_x) was chosen as our hard mask, as it has an etch rate of 0.67 nm/min in 30% KOH at 80 °C [55], which is negligible compared to the etch rates of $Al_{1-x}Sc_xN$. A 200 nm SiN_x film was deposited on top of the $Al_{1-x}Sc_xN$ film in an Oxford Plasma Lab 100 Plasma-enhanced Chemical Vapor Deposition (PECVD) machine (Oxford Instruments Plasma Technology, Bristol, UK), and on the backside of the wafer as well to protect the Si substrate during the etch. Afterwards, photoresist MICROPOSIT®S1813 was spin-coated and exposed in a Karl Süss MA6 Mask Aligner (SÜSS MicroTec SE, Garching, Germany) via contact lithography. The exposed film was developed in TMAH-0.26N developer, then transferred to an Oxford 80 plus Reactive-ion Etching (RIE) etcher (Oxford Instruments Plasma Technology, Bristol, UK). A 30 s O_2 descum was performed first to remove the remaining resist, followed by 3 min of SiN_x etch with CHF_3/O_2 mixture. Finally, the resist was stripped in MICROPOSIT® remover 1165 with ultrasonic bath at 60 °C and plasma cleaned. Figure 2 illustrates the fabrication process.

2.3. KOH Wet Etching

2.3.1. Principle

The wet etching of III–V nitrides in general involves the formation of an oxide on the surface and the subsequent dissolution of the oxide [24]. The flowing reactions occur when $AlN/Al_{1-x}Sc_xN$ is subjected to the alkaline environment [29]:

$$2AlN + 3H_2O \overset{KOH}{\rightarrow} Al_2O_3 + 2NH_3 \tag{1}$$

$$AlN + 3H_2O \overset{KOH}{\rightarrow} Al(OH)_3 + NH_3 \tag{2}$$

$$2ScN + 3H_2O \overset{KOH}{\rightarrow} Sc_2O_3 + 2NH_3 \tag{3}$$

$$ScN + 3H_2O \overset{KOH}{\rightarrow} Sc(OH)_3 + NH_3 \tag{4}$$

In this reaction, KOH acts as the catalyst that pushes the equation to the right side. Due to the origin of the reaction, N-polar $AlN/Al_{1-x}Sc_xN$ are preferred to be etched as it is difficult for OH^- to make contact with the Al/Sc atoms in the metal polar state because of repulsion from the negatively charged dangling nitrogen bonds [24].

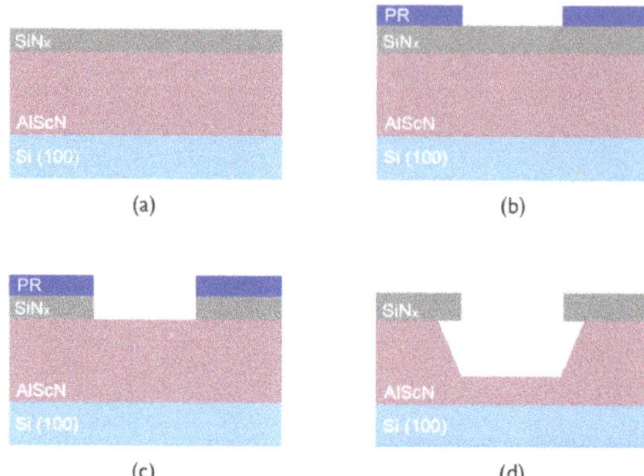

Figure 2. (**a**) $Al_{1-x}Sc_xN$ and SiN_x hard mask deposition; (**b**) photoresist patterning; (**c**) SiN_x dry etch; (**d**) resist stripping and $Al_{1-x}Sc_xN$ KOH wet etch. The opening area for the etching is $30 \times 200\ \mu m^2$.

2.3.2. Etching Process

A water bath was established for the etching to be conducted in a stable temperature environment. A trough was filled with deionized (DI) water and placed upon an Echo Thermal HP30 hotplate (Torrey Pines Scientific, Inc., Carlsbad, California). Around 300 mL of 30% KOH/diluted KOH was poured into a beaker, which was later transferred into the trough. A Teflon plate was added in between to avoid direct heating. Finally, a thermocouple was submerged into the KOH solution and connected to the hotplate through a proportional–integral–derivative (PID) control loop to adjust the solution temperature. With all these measures, the solution temperature was able to be stabilized within ±1 °C during etching. The test sample was cleaved into an $8 \times 10\ mm^2$ die and clamped by a tweezer when dipped into the solution. When submerged in the KOH, samples experienced minimal agitation, and upon removal, they were rinsed under DI water and subsequently dried with N_2. Finally, the sample was cleaved from the middle (Figure 3a) and a cross-section was imaged in a FEI Quanta 600 Environmental Scanning Electron Microscope (ESEM) (FEI Company, Hillsboro, OR, USA) (Figure 3b).

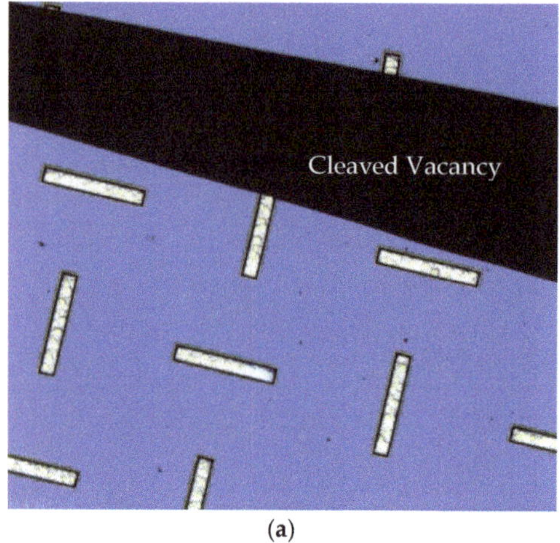

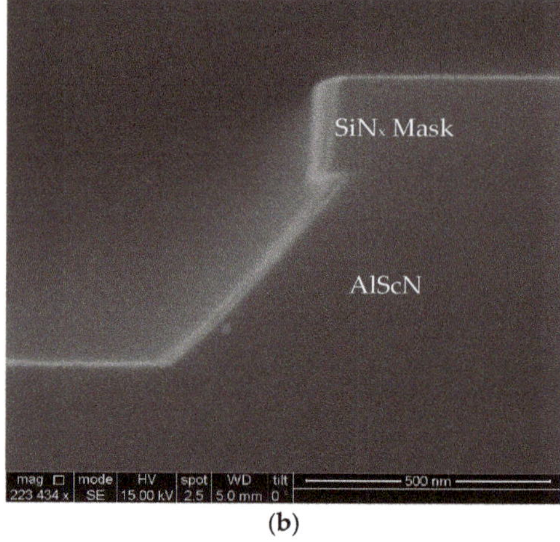

(**a**) (**b**)

Figure 3. (**a**) A cleaved sample after etching. Black background is the vacancy after cleaving. (**b**) Cross-section SEM image of an $Al_{0.72}Sc_{0.28}N$ film etched for 30 s in 30% KOH at 60 °C.

2.3.3. Data Interpretation

Three types of data were extracted from the SEM images: vertical etch depth, lateral etch length (undercutting) and the sidewall angle. The vertical etch depth is defined as the etching depth into the AlN/$Al_{1-x}Sc_xN$ film from the bottom of the SiN_x hard mask. It can be somewhat ambiguous, as hexagonal-shaped hillocks are known to form after KOH etching [29], which makes it difficult to identify the end point of the etch. Since the phenomenon is presumed to be defect-related in AlN [28,55], we conclude that the 'tip' of the pyramid acts as a mask in this process and thus the etch front should be read as the basal plane of the hexagonal pyramid. Figure 4a is a demonstration of how the etch depth was measured in case of the existence of the hillocks. The lateral etch length is defined as the etch length underneath the SiN_x hard mask (Figure 4b). Finally, the sidewall angle is simply the angle between the sidewall and base plane.

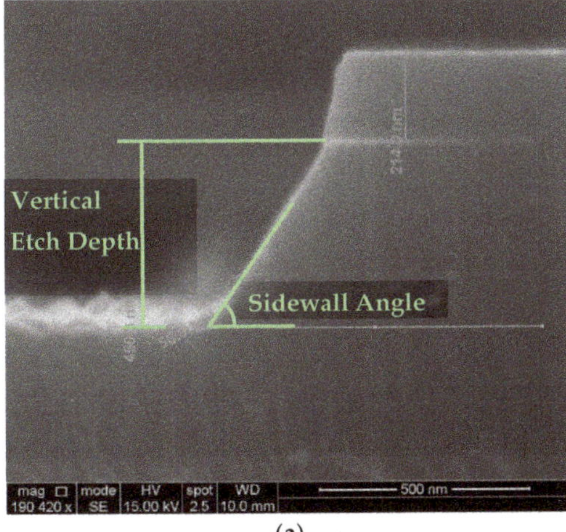

(a)

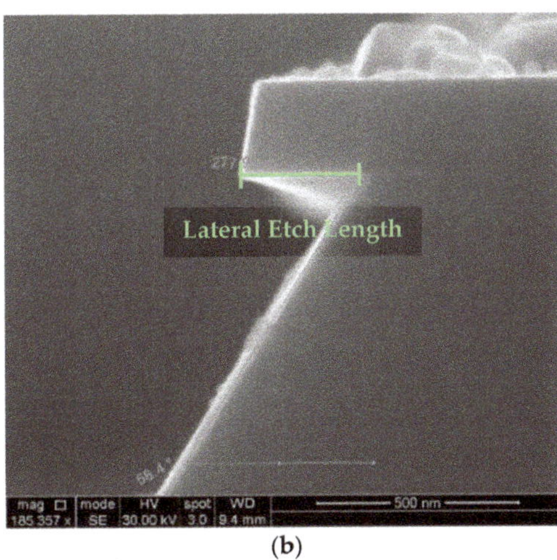

(b)

Figure 4. (**a**) $Al_{0.95}Sc_{0.05}N$ etched for 10 s in 30% KOH at 45 °C. (**b**) $Al_{0.85}Sc_{0.15}N$ etched for 2.5 min in 30% KOH at 65 °C.

3. Results and Analysis

3.1. Etch Result with 30% KOH at 45 °C

Due to the limited film thickness and the vastly different etch rates of films with varying Sc concentrations, it was not practical to use the same etch time when determining the vertical etch rate. Instead, films with lower Sc concentration were etched for a shorter amount of time. The following table (Table 4) illustrates the time used in each case:

Table 4. Scandium concentration vs. etch time (short etch time in 30% KOH at 45 °C).

Sc Alloying (%)	**0**	**5**	**10**	**15**	**20**	**25**	**28**	**32**	**36**
Etch Time (s)	5	10	20	20	60	60	60	60	60

It should be noted that not all samples fabricated, as mentioned in Section 2, had their vertical etch rate measured due to the availability of the sample at the time this experiment was conducted. For the etched samples, cross-section images were taken from several spots to avoid local variations. Figure 5 summarizes the vertical and lateral etch rates vs. Sc concentration for etching in 30% KOH at 45 °C.

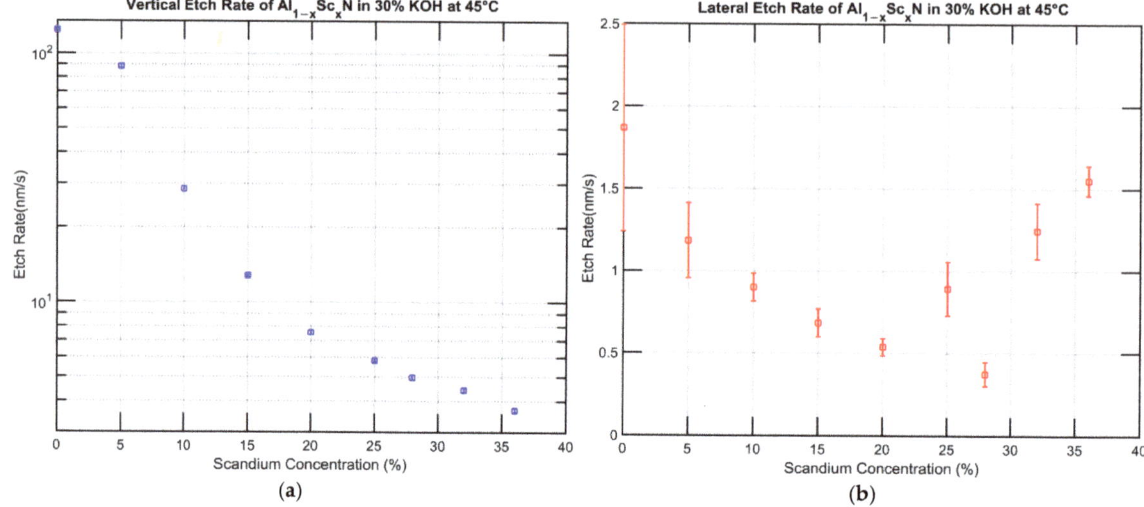

Figure 5. (**a**). Vertical etch rate of $Al_{1-x}Sc_xN$ in 30% KOH at 45 °C. (**b**) Lateral etch rate of $Al_{1-x}Sc_xN$ in 30% KOH at 45 °C (short etch time).

Intuitively, the vertical etch rate matches our expectation: a steady decline with increasing Sc concentration. The $Al_{0.80}Sc_{0.20}N$ has an average vertical etch rate of 7.58 nm/s, and the $Al_{0.64}Sc_{0.36}N$ has an average vertical etch rate of 3.68 nm/s. Compared to the values of 3.59 nm/s and 2.77 nm/s obtained by K. Bespalova et al. [25] and S. Fichtner et al. [23] respectively, they are not exact matches but in the same order of magnitude. It should be noted that as per the view of A. Ababneh et al. [56], the etch rate of AlN is a strong function of sputtering conditions; thus, this study can be used to predict the trend instead of the absolute etch rate when applied to films deposited under different conditions. The lateral etch rate has several distinctive features requiring further examination. First, the standard deviation of the lateral etch rate is considerably larger than that for the vertical etch. This is due to the finite amount of lateral etching performed. As exhibited in Figure 4a, the film has a lateral etch length of only a few dozen nanometers in some cases, which makes it difficult to accurately measure. Secondly, the etch rate of $Al_{0.72}Sc_{0.28}N$ does not fit in the line. Lastly, the lateral etch rate begins to increase for concentrations in excess of 20% Sc.

To further explore the lateral etching, a second experiment was conducted in which the lateral etch length was extracted over a longer etching period. The number of Sc concentrations in this study was also expanded for higher resolution as shown in Table 5:

Table 5. Scandium concentration vs. etch time (long etch in 30% KOH at 45 °C).

Sc Alloying in Film (%)	**0**	**5**	**10**	**12.5**	**15**	**20**	**25**	**28**	**30**	**32**	**34**	**36**	**38**	**40**	**42**
Etch Time (min)	10	10	10	10	5	5	5	5	4	5	4	5	4	4	4

Before the long-span etching was performed, a linearity check was conducted to make sure that the etch does not change during the etching process. The film tested was $Al_{0.64}Sc_{0.36}N$ and it was subjected to etches of 60 s, 80 s, 300 s, 1200 s and 2400 s, respectively. A linear regression was performed, and the lateral etching proved to be highly linear with the $R^2 = 0.9968$. The fact that the same etch rate was retained after 40 min indicates that the etch is reaction-limited, e.g., not confined by mass transferring, and matches the description of AlN etching in KOH given by Mileham et al. [26] (Figure 6).

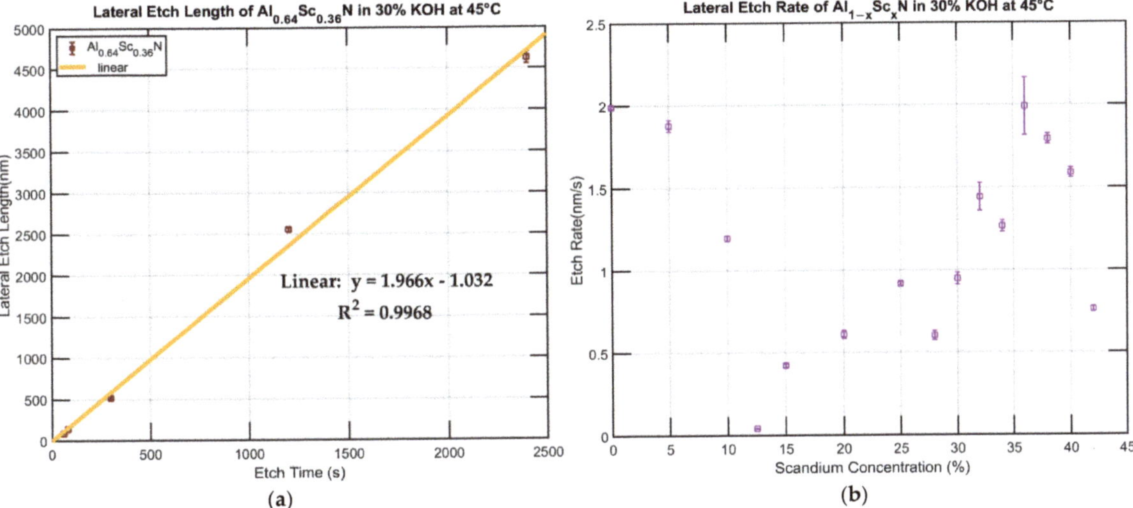

Figure 6. (**a**). Lateral etch length with respect to etch time of $Al_{0.64}Sc_{0.36}N$ in 30% KOH at 45 °C. (**b**) Lateral etch rate of $Al_{1-x}Sc_xN$ in 30% KOH at 45 °C (long etch time).

The long etch time returned similar results compared with the short etch time. The etch rate experienced a transition where it reaches the lowest at x = 0.125. However, discrepancies still exist for the etch rates of x = 0.28, 0.30, 0.34 and 0.38 when compared to other films with Sc concentration x > 0.125. One explanation might be that these films were deposited in different batches as opposed to the rest of the films. As stated above, what exhibits as different etch rates here might be the result of the different chamber condition when the films were being sputtered. It was apparent that even though they do not fit in the line made from the x = 0.125, 0.15, 0.20, 0.25, 0.32 and 0.36 films, they were able to constitute their own trend line with almost has the same slope. Deeper examination revealed that the etch rate of the x = 0.28 film was lower than expected on two different films deposited in different batches, and further studies should be carried out to better understand these subtle trends in lateral etch rate.

For most Sc concentrations, the sidewall angle of films remains invariant throughout the etching. As demonstrated in Figure 7, an extra-long submerge of the sample does not change the sidewall angle by a visible amount, and for the changes that could be measured, it can be attributed to the tilting of the sample itself during imaging.

The sidewall is a reflection of the crystal structure of the $Al_{1-x}Sc_xN$ films. As per the findings of W. Guo et al., due to the energy difference between crystalline planes, the c-plane $\{000\overline{1}\}$ will be etched first prior to the deterioration of the $\{10\overline{1}\overline{1}\}$ planes [31]. Hence, during the anisotropic etching process, the exposed $\langle\overline{1}\overline{1}23\rangle$ slip edges between the boundary of the $\{10\overline{1}1\}$ planes forms the facets that follow the $\{\overline{1}2\overline{1}2\}$ of the hexagonal crystal structure, behind which lateral etching ceases advancing (Figure 8).

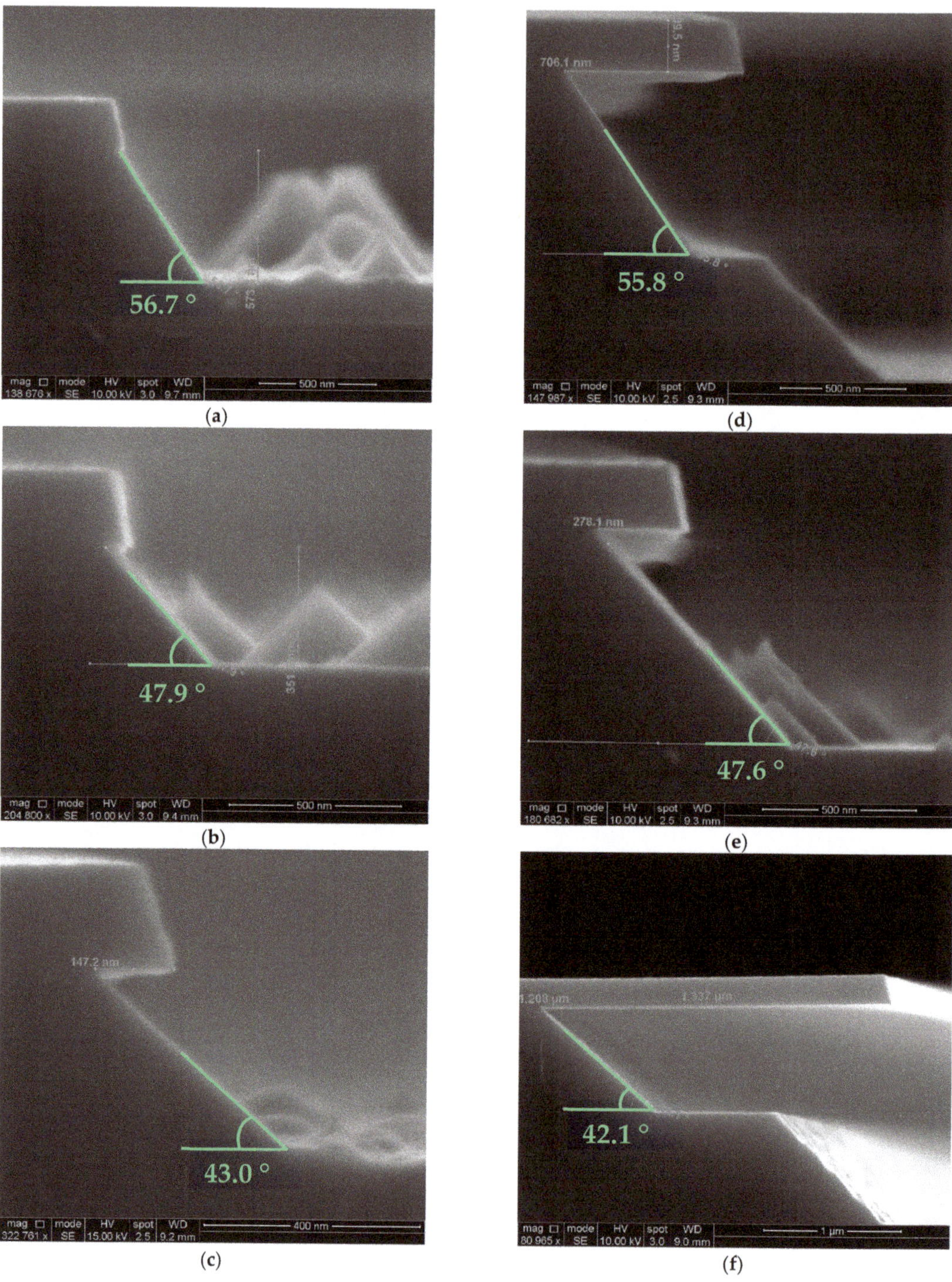

Figure 7. (**a**) $Al_{0.90}Sc_{0.10}N$ etched for 20 s; (**b**) $Al_{0.75}Sc_{0.25}N$ etched for 60 s; (**c**) $Al_{0.64}Sc_{0.36}N$ etched for 80 s; (**d**) $Al_{0.90}Sc_{0.10}N$ etched for 10 min; (**e**) $Al_{0.75}Sc_{0.25}N$ etched for 5 min; (**f**) $Al_{0.64}Sc_{0.36}N$ etched for 20 min. All etching was performed in 30 wt% KOH at 45 °C. The sidewall angle is preserved after the long etch.

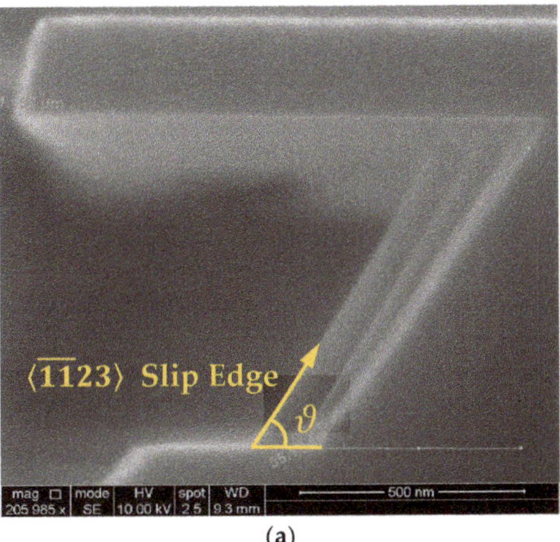

(a)

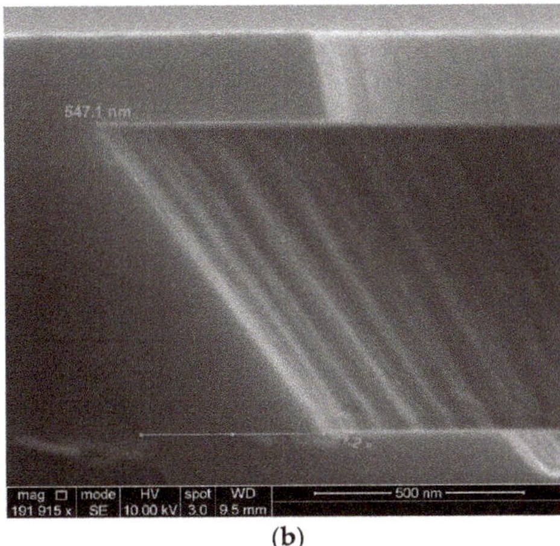

(b)

Figure 8. (**a**). Exposed ⟨$\bar{1}\bar{1}23$⟩ slip edge in an $Al_{0.95}Sc_{0.05}N$ film etched in 30% KOH at 45 °C for 10 min; (**b**) the sidewall that follows the $\{\bar{1}2\bar{1}2\}$ facet created by the boundary between the $\{10\bar{1}\bar{1}\}$ planes of the same etch condition for 5 min.

Moreover, for a Hexagonal Close-Packed (HCP) unit cell with an axis length c/a, the sidewall angle, θ (Figure 9), of the $\{\bar{1}2\bar{1}2\}$ facets as a function of lattice length can be calculated as $\theta = \arctan(c/a)$:

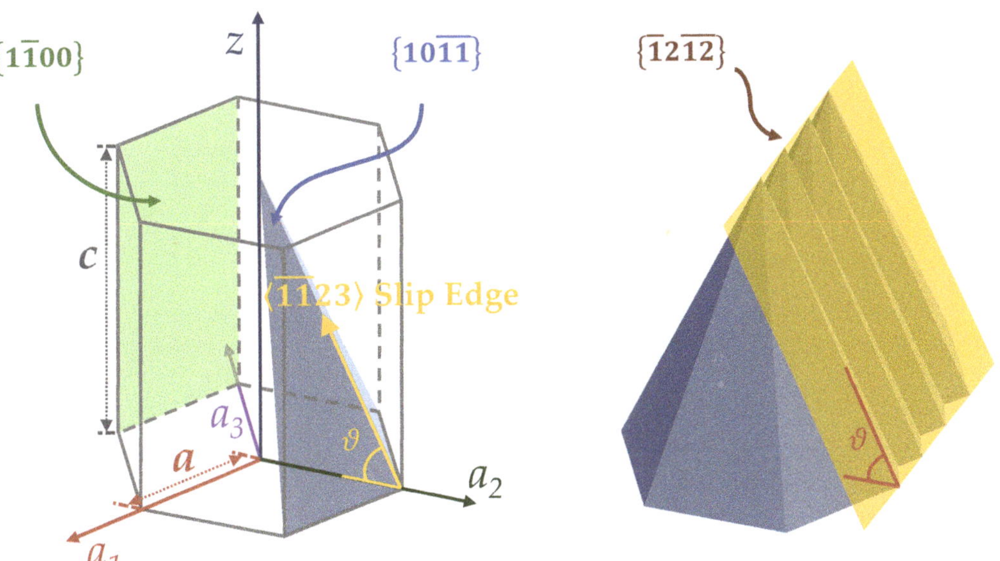

Figure 9. Crystal planes in $Al_{1-x}Sc_xN$ HCP lattice.

Using the c/a data from the work presented by Österlund et al. [45], and considering the isotropic etching of the $\{10\bar{1}\bar{1}\}$ plane, the theoretical sidewall angle φ can be plotted against the experimental value (Figure 10).

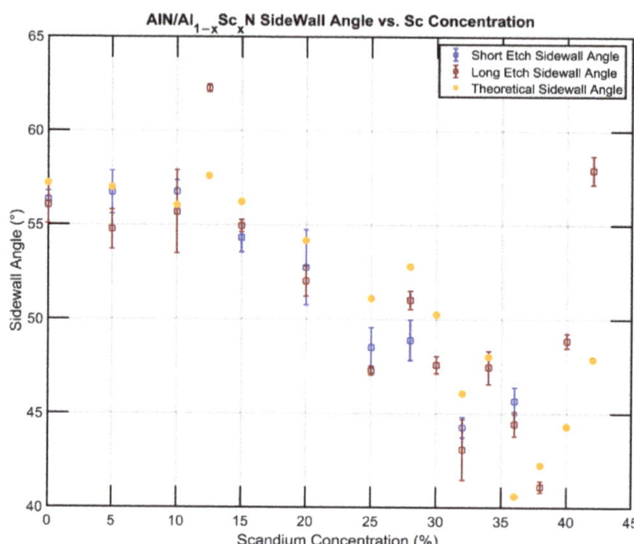

Figure 10. Comparison of experimental and theoretical sidewall angle vs. Sc concentration.

The absolute value of the experimental and theoretical angles follows the same trajectory with a parabolic downtrend with increasing Sc concentration until the Sc concentration exceeds 40%. We hypothesize that the side profile is the result of both anisotropic and isotropic etching. At lower Sc concentration, the etch rate on the c-plane $\{000\overline{1}\}$ is significantly higher, the momentum of the etch is downwards, and thus the side wall creates a facet that follows the $\{\overline{1}2\overline{1}2\}$ of the hexagonal crystal structure. At a higher Sc concentration, the low vertical etch rate slows down the descending penetration, allowing the etchant to further react with the sidewall planes already exposed. Therefore, the closer the $\{10\overline{1}\overline{1}\}$ planes are to the surface, the more they are etched away; as a result, the sidewall angle becomes lower than that predicted solely based on the anisotropic crystal etching, i.e., lower than θ. This can be partially verified by some of the abnormal points on the graph, most of which have a small lateral etch rate (e.g., 12.5% and 42% Sc), which prevents the etching of their corresponding $\{10\overline{1}\overline{1}\}$ planes.

3.2. Etch Results with 30 wt% KOH at 65 °C and 10 wt% KOH at 65 °C

The experiment was also carried out at an elevated temperature and with lower KOH concentration to rule out any possible interference to the outcome except for the intrinsic material properties. At elevated temperature, the etching was performed with a short etching time of 150 s, except for $Al_{0.875}Sc_{0.125}N$, which was etched for 450 s. Etching with lower KOH concentration was performed for 20 min for all Sc alloying concentrations. As a result of the faster etching rate, the vertical etch rate could not be measured. The lateral etch rate and sidewall angle of the experiment are presented below (Figure 11):

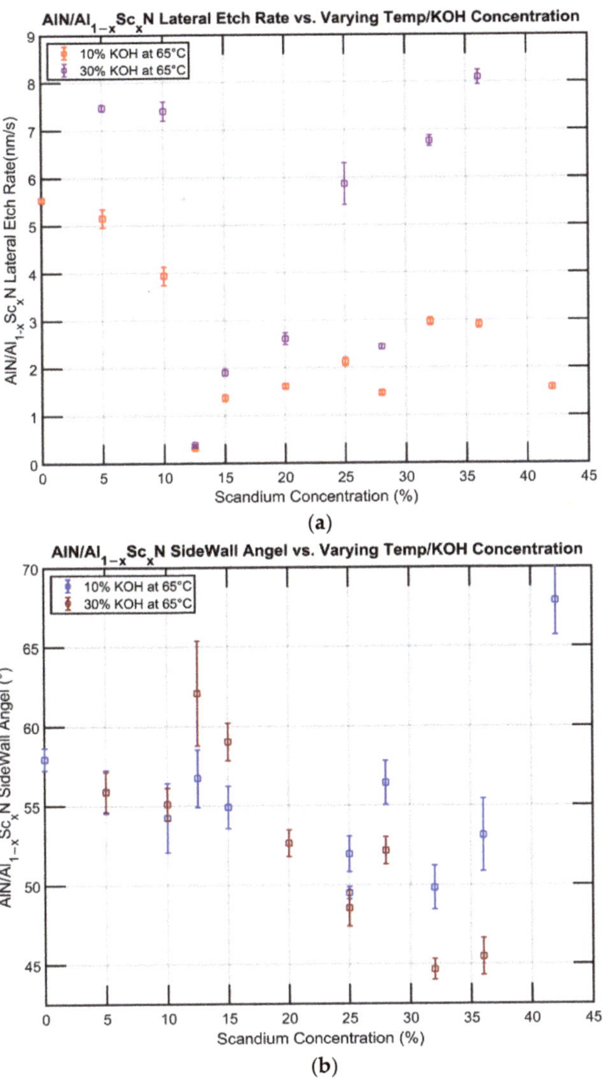

Figure 11. (**a**) Lateral etch rate at elevated temperature and lower etchant concentration. (**b**) Sidewall angle at elevated temperature and lower etchant concentration.

A trend consistent with the etching studies reported above was also observed here, where the lowest etch rate was found to be at x = 0.125 and a decreasing sidewall angle was observed with increasing Sc concentration. Under the SEM, white clusters can be seen occasionally near the etch front, which we assume to be the unsoluable reaction by-product $Sc(OH)_3$. Further research still needs to be conducted on the effect of its presence.

3.3. Formation of Vertical Sidewall in $Al_{0.875}Sc_{0.125}N$

While differences can be observed across the spectrum, the lateral etch rate unanimously reaches its lowest point when the Sc concentration is at 12.5%. An inspection of the etch results have shown that during the etching, a more vertical side wall can be formed, as shown in Figure 12.

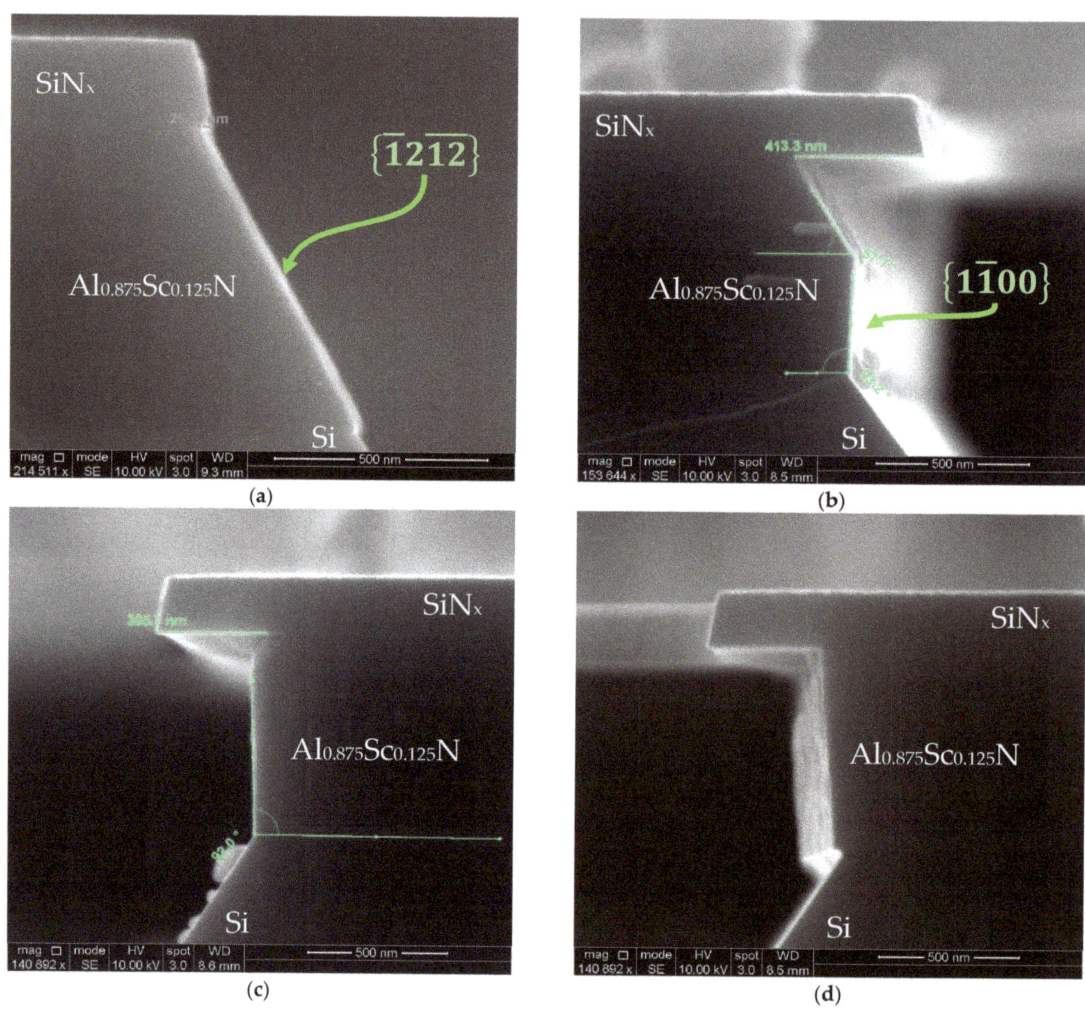

(**a**) (**b**) (**c**) (**d**)

Figure 12. Lateral etch of $Al_{0.875}Sc_{0.125}N$ in (**a**) 30% KOH at 45 °C for 10 min; (**b–d**) 10% KOH at 65 °C for 20 min.

Although (a) was etched with different parameters compared to (b–d)—plus, (b–d) were etched for almost the same time—we conjecture that these images are demonstrating the transient response of the same etching dynamics. The KOH etch is slowed significantly at the Si $\{111\}$ plane and the $Al_{0.875}Sc_{0.125}N$ $\{10\bar{1}\bar{1}\}$ plane due to the low etch rate. Because of the energy differences between removing $\{10\bar{1}\bar{1}\}$ $Al_{0.875}Sc_{0.125}N$ and $\{111\}$ Si, the KOH slowly etches Si $\{111\}$ until it comes into contact with the $Al_{0.875}Sc_{0.125}N$ $\{1\bar{1}00\}$ planes, which requires lower energy to react than the $\{10\bar{1}\bar{1}\}$ planes. As a result, the etchant begins to remove and simultaneously etch Si $\{111\}$ and $\{1\bar{1}00\}$ AlScN. By the time the entire $\{1\bar{1}00\}$ planes were exposed, a vertical sidewall was formed. It is possible that the slow lateral etch rate is necessary but not sufficient for the exposure of the $\{1\bar{1}00\}$ AlScN to occur. This has been demonstrated with the lateral etching of $Al_{0.85}Sc_{0.15}N$ and $Al_{0.72}Sc_{0.28}N$, which have slightly higher lateral etch rates than $Al_{0.875}Sc_{0.125}N$. As shown below (Figure 13), the etchant may preferentially etch Si $\{111\}$ instead of m-plane $Al_{1-x}Sc_xN$.

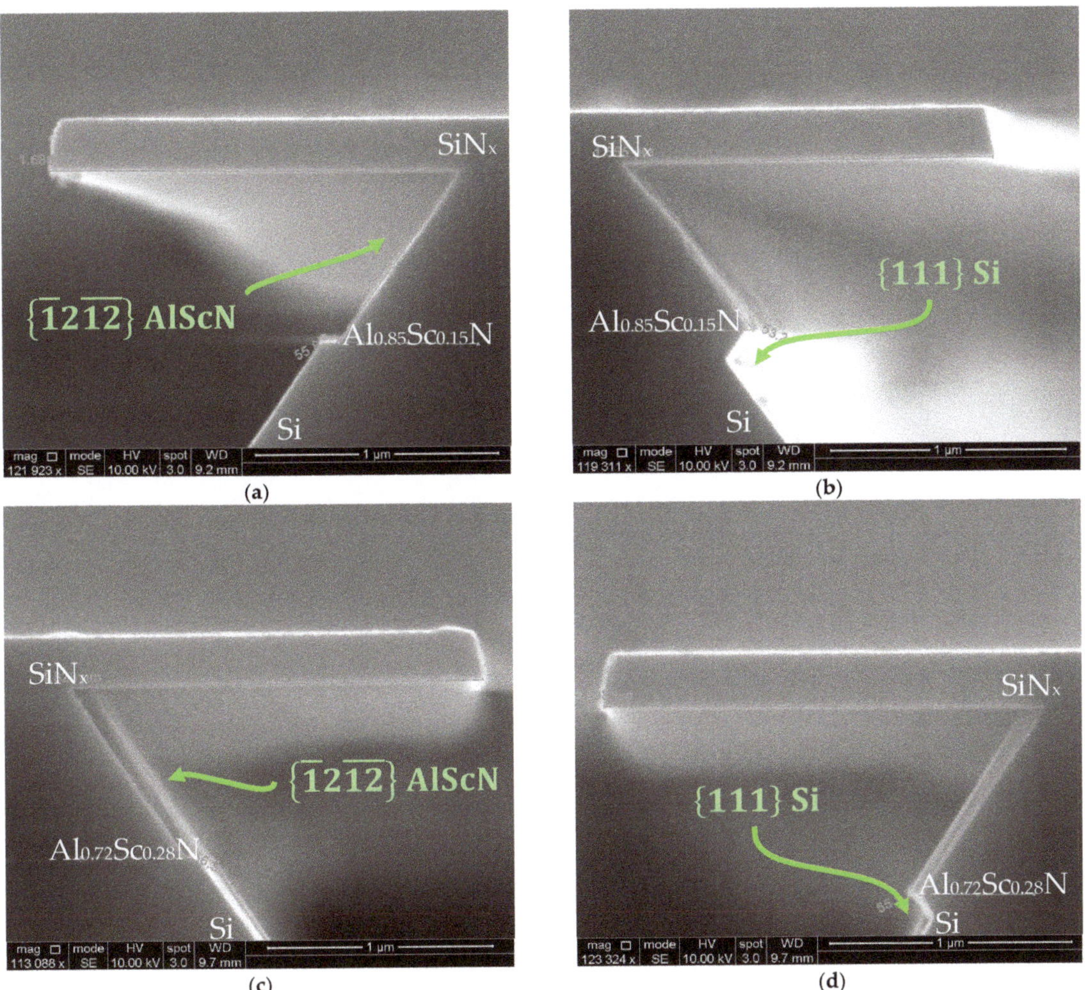

Figure 13. (**a**) $Al_{0.85}Sc_{0.15}N$ etched for 20 min in 10 wt% KOH, etch front of AlScN approaching Si {111}; (**b**) KOH preferentially etches Si {111} instead of $\{\bar{1}2\bar{1}2\}$ AlScN; (**c**) $Al_{0.72}Sc_{0.28}N$ etched for 20 min in 10 wt% KOH, etch front of AlScN aligned with Si {111}; (**d**) KOH preferential etching of Si {111} instead of $\{\bar{1}2\bar{1}2\}$ AlScN. All etches were performed at 60 °C.

The $\{1\bar{1}00\}$ plane etching has been reported before in single-crystal GaN [57] and $Al_{1-x}Ga_xN$ [58], and as per the findings of W. Chen et al. [57], the preference of its etching in GaN is a result of its smaller dangling bond density, which makes it more stable in KOH than the $\{10\bar{1}\bar{1}\}$ plane. Therefore, one explanation might be that the $Al_{0.875}Sc_{0.125}N$ film has the highest activation energy in terms of the lateral etching, which was calculated to be 23.14 kcal/mol based on the data available. Nevertheless, this is one of the limited examples where this is reported in sputtered $Al_{1-x}Sc_xN$, and more research needs to be conducted to reveal the mechanism behind the vertical sidewall formation. This method, combined with BCl_3/Cl_2 dry etching, could potentially be applied in fabricating vertical side walls using a two-step fabrication process, which will benefit the research and production of LWR, laser mirrors, UV LEDs and a variety of MEMS devices.

4. Conclusions

We extensively studied the vertical and lateral etch rate of AlN/$Al_{1-x}Sc_xN$ in aqueous KOH solutions across etch temperature, KOH concentration and a broad range of scandium alloying. It was shown that the vertical etch rate declines steadily with increased levels of Sc alloying, declining from 124.6 ± 0.68 nm/s for AlN to 3.7 ± 0.063 nm/s for $Al_{0.64}Sc_{0.36}N$ in 30 wt% KOH at 45 °C. By contrast, the resistance to lateral etching peaks at a mere 0.043 ± 0.002 nm/s when x = 0.125. This is orders of magnitude lower compared to the lateral etch rate of 1.99 ± 0.01 nm/s for AlN or 1.99 ± 0.17 nm/s for $Al_{0.64}Sc_{0.36}N$. We have also demonstrated that KOH wet etching of $Al_{1-x}Sc_xN$ is mostly anisotropic, and that the etch profile can be predicted from the crystal structure coupled with a small-scale isotropic etching of the sidewall. A technique for fabricating a vertical sidewall by exposing the $\{1\bar{1}00\}$ planes of sputtered $Al_{1-x}Sc_xN$ was also demonstrated via etching an 800 nm thick $Al_{0.875}Sc_{0.125}N$ film in 10 wt% KOH at 65 °C for 20 min. With this method, the fabrication of numerous MEMS devices such as LWRs, laser mirrors and UV-LEDs can be benefited. Future work will include detailed research on the activation energy for the lateral etching of AlN/$Al_{1-x}Sc_xN$ using Arrhenius plots formed [32] from a series of design of experiments (DOE) using the Taguchi method [59].

Author Contributions: Conceptualization, Z.T. and R.H.O.III; methodology, Z.T., G.E. and R.H.O.III; validation, Z.T. and J.Z.; formal analysis, Z.T., G.E. and J.Z.; investigation, Z.T.; resources, G.E. and R.H.O.III; data curation, Z.T.; writing—original draft preparation, Z.T.; writing—review and editing, G.E. and R.H.O.III; visualization, Z.T.; supervision, R.H.O.III; project administration, G.E. and R.H.O.III; funding acquisition, G.E. and R.H.O.III All authors have read and agreed to the published version of the manuscript.

Funding: This work was funded in part by the NSF CAREER Award (1944248). This work was carried out in part at the Singh Center for Nanotechnology at the University of Pennsylvania, a member of the National Nanotechnology Coordinated Infrastructure (NNCI) network, which is supported by the National Science Foundation (grant no. HR0011-21-9-0004). Sandia National Laboratories is a multi-mission laboratory managed and operated by National Technology & Engineering Solutions of Sandia, LLC, a wholly owned subsidiary of Honeywell International Inc., for the U.S. Department of Energy's National Nuclear Security Administration under contract DE-NA0003525. This paper describes objective technical results and analysis. Any subjective views or opinions that might be expressed in the paper do not necessarily represent the views of the U.S. Department of Energy or the United States Government.

Data Availability Statement: The data supporting the findings of this study are available from the corresponding author upon reasonable request.

Conflicts of Interest: The authors declare no conflict of interest.

References

1. Ruppel, C.C.W. Acoustic Wave Filter Technology–A Review. *IEEE Trans. Ultrason. Ferroelectr. Freq. Control* **2017**, *64*, 1390–1400. [CrossRef] [PubMed]
2. Fischerauer, G.; Ebner, T.; Kruck, P.; Morozumi, K.; Thomas, R.; Pitschi, M. SAW Filter Solutions to the Needs of 3G Cellular Phones. In Proceedings of the 2001 IEEE MTT-S International Microwave Sympsoium Digest (Cat. No. 01CH37157), Phoenix, AZ, USA, 20–24 May 2001; pp. 351–354.
3. Aigner, R.; Fattinger, G. 3G–4G–5G: How Baw Filter Technology Enables a Connected World. In Proceedings of the 2019 20th International Conference on Solid-State Sensors, Actuators and Microsystems & Eurosensors XXXIII (TRANSDUCERS & EUROSENSORS XXXIII), Berlin, Germany, 23–27 June 2019; pp. 523–526.
4. Takeuchi, N. First-Principles Calculations of the Ground-State Properties and Stability of ScN. *Phys. Rev. B* **2002**, *65*, 045204. [CrossRef]
5. Akiyama, M.; Kamohara, T.; Kano, K.; Teshigahara, A.; Takeuchi, Y.; Kawahara, N. Enhancement of Piezoelectric Response in Scandium Aluminum Nitride Alloy Thin Films Prepared by Dual Reactive Cosputtering. *Adv. Mater.* **2009**, *21*, 593–596. [CrossRef] [PubMed]
6. Akiyama, M.; Kano, K.; Teshigahara, A. Influence of Growth Temperature and Scandium Concentration on Piezoelectric Response of Scandium Aluminum Nitride Alloy Thin Films. *Appl. Phys. Lett.* **2009**, *95*, 162107. [CrossRef]

7. Teshigahara, A.; Hashimoto, K.; Akiyama, M. Scandium Aluminum Nitride: Highly Piezoelectric Thin Film for RF SAW Devices in Multi GHz Range. In Proceedings of the 2012 IEEE International Ultrasonics Symposium, Dresden, Germany, 7–10 October 2012; pp. 1–5.
8. Ansari, A. Single Crystalline Scandium Aluminum Nitride: An Emerging Material for 5G Acoustic Filters. In Proceedings of the 2019 IEEE MTT-S International Wireless Symposium (IWS), 19–22 May 2019; pp. 1–3.
9. Park, M.; Hao, Z.; Dargis, R.; Clark, A.; Ansari, A. Epitaxial Aluminum Scandium Nitride Super High Frequency Acoustic Resonators. *J. Microelectromech. Syst.* **2020**, *29*, 490–498. [CrossRef]
10. Green, A.J.; Moser, N.; Miller, N.C.; Liddy, K.J.; Lindquist, M.; Elliot, M.; Gillespie, J.K.; Fitch, R.C.; Gilbert, R.; Walker, D.E.; et al. RF Power Performance of Sc(Al,Ga)N/GaN HEMTs at Ka-Band. *IEEE Electron Device Lett.* **2020**, *41*, 1181–1184. [CrossRef]
11. Pinto, R.M.R.; Gund, V.; Calaza, C.; Nagaraja, K.K.; Vinayakumar, K.B. Piezoelectric Aluminum Nitride Thin-Films: A Review of Wet and Dry Etching Techniques. *Microelectron. Eng.* **2022**, *257*, 111753. [CrossRef]
12. Khan, F.A.; Zhou, L.; Kumar, V.; Adesida, I.; Okojie, R. High Rate Etching of AlN Using BCl3/Cl2/Ar Inductively Coupled Plasma. *Mater. Sci. Eng. B* **2002**, *95*, 51–54. [CrossRef]
13. Wang, Q.; Lu, Y.; Mishin, S.; Oshmyansky, Y.; Horsley, D.A. Design, Fabrication, and Characterization of Scandium Aluminum Nitride-Based Piezoelectric Micromachined Ultrasonic Transducers. *J. Microelectromech. Syst.* **2017**, *26*, 1132–1139. [CrossRef]
14. Mayrhofer, P.M.; Wistrela, E.; Kucera, M.; Bittner, A.; Schmid, U. Fabrication and Characterisation of ScAlN -Based Piezoelectric MEMS Cantilevers. In Proceedings of the 2015 Transducers—2015 18th International Conference on Solid-State Sensors, Actuators and Microsystems (TRANSDUCERS), Anchorage, AK, USA, 2–25 June 2015; pp. 2144–2147.
15. Kusano, Y.; Ishii, I.; Kamiya, T.; Teshigahara, A.; Luo, G.-L.; Horsley, D.A. High-SPL Air-Coupled Piezoelectric Micromachined Ultrasonic Transducers Based on 36% ScAlN Thin-Film. *IEEE Trans. Ultrason. Ferroelectr. Freq. Control* **2019**, *66*, 1488–1496. [CrossRef]
16. Hardy, M.T.; Downey, B.P.; Meyer, D.J.; Nepal, N.; Storm, D.F.; Katzer, D.S. Epitaxial ScAlN Etch-Stop Layers Grown by Molecular Beam Epitaxy for Selective Etching of AlN and GaN. *IEEE Trans. Semicond. Manuf.* **2017**, *30*, 475–479. [CrossRef]
17. James, R.; Pilloux, Y.; Hegde, H. Reactive Ion Beam Etching of Piezoelectric ScAlN for Bulk Acoustic Wave Device Applications. *J. Phys. Conf. Ser.* **2019**, *1407*, 012083. [CrossRef]
18. Henry, M.D.; Timon, R.; Young, T.R.; Nordquist, C.; Griffin, B. (Invited) AlN and ScAlN Contour Mode Resonators for RF Filters. *ECS Trans.* **2017**, *77*, 23. [CrossRef]
19. Saravanan, S.; Berenschot, E.; Krijnen, G.; Elwenspoek, M. Surface Micromachining Process for the Integration of AlN Piezoelectric Microstructures. In Proceedings of the Annual Workshop on Semiconductor Advances for Future Electronics, SAFE 2004, Veldhoven, the Netherlands, 25–26 November 2004.
20. Yasue, S.; Sato, K.; Kawase, Y.; Ikeda, J.; Sakuragi, Y.; Iwayama, S.; Iwaya, M.; Kamiyama, S.; Takeuchi, T.; Akasaki, I. The Dependence of AlN Molar Fraction of AlGaN in Wet Etching by Using Tetramethylammonium Hydroxide Aqueous Solution. *Jpn. J. Appl. Phys.* **2019**, *58*, SCCC30. [CrossRef]
21. Marauska, S.; Dankwort, T.; Quenzer, H.J.; Wagner, B. Sputtered Thin Film Piezoelectric Aluminium Nitride as a Functional MEMS Material and CMOS Compatible Process Integration. *Procedia Eng.* **2011**, *25*, 1341–1344. [CrossRef]
22. Ababneh, A.; Kreher, H.; Schmid, U. Etching Behaviour of Sputter-Deposited Aluminium Nitride Thin Films in H_3PO_4 and KOH Solutions. *Microsyst. Technol.* **2008**, *14*, 567–573. [CrossRef]
23. Fichtner, S.; Wolff, N.; Lofink, F.; Kienle, L.; Wagner, B. AlScN: A III-V Semiconductor Based Ferroelectric. *J. Appl. Phys.* **2019**, *125*, 114103. [CrossRef]
24. Zhuang, D.; Edgar, J.H. Wet Etching of GaN, AlN, and SiC: A Review. *Mater. Sci. Eng. R Rep.* **2005**, *48*, 1–46. [CrossRef]
25. Bespalova, K.; Österlund, E.; Ross, G.; Paulasto-Kröckel, M.; Sebastian, A.T.; Karuthedath, C.B.; Mertin, S.; Pensala, T. Characterization of AlScN-Based Multilayer Systems for Piezoelectric Micromachined Ultrasound Transducer (PMUT) Fabrication. *J. Microelectromech. Syst.* **2021**, *30*, 290–298. [CrossRef]
26. Mileham, J.R.; Pearton, S.J.; Abernathy, C.R.; MacKenzie, J.D.; Shul, R.J.; Kilcoyne, S.P. Wet Chemical Etching of AlN. *Appl. Phys. Lett.* **1995**, *67*, 1119–1121. [CrossRef]
27. Vartuli, C.B.; Pearton, S.J.; Lee, J.W.; Abernathy, C.R.; Mackenzie, J.D.; Zolper, J.C.; Shul, R.J.; Ren, F. Wet Chemical Etching of AlN and InAlN in KOH Solutions. *J. Electrochem. Soc.* **1996**, *143*, 3681. [CrossRef]
28. Zhuang, D.; Edgar, J.H.; Liu, L.; Liu, B.; Walker, L. Wet Chemical Etching of AlN Single Crystals. *MRS Internet J. Nitride Semicond. Res.* **2020**, *7*, 4. [CrossRef]
29. Guo, W.; Kirste, R.; Bryan, I.; Bryan, Z.; Hussey, L.; Reddy, P.; Tweedie, J.; Collazo, R.; Sitar, Z. KOH Based Selective Wet Chemical Etching of AlN, AlxGa1−xN, and GaN Crystals: A Way towards Substrate Removal in Deep Ultraviolet-Light Emitting Diode. *Appl. Phys. Lett.* **2015**, *106*, 082110. [CrossRef]
30. Dunk, I. Robust Scandium Aluminum Nitride MEMS Resonators for L Band Operation in Orbital Environments. *Theses Diss.* **2021**.
31. Guo, W.; Xie, J.; Akouala, C.; Mita, S.; Rice, A.; Tweedie, J.; Bryan, I.; Collazo, R.; Sitar, Z. Comparative Study of Etching High Crystalline Quality AlN and GaN. *J. Cryst. Growth* **2013**, *366*, 20–25. [CrossRef]
32. Airola, K.; Mertin, S.; Likonen, J.; Hartikainen, E.; Mizohata, K.; Dekker, J.; Thanniyil Sebastian, A.; Pensala, T. High-Fidelity Patterning of AlN and ScAlN Thin Films with Wet Chemical Etching. *Materialia* **2022**, *22*, 101403. [CrossRef]

33. Cimalla, I.; Foerster, C.; Cimalla, V.; Lebedev, V.; Cengher, D.; Ambacher, O. Wet Chemical Etching of AlN in KOH Solution. *Phys. Status Solidi C* **2006**, *3*, 1767–1770. [CrossRef]
34. Fu, D.; Lei, D.; Li, Z.; Zhang, G.; Huang, J.; Sun, X.; Wang, Q.; Li, D.; Wang, J.; Wu, L. Toward Φ56 Mm Al-Polar AlN Single Crystals Grown by the Homoepitaxial PVT Method. *Cryst. Growth Des.* **2022**, *22*, 3462–3470. [CrossRef]
35. Wang, Q.; Lei, D.; He, G.; Gong, J.; Huang, J.; Wu, J. Characterization of 60 Mm AlN Single Crystal Wafers Grown by the Physical Vapor Transport Method. *Phys. Status Solidi A* **2019**, *216*, 1900118. [CrossRef]
36. Wu, H.; Zhang, K.; He, C.; He, L.; Wang, Q.; Zhao, W.; Chen, Z. Recent Advances in Fabricating Wurtzite AlN Film on (0001)-Plane Sapphire Substrate. *Crystals* **2021**, *12*, 38. [CrossRef]
37. Lee, K.; Cho, Y.; Schowalter, L.J.; Toita, M.; Xing, H.G.; Jena, D. Surface Control and MBE Growth Diagram for Homoepitaxy on Single-Crystal AlN Substrates. *Appl. Phys. Lett.* **2020**, *116*, 262102. [CrossRef]
38. Hardy, M.T.; Jin, E.N.; Nepal, N.; Katzer, D.S.; Downey, B.P.; Gokhale, V.J.; Storm, D.F.; Meyer, D.J. Control of Phase Purity in High Scandium Fraction Heteroepitaxial ScAlN Grown by Molecular Beam Epitaxy. *Appl. Phys. Express* **2020**, *13*, 065509. [CrossRef]
39. Zukauskaite, A.; Wingqvist, G.; Palisaitis, J.; Jensen, J.; Persson, P.O.Å.; Matloub, R.; Muralt, P.; Kim, Y.; Birch, J.; Hultman, L. Microstructure and Dielectric Properties of Piezoelectric Magnetron Sputtered W-ScxAl1−xN Thin Films. *J. Appl. Phys.* **2012**, *111*, 093527. [CrossRef]
40. Ligl, J.; Leone, S.; Manz, C.; Kirste, L.; Doering, P.; Fuchs, T.; Prescher, M.; Ambacher, O. Metalorganic Chemical Vapor Phase Deposition of AlScN/GaN Heterostructures. *J. Appl. Phys.* **2020**, *127*, 195704. [CrossRef]
41. Manz, C.; Leone, S.; Kirste, L.; Ligl, J.; Frei, K.; Fuchs, T.; Prescher, M.; Waltereit, P.; Verheijen, M.A.; Graff, A.; et al. Improved AlScN/GaN Heterostructures Grown by Metal-Organic Chemical Vapor Deposition. *Semicond. Sci. Technol.* **2021**, *36*, 034003. [CrossRef]
42. Fichtner, S.; Reimer, T.; Chemnitz, S.; Lofink, F.; Wagner, B. Stress Controlled Pulsed Direct Current Co-Sputtered Al1−xScxN as Piezoelectric Phase for Micromechanical Sensor Applications. *APL Mater.* **2015**, *3*, 116102. [CrossRef]
43. Mertin, S.; Pashchenko, V.; Parsapour, F.; Nyffeler, C.; Sandu, C.S.; Heinz, B.; Rattunde, O.; Christmann, G.; Dubois, M.-A.; Muralt, P. Enhanced Piezoelectric Properties of C-Axis Textured Aluminium Scandium Nitride Thin Films with High Scandium Content: Influence of Intrinsic Stress and Sputtering Parameters. In Proceedings of the 2017 IEEE International Ultrasonics Symposium (IUS), Washington, DC, USA, 1–25 September 2017; pp. 1–4.
44. Clement, M.; Felmetsger, V.; Mirea, T.; Iborra, E. Reactive Sputtering of AlScN Thin Ulms with Variable Sc Content on 200 mm Wafers. In Proceedings of the 2018 European Frequency and Time Forum (EFTF), Turin, Italy, 10–12 April 2018; pp. 13–16.
45. Österlund, E.; Ross, G.; Caro, M.A.; Paulasto-Kröckel, M.; Hollmann, A.; Klaus, M.; Meixner, M.; Genzel, C.; Koppinen, P.; Pensala, T.; et al. Stability and Residual Stresses of Sputtered Wurtzite AlScN Thin Films. *Phys. Rev. Mater.* **2021**, *5*, 035001. [CrossRef]
46. Zywitzki, O.; Modes, T.; Barth, S.; Bartzsch, H.; Frach, P. Effect of Scandium Content on Structure and Piezoelectric Properties of AlScN Films Deposited by Reactive Pulse Magnetron Sputtering. *Surf. Coat. Technol.* **2017**, *309*, 417–422. [CrossRef]
47. Lu, Y. Development and Characterization of Piezoelectric AlScN-Based Alloys for Electroacoustic Applications. PhD Thesis, Albert-Ludwigs-Universität Freiburg im Breisgau, Freiburg im Breisgau, Germany, 2019.
48. Beaucejour, R.; Roebisch, V.; Kochhar, A.; Moe, C.G.; Hodge, M.D.; Olsson, R.H. Controlling Residual Stress and Suppression of Anomalous Grains in Aluminum Scandium Nitride Films Grown Directly on Silicon. *J. Microelectromech. Syst.* **2022**, 1–8. [CrossRef]
49. Sandu, C.S.; Parsapour, F.; Mertin, S.; Pashchenko, V.; Matloub, R.; LaGrange, T.; Heinz, B.; Muralt, P. Abnormal Grain Growth in AlScN Thin Films Induced by Complexion Formation at Crystallite Interfaces. *Phys. Status Solidi A* **2018**, *216*, 1800569. [CrossRef]
50. Fichtner, S.; Wolff, N.; Krishnamurthy, G.; Petraru, A.; Bohse, S.; Lofink, F.; Chemnitz, S.; Kohlstedt, H.; Kienle, L.; Wagner, B. Identifying and Overcoming the Interface Originating C-Axis Instability in Highly Sc Enhanced AlN for Piezoelectric Micro-Electromechanical Systems. *J. Appl. Phys.* **2017**, *122*, 035301. [CrossRef]
51. Henry, M.D.; Young, T.R.; Douglas, E.A.; Griffin, B.A. Reactive Sputter Deposition of Piezoelectric Sc0.12Al0.88N for Contour Mode Resonators. *J. Vac. Sci. Technol. B* **2018**, *36*, 03E104. [CrossRef]
52. Sandu, C.S.; Parsapour, F.; Xiao, D.; Nigon, R.; Riemer, L.M.; LaGrange, T.; Muralt, P. Impact of Negative Bias on the Piezoelectric Properties through the Incidence of Abnormal Oriented Grains in Al0.62Sc0.38N Thin Films. *Thin Solid Films* **2020**, *697*, 137819. [CrossRef]
53. Li, M.; Xie, J.; Chen, B.; Wang, N.; Zhu, Y. Microstructural Evolution of the Abnormal Crystallite Grains in Sputtered ScAlN Film for Piezo-MEMS Applications. In Proceedings of the 2019 IEEE International Ultrasonics Symposium (IUS), Glasgow, UK, 6–9 October 2019; pp. 1124–1126.
54. Liu, C.; Chen, B.; Li, M.; Zhu, Y.; Wang, N. Evaluation of the Impact of Abnormally Orientated Grains on the Performance of ScAlN-Based Laterally Coupled Alternating Thickness (LCAT) Mode Resonators and Lamb Wave Mode Resonators. In Proceedings of the 2020 IEEE International Ultrasonics Symposium (IUS), Las Vegas, NV, USA, 1–25 September 2020; pp. 1–3.
55. Williams, K.R.; Gupta, K.; Wasilik, M. Etch Rates for Micromachining Processing-Part II. *J. Microelectromech. Syst.* **2003**, *12*, 761–778. [CrossRef]
56. Ababneh, A.; Kreher, H.; Seidel, H.; Schmid, U. The Influence of Varying Sputter Deposition Conditions on the Wet Chemical Etch Rate of AlN Thin Films. In Proceedings of the Smart Sensors, Actuators, and MEMS III, SPIE, Maspalomas, Spain, 15 May 2007; Volume 6589, pp. 1–7.

57. Chen, W.; Lin, J.; Hu, G.; Han, X.; Liu, M.; Yang, Y.; Wu, Z.; Liu, Y.; Zhang, B. GaN Nanowire Fabricated by Selective Wet-Etching of GaN Micro Truncated-Pyramid. *J. Cryst. Growth* **2015**, *426*, 168–172. [CrossRef]
58. Tian, Y.; Zhang, Y.; Yan, J.; Chen, X.; Wang, J.; Li, J. Stimulated Emission at 272 Nm from an AlxGa1−xN-Based Multiple-Quantum-Well Laser with Two-Step Etched Facets. *RSC Adv.* **2016**, *6*, 50245–50249. [CrossRef]
59. Wang, H.; Godara, M.; Chen, Z.; Xie, H. A One-Step Residue-Free Wet Etching Process of Ceramic PZT for Piezoelectric Transducers. *Sens. Actuators Phys.* **2019**, *290*, 130–136. [CrossRef]

micromachines

Article

Characterization of Ferroelectric $Al_{0.7}Sc_{0.3}N$ Thin Film on Pt and Mo Electrodes

Ran Nie [1,†], Shuai Shao [1,2,3,†], Zhifang Luo [1,2,3], Xiaoxu Kang [4] and Tao Wu [1,2,3,*]

1 Shanghai Engineering Research Center of Energy Efficient and Custom AI IC, School of Information Science and Technology, ShanghaiTech University, Shanghai 201210, China
2 Shanghai Institute of Microsystem and Information Technology, Chinese Academy of Sciences, Shanghai 200050, China
3 University of Chinese Academy of Sciences, Beijing 100049, China
4 Process Technologies Department, Shanghai IC R&D Center, Shanghai 201203, China
* Correspondence: wutao@shanghaitech.edu.cn
† These authors contributed equally to this work.

Abstract: In the past decade, aluminum scandium nitride (AlScN) with a high Sc content has shown ferroelectric properties, which provides a new option for CMOS-process-compatible ferroelectric memory, sensors and actuators, as well as tunable devices. In this paper, the ferroelectric properties of $Al_{0.7}Sc_{0.3}N$ grown on different metals were studied. The effect of metal and abnormal orientation grains (AOGs) on ferroelectric properties was observed. A coercive field of approximately 3 MV/cm and a large remanent polarization of more than 100 μC/cm^2 were exhibited on the Pt surface. The $Al_{0.7}Sc_{0.3}N$ thin film grown on the Mo metal surface exhibited a large leakage current. We analyzed the leakage current of $Al_{0.7}Sc_{0.3}N$ during polarization with the polarization frequency, and found that the $Al_{0.7}Sc_{0.3}N$ films grown on either Pt or Mo surfaces have large leakage currents at frequencies below 5 kHz. The leakage current decreases significantly as the frequency approaches 10 kHz. The positive up negative down (PUND) measurement was used to obtain the remanent polarization of the films, and it was found that the remanent polarization values were not the same in the positive and negative directions, indicating that the electrode material has an effect on the ferroelectric properties.

Keywords: AlScN; ferroelectric; thin film; leakage current; PUND test

Citation: Nie, R.; Shao, S.; Luo, Z.; Kang, X.; Wu, T. Characterization of Ferroelectric $Al_{0.7}Sc_{0.3}N$ Thin Film on Pt and Mo Bottom Electrodes. *Micromachines* **2022**, *13*, 1629. https://doi.org/10.3390/mi13101629

Academic Editor: Agnė Žukauskaitė

Received: 20 June 2022
Accepted: 19 July 2022
Published: 28 September 2022

1. Introduction

Ferroelectrics are materials that possess spontaneous polarization in the absence of an applied electric field, and the direction of its polarization vector can be flipped by the applied electric field [1,2]. It is an essential component in a wide range of applications, such as non-volatile memories and radio frequency (RF) devices [3–5]. Components based on ferroelectric thin films are also being developed for a variety of sensor and actuator applications, as well as tunable microwave circuits [6,7]. Many ferroelectric materials are perovskites with drawbacks, such as low quasi-electric transition temperatures, non-linear shifts or limited compatibility with complementary metal oxide semiconductors (CMOSs) or III-nitride technologies. These issues have so far prevented the popularization of ferroelectric functionality in microelectronics [8].

Aluminum nitride (AlN) thin films have a relatively high acoustic phase velocity, low acoustic wave loss, considerable piezoelectric coupling constant, and a coefficient of thermal expansion similar to that of Si and GaAs. These unique properties of AlN films make them widely used in mechanical, microelectronic, optical, MEMS transducers, surface wave devices (SAWs) and high-frequency broadband RF filters in the communication front-end [9,10]. AlN thin film is a III-V group semiconductor with a Wurtzite-type structure, and possesses polarization (N-polar and metal-polar) along the c-axis due to the separation of aluminum and nitrogen atoms in each plane under certain stress conditions [8]. However, AlN does not possess ferroelectricity because its polarization direction cannot be switched in an electric field below its own dielectric breakdown limit. In recent years, aluminum scandium nitride (AlScN) has become a hot research topic [11,12]. The significantly higher piezoelectric coefficient of AlScN compared to AlN has led to piezoelectric devices based on AlScN with high electromechanical coupling coefficients [13]. More interestingly, AlScN has been approved to be ferroelectric with a high Sc ratio, and its ferroelectric switching voltage can be flexibly adjusted depending on the Remanent stress and Sc content to meet the needs of ferroelectric thin films in a wide range of application scenarios [14,15]. The two polarization states of AlScN are shown in Figure 1.

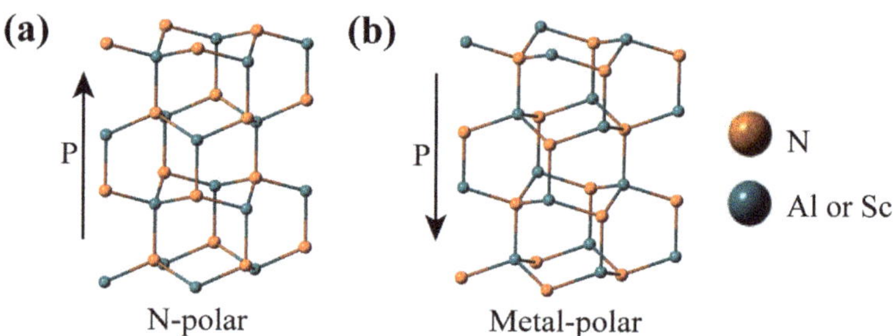

Figure 1. The two polarization states formed when the scandium atom occupies the position of the aluminum atom: (**a**) N-polar and (**b**) metal-polar.

In this work, the ferroelectric properties of AlScN with a 30% Sc content on Pt and Mo bottom electrodes were studied. It was found that the ferroelectric properties of the films grown on Mo and Pt are significantly different. The leakage current of the Mo sample is quite high and the polarization value obtained from the hysteresis curve is almost double that of the Pt sample. In the PUND test, this gap was maintained and it was observed that the remanent polarizations in the positive and negative directions were not equal. Given that the top and bottom electrodes of the device are not of the same material, it can be concluded that different metal electrodes play a role in the ferroelectricity of the film [16].

2. Fabrication and Experiment Setup

According to previous studies, the AlScN films with Sc contents below 27% are prone to break down near the coercive field, and a distinct ferroelectric polarization occurs with a Sc content of more than 27%. As the Sc contents increase, the coercivity, saturation polarization and remanent polarization all decrease [8,11,14]. Therefore, we prepared AlScN thin films at a specific Sc content of 30% in this study. In the study, 200 nm thick $Al_{0.7}Sc_{0.3}N$ films were deposited using a pulsed DC magnetron reactive sputtering (EVATEC CLUSTERLINE ® 200 MSQ) with a single 4-inch $Al_{0.7}Sc_{0.3}N$ alloy target, as shown in Figure 2a. The films deposited in this way grow along the *c*-axis direction [17,18]. In order to apply the electric field across the film thickness direction, 100 nm Pt and 200 nm Mo were used as the bottom electrodes, respectively. Then, 100 nm Al was used as the top electrodes for a simplified process flow. The final device structure is shown in Figure 2b.

X-ray diffraction (XRD) was used to characterize the crystalline quality of $Al_{0.7}Sc_{0.3}N$. A comparison and evaluation of the number of abnormal orientation grains on the film surface was obtained from SEM images. Then, the dielectric properties were measured using Keysight B1500 to test the I−V and C−V curves of the samples. To characterize the ferroelectric polarization of the films in Pt and Mo, hysteresis tests with different polarization voltage and frequencies were performed using a Radiant Multiferroic II II tester. However, other components of the system, including electrodes, leads and interfaces, could dominate the electrical response rather than the intrinsic properties of the material of interest [19]. Therefore, PUND measurements were used to separate the different components of the electrical response of a ferroelectric film. In this measurement, a sequence of five pulses was introduced. The first pulse (pulse 1) flips the polarization of the sample to a defined state. The second pulse is in the opposite direction of the first pulse and V_{max} is maintained for one pulse width to ensure that the sample is saturated with polarization, at which point, the polarization value P_1 is recorded. After the second pulse, their is a wait of one pulse width and the second polarization value P_2 is recorded. After certain pulse delay, a third pulse is applied and the third polarization value P_3 is recorded at its end. Then, there is a wait of one pulse width to record the polarization value P_4 and subtract P_1 from P_3 to obtain dP, since P_1 contains both switching and non-switching components, whereas P_3 contains only non-switching components, so dP can represent the correct remanent polarization. Typically, P_2-P_4 is written as dP_r, since P_2 and P_4 are the polarization values recorded after waiting for a pulse width and losing a certain polarization, and dP_r should be equal to dP. Pulse 4 and pulse 5 are similar to pulse 2 and pulse 3, only in the opposite direction, in order to obtain $-dP$ and $-dP_r$. Figure 3 shows the pulse sequence of the PUND test.

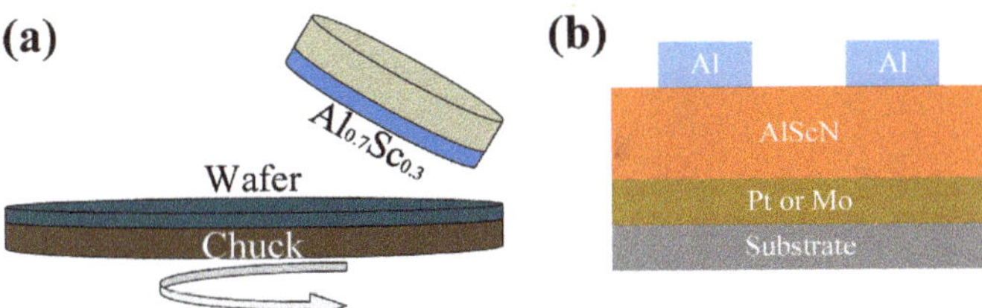

Figure 2. (**a**) Schematic diagram of magnetron sputtering deposition of $Al_{0.7}Sc_{0.3}N$. (**b**) Stacking schematic of Pt/Mo-AlScN-Al structure.

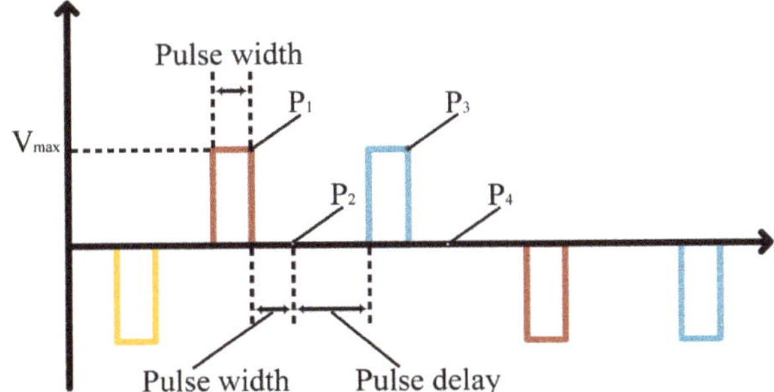

Figure 3. Schematic of the pulse sequence for the PUND test. The pulse intensity is V_{max}, and each pulse maintains a pulse width. When a pulse ends, a pulse width plus pulse delay is waited for in order to input the next pulse. The polarization values are recorded twice for each pulse cycle.

3. Results and Discussions

3.1. Film Quality Characterization

As shown in Figure 4a, the full width half maximum (FWHM) of the $2\theta/\theta$ scans of the 200 nm $Al_{0.7}Sc_{0.3}N$ films grown on Pt and Mo surfaces were 0.36° and 0.50°, respectively. The FWHM of the ω-rocking curve of $Al_{0.7}Sc_{0.3}N$ (0002) peak is below 2.5° on Pt samples, and approximately 3° on Mo samples. The absence of other peaks near the $Al_{0.7}Sc_{0.3}N$ (0002) peak indicates that the film has a good c-axis orientation [20]. The diffraction peak of the sample on the Pt surface is stronger and has a better grain orientation [21]. Comparing SEM images Figure 4b,c, there are a large number of abnormal grains of $Al_{0.7}Sc_{0.3}N$ grown on the Mo surface. These AOGs lead to a partial or total loss of the c-axis texture in the surface layer of the films. Since the polarization direction of AlScN is along the c-axis, it can be assumed that the ferroelectric properties of the Mo sample will be worse than those of the Pt sample, which will be further verified in the later measurement.

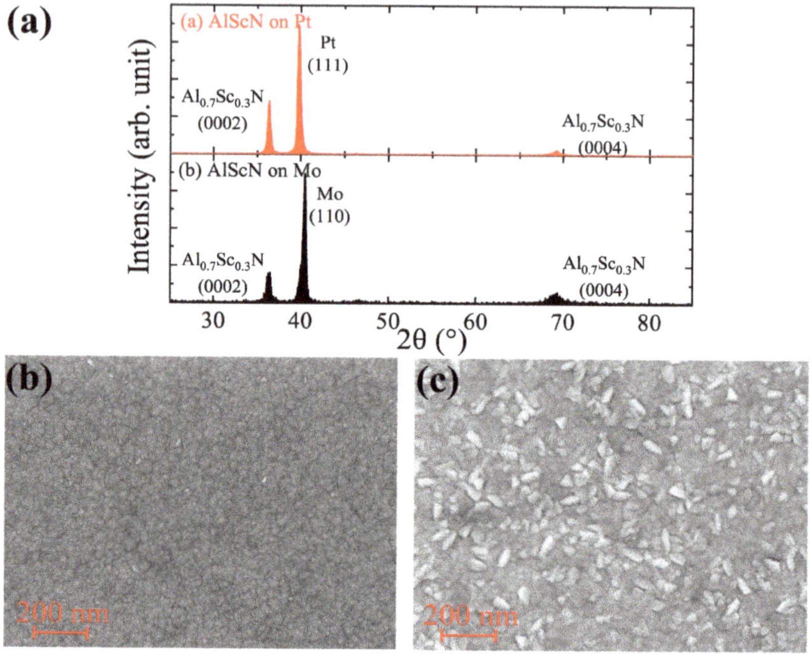

Figure 4. (**a**) X-ray diffraction $2\theta/\theta$ scan $Al_{0.7}Sc_{0.3}N$ on (**a**) Pt and (**b**) Mo. The peak of $Al_{0.7}Sc_{0.3}N$ (0002) on the Pt sample is stronger than that of the Mo sample. SEM image of $Al_{0.7}Sc_{0.3}N$ grown on: (**b**) Pt with a good crystal orientation; (**c**) Mo with a large number of abnormal grains.

3.2. Dielectric Properties Measurement

We performed I−V and C−F tests on Pt samples. The scanned voltage from the I-V test does not exceed the coercivity field of the film in order to observe the leakage current of the device. The leakage current of the Pt sample is small as can be seen in Figure 5a. The currents in the positive and negative directions are not symmetrical, which is caused by the Schottky contact between Pt and AlScN and the ohmic contact between Al and AlScN. [22] As shown in Figure 5b, in the C-F test, the frequency was scanned from 1 kHz to 5 MHz, the measured capacitance value increased from 10 pF to 18 pF, and the area of the device under the test was 4×10^{-4} cm^2.

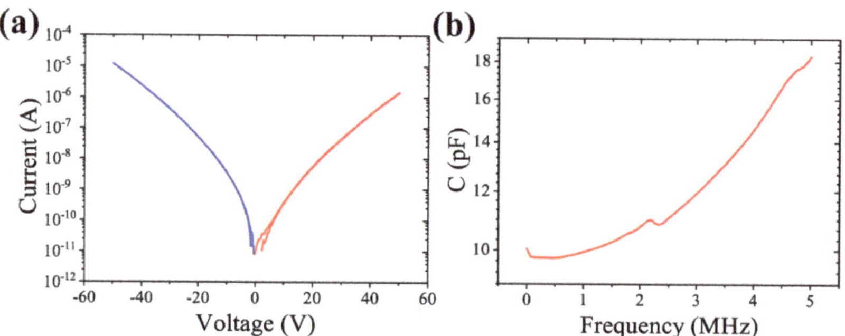

Figure 5. $Al_{0.7}Sc_{0.3}N$ with Pt as the bottom electrode. (**a**) I−V curve with voltage scanning from 0 V to 50 V and then returning to 0 V, similar in the negative direction. (**b**) The C−F scan goes from 1 kHz to 5 MHz and back.

3.3. Effect of Different Electrodes on P-E Ferroelectric Hysteresis

As shown in Figure 6a,b, AlScN has typical ferroelectric properties. The maximum applied drive voltage ranges from 50 V to 82 V in steps of 4 V. The maximum voltage of the Pt sample was only up to 78 V. The sample was broken down after the voltage was increased to 84 V, whereas the Mo sample was broken down at 78 V. As can be seen from the hysteresis loop, the coercive field is approximately 3 MV/cm. However, the samples with two different bottom electrodes show completely different leakage behavior. The $Al_{0.7}Sc_{0.3}N$ films grown on Mo have a large leakage current during the negative polarization. We speculate that this is due to the poor crystal quality and the huge amount of abnormal orientation grains, as the SEM image shows. Comparatively, the Pt sample also exhibits current asymmetry, but very weakly. The remanent polarizations of the two samples obtained from the hysteresis line test were approximately 100 μC/cm^2 and 350 μC/cm^2, respectively.

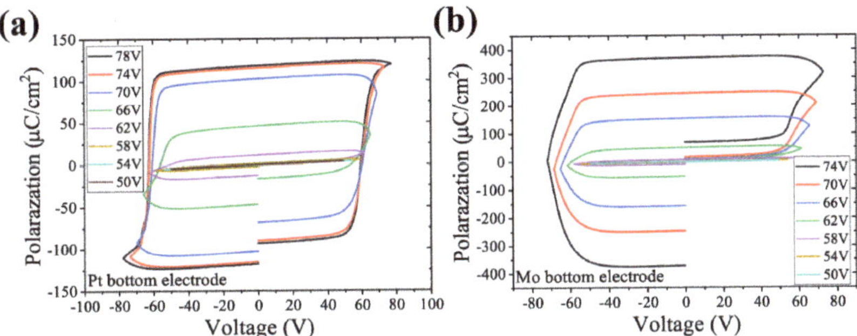

Figure 6. Hysteresis lines of $Al_{0.7}Sc_{0.3}N$ on two electrodes at different voltages. (**a**) Pt samples, with maximum remanent polarization, were approximately 100 μC/cm^2 and (**b**) Mo samples were approximately 350 μC/cm^2.

In order to further analyze this leakage behavior, we measured the hysteresis loops at different frequencies and set the drive voltage just beyond the coercive field. For the Pt sample, ignoring the "gap" caused by electrode asymmetry, the P-E curves in the range of 1 K to 10 K show near standard ferroelectricity, as shown in Figure 7. When the frequency reaches 10 K, there is almost symmetry. However, the P-E curve of the Mo sample is not so good. Although the frequency is increased to a relatively high level so that the polarization does not switch repeatedly, it still exhibits significant asymmetry.

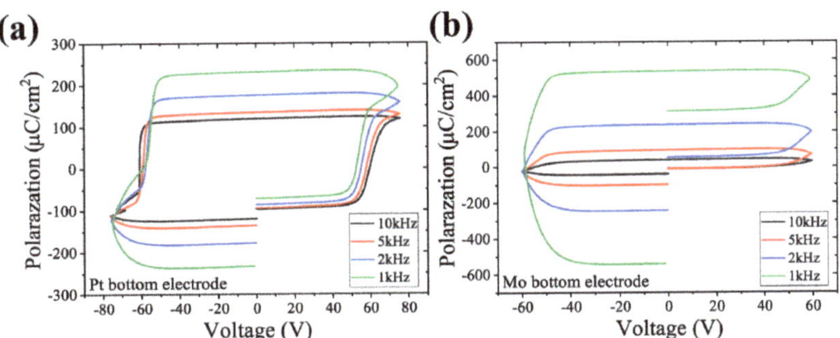

Figure 7. The hysteresis P-E loops with different frequencies of the $Al_{0.7}Sc_{0.3}N$ on (**a**) Pt—as the frequency increases, the polarization flips incompletely and gradually becomes symmetrical—and (**b**) Mo—always asymmetrical.

Since the Mo sample exhibited asymmetry, we swapped the drive and sense terminals and obtained the electrical response as shown in Figure 8a. The maximum voltage applied was 50 V, which does not exceed the coercivity field, so the resulting current contains only the leakage current component and not the polarization current. The results show that there is a large leakage current of 1 mA in the negative direction only. Such a phenomenon could be attributed to different electrode materials, as well as asymmetric polarization hysteresis. Once the driving voltage exceeds the coercive field, the large leakage current on the Mo sample causes the hysteresis loop to completely deform, as illustrated in Figure 7b.

In Figure 8b, the peak currents of negative polarization at different frequencies are compared by extracting at 60 V. The measured current of the Mo sample is around 8 mA at a 1 kHz frequency, while showing currents over two times that of the Pt at all frequencies. Such a large leakage current will make the Mo electrode sample easier to break down when its polarization is reversed. On the other hand, the large leakage current makes it possible to output a stronger signal during polarization reversal, which greatly reduces the possibility of a loss of reading.

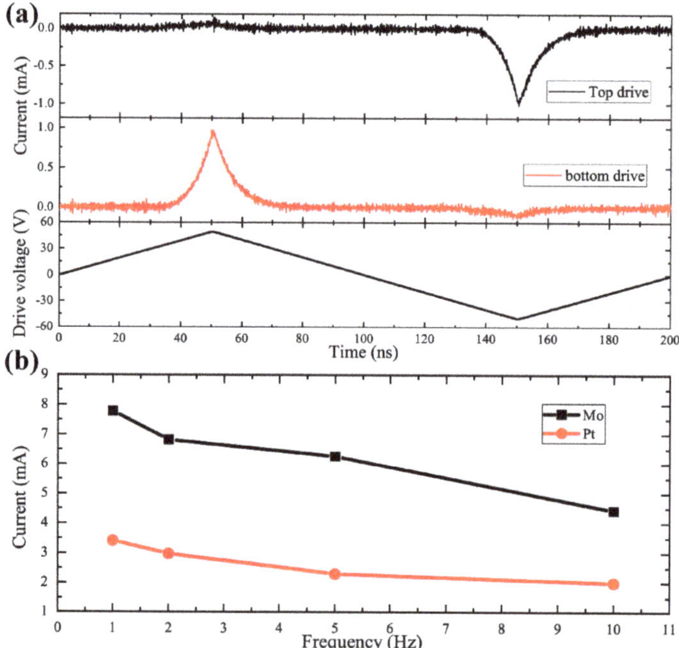

Figure 8. (**a**) Switching current with top and bottom drive under maximum 50 V triangle driving voltage. (**b**) The maximum voltage current as a function of driving frequency with the maximum voltage set to 60 V.

3.4. PUND Test to Obtain the Remanent Polarization

A PUND test was performed to further analyze the ferroelectric properties of $Al_{0.7}Sc_{0.3}N$ on both metals. As mentioned earlier, the main conditions that can be changed in the PUND test are V_{max}, pulse width and pulse delay. V_{max} just needs to be large enough to ensure that the polarization can switch. Therefore, we only changed the pulse width and pulse delay to see how the remanent polarization of the device changes.

First, V_{max} was set to 60 V, pulse width to 0.5 ms, pulse delay to 10 ms and each test was subjected to 20 repetitions of the experiment, as shown in Figure 9a,b. An interesting phenomenon appears here: the remanent polarization in the negative direction of the Pt sample is larger for the first few times of the power-up test after resting at one end of the time, and then gradually decreases and stabilizes. This may be due to some parasitic parameters, which are subject to further analysis. In addition, the remanent polarization of the Pt sample is around 200 μC/cm² in the positive direction and 260 μC/cm² in the negative direction, a difference brought about by the Schottky contact between Pt and AlScN. The polarization of the Mo sample is very large, more than twice that of the Pt sample in both the positive and negative directions, which is in agreement with the polarization current pattern recorded earlier.

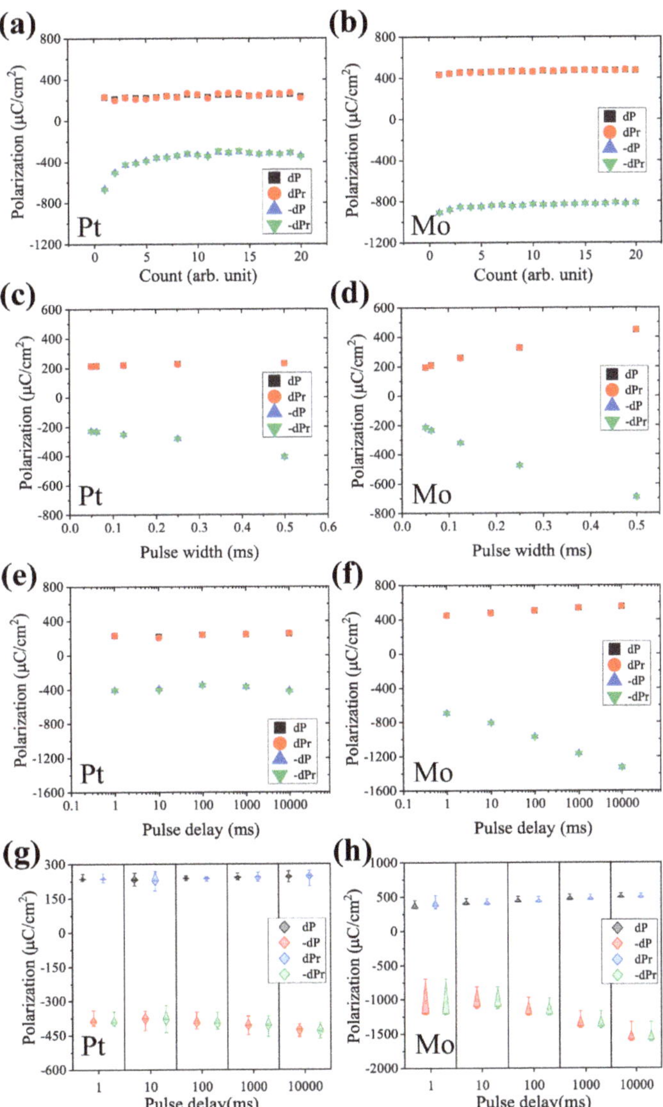

Figure 9. PUND measurements under different conditions. Repeat test at fixed V_{max} = 60, V, pulse width = 0.5 ms and pulse delay = 10 ms: (**a**) Pt and (**b**) Mo. V_{max} and pulse width were kept constant: 60 V and 0.5 ms, pulse delay set to 1, 10, 100, 1000 and 10,000 ms. (**c**) Pt and (**d**) Mo. Pulse width was taken as 0.5, 0.25, 0.125, 0.625 and 0.05 ms. In addition, V_{max} = 60 V, pulse delay = 1 ms. (**e**) Pt and (**f**) Mo. Distribution of PUND results for different samples at pulse wide was 0.5 ms, pulse delay changed from 1 ms to 10,000 ms, (**g**) Pt and (**h**) Mo.

Then, the test was performed with different pulse widths. The V_{max} and pulse delay were set to 60 V and 1 ms, respectively, and the pulse width was taken as 0.5, 0.25, 0.125, 0.625 and 0.05 ms. It can be seen in Figure 9c,d that the remanent polarization increases with increasing pulse width for both samples. At small pulse widths, it is not enough to support a complete flip of polarization, resulting in a decrease in the remanent polarization value. Therefore, devices utilizing the ferroelectricity of AlScN thin films require a special design when setting the operating frequency. The remanent polarization, leakage currents and breakdown voltage, as well as the retention of the ferroelectricity, should be taken into account during the device and architecture design.

Finally, the V_{max} and pulse width were kept constant and tested at pulse delays of 1, 10, 100, 1000 and 10,000 ms, respectively. It can be seen that the Pt sample results are smooth with no significant change, which is basically the same as the previous test. The remaining polarization value in the positive direction of the Mo sample also has no significant change, whereas the value in the negative direction gradually increases, as shown in Figure 9e,f. This means that the Mo sample has a larger polarization loss in the negative direction, which also corresponds to a larger leakage current in the negative direction. Therefore, AOGs on the film surface perpendicular to the c-axis can greatly compromise the ferroelectric properties. Moreover, multiple samples were tested to observe whether there is good consistency, as shown in Figure 9g,h. It can be seen that the negative polarization fluctuation of the Mo sample is slightly larger, and other points float in a small range.

Therefore, the abnormal grain orientation may seriously affect the ferroelectric properties of the films. It is possible that the abnormal grain orientation changes the original wurtzite structure near the interface between the metal and dielectric, and then affects the polarization properties of the film. From the test results, this effect is unidirectional and will greatly change the polarization characteristics in one direction. It can be reasonably speculated that, in addition to Mo, other metal materials may also bring different effects, which is worthy of further experimental verification.

4. Conclusions

In this paper, we analyzed the correlation between the ferroelectricity of aluminum scandium nitride and bottom metal electrodes. On one hand, the difference in crystal orientation of films grown on Pt or Mo metals will affect the ferroelectric properties. On the other hand, the inherent contact barrier between metal and dielectric materials will also affect ferroelectricity. The direction of driving voltage and frequency will also lead to different phenomena. After solving the problem of the leakage current, the high remnant polarization of more than 100 $\mu C/cm^2$ and coercive field of 3 MV/cm exhibited by $Al_{0.7}Sc_{0.3}N$ films with good quality are suitable for ferroelectric memory devices. In FeRAM or FeFET, a high remnant polarization value can increase the storage density, while a suitable coercivity field can meet the storage window at a thin thickness. With further development, AlScN is expected to be widely used in commercial memory devices, as well as tunable RF applications.

Author Contributions: R.N.: Conceptualization, validation, graphics, original draft preparation; S.S.: Methodology, formal analysis, review and editing; Z.L.: Investigation, data curation; X.K.: Supervision; T.W.: Supervision, project administration, editing. All authors have read and agreed to the published version of the manuscript.

Funding: This research was funded by National Natural Science Foundation of China (61874073) and Lingang Laboratory under Grant LG-QS-202202-05.

Data Availability Statement: Data is available upon request.

Acknowledgments: The authors appreciate the device fabrication support from the ShanghaiTech Quantum Device Lab (SQDL), and Analytical Instrumentation Center (SPSTAIC10112914) XRD Lab.

Conflicts of Interest: The authors declare no conflict of interest.

References

1. Li, S.; Yilin, W.; Yang, M.; Miao, J.; Lin, K.; Li, Q.; Chen, X.; Deng, J.; Xing, X. Ferroelectric thin films: Performance modulation and application. *Mater. Adv.* **2022**, *3*, 5735–5752. [CrossRef]
2. Martin, L.W.; Rappe, A.M. Thin-film ferroelectric materials and their applications. *Nat. Rev. Mater.* **2017**, *2*, 16087. [CrossRef]
3. Setter, N.; Damjanovic, D.; Eng, L.; Fox, G.; Gevorgian, S.; Hong, S.; Kingon, A.; Kohlstedt, H.; Park, N.Y.; Stephenson, G.B.; et al. Ferroelectric thin films: Review of materials, properties, and applications. *J. Appl. Phys.* **2006**, *100*, 051606. [CrossRef]
4. Wang, D.; Zheng, J.; Musavigharavi, P.; Zhu, W.; Foucher, A.C.; Trolier-McKinstry, S.E.; Stach, E.A.; Olsson, R.H. Ferroelectric Switching in Sub-20 nm Aluminum Scandium Nitride Thin Films. *IEEE Electron. Device Lett.* **2020**, *41*, 1774–1777. [CrossRef]
5. Schenk, T.; Mueller, S. A New Generation of Memory Devices Enabled by Ferroelectric Hafnia and Zirconia. In Proceedings of the International Symposium on Applications of Ferroelectrics (ISAF), Sydney, Australia, 16–21 May 2021; pp. 1–11, ISSN 2375-0448. [CrossRef]
6. Kreutzer, T.N.; Fichtner, S.; Wagner, B.; Lofink, F. A double-layer MEMS actuator based on ferroelectric polarization inversion in AlScN. In Proceedings of the International Symposium on Applications of Ferroelectrics (ISAF), Sydney, Australia, 16–21 May 2021; pp. 1–3, ISSN 2375-0448. [CrossRef]
7. Wang, J.; Park, M.; Mertin, S.; Pensala, T.; Ayazi, F.; Ansari, A. A Film Bulk Acoustic Resonator Based on Ferroelectric Aluminum Scandium Nitride Films. *J. Microelectromech. Syst.* **2020**, *29*, 741–747. [CrossRef]
8. Fichtner, S.; Wolff, N.; Lofink, F.; Kienle, L.; Wagner, B. AlScN: A III-V semiconductor based ferroelectric. *J. Appl. Phys.* **2019**, *125*, 114103. [CrossRef]
9. Shao, S.; Luo, Z.; Lu, Y.; Mazzalai, A.; Tosi, C.; Wu, T. High Quality Co-Sputtering AlScN Thin Films for Piezoelectric Lamb-Wave Resonators. *J. Microelectromech. Syst.* **2022**, *31*, 328–337. [CrossRef]
10. Shao, S.; Luo, Z.; Lu, Y.; Mazzalai, A.; Tosi, C.; Wu, T. Low Loss Al0.7Sc0.3N Thin Film Acoustic Delay Lines. *IEEE Electron. Device Lett.* **2022**, *43*, 647–650. [CrossRef]
11. Olsson, R.H.; Tang, Z.; D'Agati, M. Doping of Aluminum Nitride and the Impact on Thin Film Piezoelectric and Ferroelectric Device Performance. In Proceedings of the Custom Integrated Circuits Conference (CICC), Boston, MA, USA, 22–25 March 2020; pp. 1–6, ISSN 2152-3630. [CrossRef]
12. Shao, S.; Luo, Z.; Wu, T. High Figure-of-Merit Lamb Wave Resonators Based on Al0.7Sc0.3N Thin Film. *IEEE Electron. Device Lett.* **2021**, *42*, 1378–1381. [CrossRef]
13. Wang, J.; Park, M.; Mertin, S.; Pensala, T.; Ayazi, F.; Ansari, A. A High-k_t^2 Switchable Ferroelectric Al0.7Sc0.3N Film Bulk Acoustic Resonator. In Proceedings of the Joint Conference of the IEEE International Frequency Control Symposium and International Symposium on Applications of Ferroelectrics (IFCS-ISAF), Keystone, CO, USA, 19–23 July 2020; pp. 1–3, ISSN 2375-0448. [CrossRef]
14. Pirro, M.; Herrera, B.; Assylbekova, M.; Giribaldi, G.; Colombo, L.; Rinaldi, M. Characterization of Dielectric and Piezoelectric Properties of Ferroelectric AlScN Thin Films. In Proceedings of the 34th International Conference on Micro Electro Mechanical Systems (MEMS), Gainesville, FL, USA, 25–29 January 2021; pp. 646–649. [CrossRef]
15. Luo, Z.; Shao, S.; Wu, T. Al0.78Sc0.22N Lamb Wave Contour Mode Resonators. *IEEE Trans. Ultrason. Ferroelectr. Freq. Control.* **2021**, 1. [CrossRef] [PubMed]
16. Fichtner, S.; Lofink, F.; Wagner, B.; Schönweger, G.; Kreutzer, T.N.; Petraru, A.; Kohlstedt, H. Ferroelectricity in AlScN: Switching, Imprint and sub-150 nm Films. In Proceedings of the Joint Conference of the IEEE International Frequency Control Symposium and International Symposium on Applications of Ferroelectrics (IFCS-ISAF), Keystone, CO, USA, 19–23 July 2020; pp. 1–4, ISSN 2375-0448. [CrossRef]
17. Pirro, M.; Zhao, X.; Herrera, B.; Simeoni, P.; Rinaldi, M. Effect of Substrate-RF on Sub-200 nm Al0.7Sc0.3N Thin Films. *Micromachines* **2022**, *13*, 877. [CrossRef] [PubMed]
18. Yasuoka, S.; Shimizu, T.; Tateyama, A.; Uehara, M.; Yamada, H.; Akiyama, M.; Hiranaga, Y.; Cho, Y.; Funakubo, H. Effects of deposition conditions on the ferroelectric properties of $(Al_{1-x}Sc_x)N$ thin films. *J. Appl. Phys.* **2020**, *128*, 114103. [CrossRef]
19. Rabe, K.M.; Dawber, M.; Lichtensteiger, C.; Ahn, C.H.; Triscone, J.M. Modern Physics of Ferroelectrics:Essential Background. In *Physics of Ferroelectrics: A Modern Perspective*; Topics in Applied Physics, Springer: Berlin/Heidelberg, Germany, 2007; pp. 1–30. [CrossRef]
20. Wang, D.; Zheng, J.; Tang, Z.; D'Agati, M.; Gharavi, P.S.M.; Liu, X.; Jariwala, D.; Stach, E.A.; Olsson, R.H.; Roebisch, V.; et al. Ferroelectric C-Axis Textured Aluminum Scandium Nitride Thin Films of 100 nm Thickness. In Proceedings of the Joint Conference of the IEEE International Frequency Control Symposium and International Symposium on Applications of Ferroelectrics (IFCS-ISAF), Keystone, CO, USA, 19–23 July 2020; pp. 1–4, ISSN 2375-0448. [CrossRef]
21. Schönweger, G.; Petraru, A.; Islam, M.R.; Wolff, N.; Haas, B.; Hammud, A.; Koch, C.; Kienle, L.; Kohlstedt, H.; Fichtner, S. From Fully Strained to Relaxed: Epitaxial Ferroelectric Al1-xScxN for III-N Technology. *Adv. Funct. Mater.* **2022**, *32*, 2109632. Available online: https://onlinelibrary.wiley.com/doi/pdf/10.1002/adfm.202109632 (accessed on 12 February 2022). [CrossRef]
22. Giribaldi, G.; Pirro, M.; Soukup, B.H.; Assylbekova, M.; Colombo, L.; Rinaldi, M. Compensation of Contact Nature-Dependent Asymmetry in The Leakage Current of Ferroelectric ScxAl1-x N Thin-Film Capacitors. In Proceedings of the 34th International Conference on Micro Electro Mechanical Systems (MEMS), Gainesville, FL, USA, 25–29 January 2021; pp. 650–653, ISSN 2160-1968. [CrossRef]

Article

Compensation of the Stress Gradient in Physical Vapor Deposited $Al_{1-x}Sc_xN$ Films for Microelectromechanical Systems with Low Out-of-Plane Bending

Rossiny Beaucejour [1], Michael D'Agati [2], Kritank Kalyan [3] and Roy H. Olsson III [2,*]

1 Department of Mechanical Engineering and Applied Mechanics, University of Pennsylvania, 220 S. 33rd St., Philadelphia, PA 19104, USA; rossiny@seas.upenn.edu
2 Department of Electrical and Systems Engineering, University of Pennsylvania, 3205 Walnut St., Philadelphia, PA 19104, USA; mdagati@seas.upenn.edu
3 Singh Center for Nanotechnology, University of Pennsylvania, 3205 Walnut St., Philadelphia, PA 19104, USA; kritank@seas.upenn.edu
* Correspondence: rolsson@seas.upenn.edu

Abstract: Thin film through-thickness stress gradients produce out-of-plane bending in released microelectromechanical systems (MEMS) structures. We study the stress and stress gradient of $Al_{0.68}Sc_{0.32}N$ thin films deposited directly on Si. We show that $Al_{0.68}Sc_{0.32}N$ cantilever structures realized in films with low average film stress have significant out-of-plane bending when the $Al_{1-x}Sc_xN$ material is deposited under constant sputtering conditions. We demonstrate a method where the total process gas flow is varied during the deposition to compensate for the native through-thickness stress gradient in sputtered $Al_{1-x}Sc_xN$ thin films. This method is utilized to reduce the out-of-plane bending of 200 μm long, 500 nm thick $Al_{0.68}Sc_{0.32}N$ MEMS cantilevers from greater than 128 μm to less than 3 μm.

Keywords: aluminum scandium nitride; physical vapor deposition; stress; stress gradient; fabrication; cantilever beams; MEMS

Citation: Beaucejour, R.; D'Agati, M.; Kalyan, K.; Olsson, R.H., III Compensation of the Stress Gradient in Physical Vapor Deposited $Al_{1-x}Sc_xN$ Films for Microelectromechanical Systems with Low Out-of-Plane Bending. *Micromachines* **2022**, *13*, 1169. https://doi.org/10.3390/mi13081169

Academic Editor: Nam-Trung Nguyen

Received: 2 June 2022
Accepted: 21 July 2022
Published: 24 July 2022

1. Introduction

Microelectromechanical System (MEMS) structures are utilized to form cantilevers for sensors [1,2], resonators [3], and piezoelectric energy harvesters [4]. In piezoelectric energy harvesters [4] and micro-accelerometers [5], the properties of the cantilevers and their geometric limits directly impact device performance. The amount of deformation in cantilever beams and other structures depends on the growth technique utilized and material selection.

Aluminum Nitride (AlN) is a piezoelectric material commonly selected for the fabrication of vibrating MEMS structures. In 2009, it was determined that alloying AlN with scandium (Sc) can increase the piezoelectric coefficients by 500% [6]. AlN and Aluminum Scandium Nitride ($Al_{1-x}Sc_xN$) films deposited using Molecular Beam Epitaxy (MBE) or metal-organic Chemical Vapor Deposition (MOCVD) demonstrate exceptional film quality and high electromechanical coupling (k_t^2) but require processing at high temperature [7,8] and exhibit challenges due to stress gradients in the epi layers and complications in the releasing of structures at the end of the fabrication process [8]. For example, Park et al. [8] developed MBE $Al_{1-x}Sc_xN$ Lamb wave resonators at super high frequencies directly on Silicon with acoustic velocities of 14,000 m/s and k_t^2 of 7.2%, but compressive film stresses produced buckling in the released films [9].

Sputtering is a Physical Vapor Deposition (PVD) method used to produce semi-conformal films of controlled thickness with high deposition rates at low substrate temperatures for realizing MEMS devices. During sputter deposition, ions accelerate and bombard a target removing atoms to be deposited on the wafer. There are numerous collisions that impact the mean free path of atoms and which control coverage on all surfaces. These processes also introduce stresses within the resulting film [10,11]. Reush et al. grew 001 polycrystalline AlN films using reactive RF magnetron sputtering directly on Silicon [7]. During growth, differences in grain size induced forces between grains at the grain boundaries. The increase in grain size with increasing thickness produced an intrinsic stress gradient through the thickness caused by zone T growth [12] and self-shadowing [6,13]. Knisely et al. determined that stress gradients in AlN film growth were a result of film growth stress, nucleation, and coalescence [14]. Knisely [14] further demonstrated a method to control both the stress and through-thickness stress gradient in AlN thin films using radio frequency (RF) substrate bias power modeled using a power law function. Introduction and further increases in RF substrate bias power during the sputtering of AlN based materials results in films with more compressive stress [14]. For $Al_{1-x}Sc_xN$ films with high Sc substitution for Al, the resulting average film stress is often compressive [3,15] without the introduction of an RF substrate bias, especially when optimizing for low anomalously oriented grain (AOG) growth. Thus, it may not be possible to utilize RF substrate bias to simultaneously optimize for low average film stress and low through-thickness stress gradient in highly Sc alloyed AlN thin films.

Several stress and stress control techniques have been produced for MEMS devices. Pulskamp et al. mitigated the stress and stress gradient in sol-gel Lead Zirconium Titanate (PZT) multi-layer MEMS devices by altering the silane flow rates, annealing the top layer Platinum (Pt), and adjusting the film thickness [16]. For the 0.5 μm devices, a Titanium/Platinum 651 MPa (Ti/Pt) adhesive layer produced a stress of 35 MPa in the PZT with displacements of −250 μm [16]. Characterizing the released structures with analytical modeling and controlling process variability during growth further characterized the stress within these films [16]. Devices such as RF switches, accelerometers, and optical mirrors are very sensitive to the stress and stress gradient [9,17]. Sedky et al. developed these MEMS devices with p-type Silicon Germanium (Si_xGe_{1-x}) using low pressure chemical vapor deposition (LPCVD) [17]. The stress was optimized by modifying the Se concentration from 40% to 90%, tuning the pressure from 650 mtorr to 800 mtorr, and controlling the deposition temperature to be between 400 and 450 °C. The stress gradient was reduced with a laser annealing technique which reduced stress from 125 MPa to 25 MPa and deflections from 118 μm to 5 μm [17]. Zhu determined the stress was caused by variations in film growth resulting in uneven grains and nonuniform boundaries [18]. When Aluminum Nitride (AlN) and Ruthenium (Ru) is sputtered to fabricate MEMS resonators, a folded beam design was used to limit deflection [18]. Mulloni et al. used similar clamped-clamped and single-clamped cantilever designs for Gold microstructure using electrodepositing [19]. The clamped-clamped released structures were more sensitive to stress variations than the single-clamped design [19]. Depending on the desired bending, the release temperature was reduced to limit stress which increased fabrication time [19]. These methods including temperature control, thickness variation, process control, laser annealing, design alterations, and pressure tuning show promise in reducing stress and stress gradients in other material systems.

Previous studies have explored methods using total N_2 process gas flow to control average $Al_{1-x}Sc_xN$ film stress [3,20,21]. These studies show that total process gas flow and the process gas mixture have a significant impact on the crystallinity, defects, and stress of sputter-deposited $Al_{1-x}Sc_xN$ films [3,20,21]. $Al_{1-x}Sc_xN$ films with a high degree of crystallinity and few defects require long mean free path and thus low process pressures. Sputtering $Al_{1-x}Sc_xN$ at low process pressure in a pure N_2 environment improves c-axis orientation and decreases surface roughness [3]. Increases in N_2 flow increases ion scattering, reduces ion mean free path (MFP) and increases average film stress [20–22]. A method to suppress the stress gradients which produce out-of-plane bending while holding Sc concentration constant, maintaining film quality, and minimizing average stress has not been established for $Al_{1-x}Sc_xN$ thin film materials. We previously demonstrated that 500 nm thick $Al_{0.68}Sc_{0.32}N$ materials, free of AOGs, with low surface roughness and strong c-axis orientation, could be sputter-deposited on Si with a controlled average film stress ranging from −458 to 287 MPa, by controlling N_2 flow between 20–30 sccm [3]. Here we provide a novel method that demonstrates both low average film stress and low through-thickness film stress gradient simultaneously by varying the N_2 flow within the 20–30 sccm range during the deposition of $Al_{0.68}Sc_{0.32}N$. Furthermore, we demonstrate the utility of this method by fabricating $Al_{0.68}Sc_{0.32}N$ MEMS cantilevers with low out-of-plane bending.

$Al_{1-x}Sc_xN$ with high Sc alloying is a promising piezoelectric material for MEMS radio frequency (RF) filtering [3], energy harvesting [4], and sensing [23,24] devices because of its demonstrated high figure of merit [2] for each of these application areas. Piezoelectric RF bulk acoustic wave (BAW) resonator filters can often be implemented as anchored plates with extremely high out-of-plane bending stiffness. Such an implementation allows the realization of high-performance BAW resonators and filters using $Al_{1-x}Sc_xN$ materials with high through-thickness stress gradient [3]. By contrast, high performance energy harvesters [4] and piezoelectric sensors [24,25] often require implementations using compliant cantilever [24] and clamped-guided beams [25] with low out-of-plane bending stiffness. In these devices, large film through-thickness stress gradients and the resulting out-of-plane bending can cause a significant degradation of the on-axis sensitivity and a corresponding increase in the cross-axis sensitivity. In addition, high out-of-plane bending complicates wafer level packaging (WLP) of MEMS devices because the packaging cavity must be able to accommodate the out-of-plane bending displacement. Therefore, if high performance $Al_{1-x}Sc_xN$ materials are to be utilized in sensing and energy harvesting applications it is imperative to achieve low through-thickness stress gradient for the $Al_{1-x}Sc_xN$ film and low out-of-plane bending for the MEMS structures implemented with the $Al_{1-x}Sc_xN$ film.

2. Background and Theory

2.1. Average Film Stress Measurement

Once released from a substrate, as is common in MEMS processing, films will relax their built-in stress, which can lead to undesired deformation and/or buckling. Thus, knowing the stress in each layer of a MEMS process is vitally important. The average film stress is computed by measuring the wafer's radius of curvature with a profilometer and subsequently using the Stoney equation [3,14,20,26,27]. To separate the various components leading to bending of the substrate, the radius of curvature is measured both before, R_0, and after, R, the deposition of each film, allowing the built-in stress of each layer, $T_{f,BI}$, [3,14,20,26,27] in a process to be isolated

$$T_{f,BI} = \frac{1}{6}\frac{Y_s}{(1-v_s)}\frac{t_s^2}{t_f}\left(\frac{1}{R}-\frac{1}{R_0}\right) \quad (1)$$

where Y_s, v_s and t_s are the Young's Modulus, Poisson's ratio, and thickness of the substrate respectively, and t_f is the thickness of the film.

2.2. Relationship between the Total Process Gas Flow and the Resulting Film Stress

Knisely [14] introduced a model describing the power law relation between the RF substrate bias and the resulting AlN film stress. For a fixed Sc alloy this power law equation can be adapted to instead describe the total $Al_{1-x}Sc_xN$ films stress T_{ave} as a function of N_2 process gas flow

$$T_{ave} = \alpha\left(\beta(t_f)^{\gamma} + F_{N_2}\right) \quad (2)$$

where α is determined from the slope of the stress versus N_2 gas flow, F_{N_2} is the N_2 gas flow in the chamber, and β and γ are empirical fits based on the deposition parameters and environment. Similar to Knisely [14] we implement a film that utilizes multiple layers deposited under different sputtering gas conditions where the average film stress of each layer is used to compensate for the through-thickness stress gradient. Where Knisely [14] utilizes a different RF substrate bias to control the stress of each AlN layer, we utilize a different N_2 flow to control the stress of each $Al_{0.68}Sc_{0.32}N$ layer. The layer film stress, t_f, based on the thickness and N_2 flow of the layer is derived from integrating the average film stress [14] and is given by

$$T_f(t, F_{N_2}) = \alpha(\beta(1+\gamma)(t_f)^{\gamma} - F_{N_2,n}) \tag{3}$$

where F_{N_2} is the constant flow applied to the layer and t is the thickness of the layer. Equation (3) can be used to calculate the layer stresses needed to compensate for the through-thickness stress gradient within the film. The stress gradient is calculated from the average stress measurements taken from different wafers deposited using identical deposition process parameters and at varying AlScN thicknesses. Equation (3) is then used to interpolate between the experimental measurements of films with different thicknesses to find the through-thickness stress gradient.

3. Experimentation Details

The average stress through the thickness of the film is evaluated using the methods described in Section 2.1. The slope of the stress versus F_{N_2} curve, α, is determined using a linear fit to the measured stress data. The local stress of each individual layer required to compensate for the through-thickness stress gradient is calculated using Equation (3). Cantilevers with a width of 50 µm and a length of 200 µm are realized in 500 nm thick $Al_{0.68}Sc_{0.32}N$ films with out-of-plane displacements measured from the difference between the anchor and beam tip heights. The $Al_{1-x}Sc_xN$ cantilever fabrication is shown in Figure 1 and begins with deposition of $Al_{0.68}Sc_{0.32}N$ on 100 mm p-type (100) Si wafers in an Evatec CLUSTERLINE® 200 II Physical Vapor Deposition System in steps (a) and (b). Table 1 summarizes the DC reactive co-sputtering parameters used to deposit the $Al_{0.68}Sc_{0.32}N$ films. The $Al_{1-x}Sc_xN$ films are deposited on a 15 nm thick AlN seed layer and a 35 nm thick gradient seed layer ($Al_{1\rightarrow0.68}Sc_{0\rightarrow0.32}N$) where the Sc alloying ratio is linearly varied through the thickness from 0 to 32%. This seed and gradient layer was previously demonstrated to suppress anomalously oriented grains while maintaining crystal quality [3]. The crystal quality characterized in a previous study using the full width half maximum (FWHM) of a rocking curve omega-scan showed a FWHM of 2.18° with the seed layer and 2.23° without the seed layer for films of 500 nm total (film plus seed) thickness. In step (c) Plasma Enhanced Chemical Vapor Deposition (PECVD) Silicon Nitride (SiN) is deposited and patterned using CF_4 reactive ion etching (RIE) to form a hard mask for $Al_{1-x}Sc_xN$ etching. In step (d) aqueous Potassium Hydroxide (KOH in 45% H_2O) at 45 °C for 100 s is used to etch the $Al_{1-x}Sc_xN$ and define the cantilever dimensions with SiN protecting the $Al_{1-x}Sc_xN$ film where etching is undesired. Finally, in step (e) the SiN hard mask is stripped using CF_4 RIE and the $Al_{1-x}Sc_xN$ cantilevers are released from the substrate using isotropic XeF_2 dry etching. After release, the cantilever out-of-plane deflection is measured using a VHX-5000 Digital Microscope Multiscan.

Table 1. Summary of sputter deposition parameters for $Al_{0.68}Sc_{0.32}N$.

Process Parameter	Value
Temperature	350 °C
Sputter Power Al Cathode	1000 W
Sputter Power Sc Cathode	555 W
DC Pulsing Frequency	150 kHz
N_2 Flow	20–30 sccm
Film Thickness	100–1000 nm
Base Pressure	$<3 \times 10^{-7}$ mbar

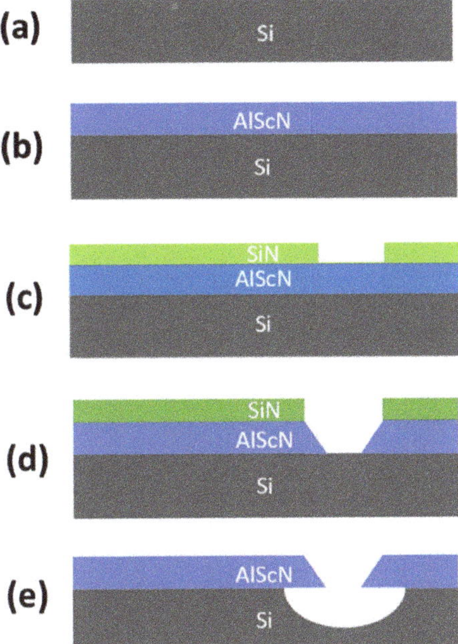

Figure 1. Fabrication process for realizing $Al_{1-x}Sc_xN$ cantilevers with (**a**) p-type (100) Si wafer (**b**) $Al_{1-x}Sc_xN$ deposition using Evatec CLUSTERLINE® 200 II PVD system (**c**) PECVD SiN deposition and patterning using CF_4 RIE (**d**) KOH in 45% H_2O etch of $Al_{1-x}Sc_xN$ (**e**) SiN hard mask stripped using CF_4 RIE and $Al_{1-x}Sc_xN$ cantilevers released using isotropic XeF_2 dry etching.

4. Results and Discussion

4.1. Through-Thickness Stress Gradients in Sputtered $Al_{1-x}Sc_xN$ films

During film growth, stresses due to lattice mismatch, intrinsic strains and microstructure produces tensile and compressive stresses. The average stress of a sputtered $Al_{1-x}Sc_xN$ film is strongly correlated to the value of the chamber pressure during deposition [3]. At 25 sccm process gas flow, the sputtering chamber will be at a near constant pressure of 1.09×10^{-3} mbar. At 25 sccm pure N_2 flow, when $Al_{0.68}Sc_{0.32}N$ is deposited to a final thickness of 500 nm, the average stress within the film will be approximately 137 MPa. The average film stress vs. thickness for $Al_{0.68}Sc_{0.32}N$ films deposited under 25 sccm pure N_2 flow using the process conditions in Table 1 is provided in Figure 2a. The average film stress is a strong function of the final film thickness due to the through-thickness stress gradient of the films. The through-thickness stress gradient can be modeled using Equation (2) in conjunction with the data in in Figure 2a where α is 41.12, β is 6.6522, and γ is 0.2194. α is determined using a linear fit of the measured average stress versus flow data shown in Figure 2b. In Figure 2a, the stress starts highly compressive and becomes more tensile as the thickness of the film increases. At lower film thicknesses, the microstructure of the film continuously changes and the grain size increases with increasing film thickness. As the thickness increases, the columnar growth of the film is more stable causing the through-thickness stress gradient to reduce and the average film stress to asymptote towards a constant value with further increases in thickness. These trends are clearly observable in Figure 2a.

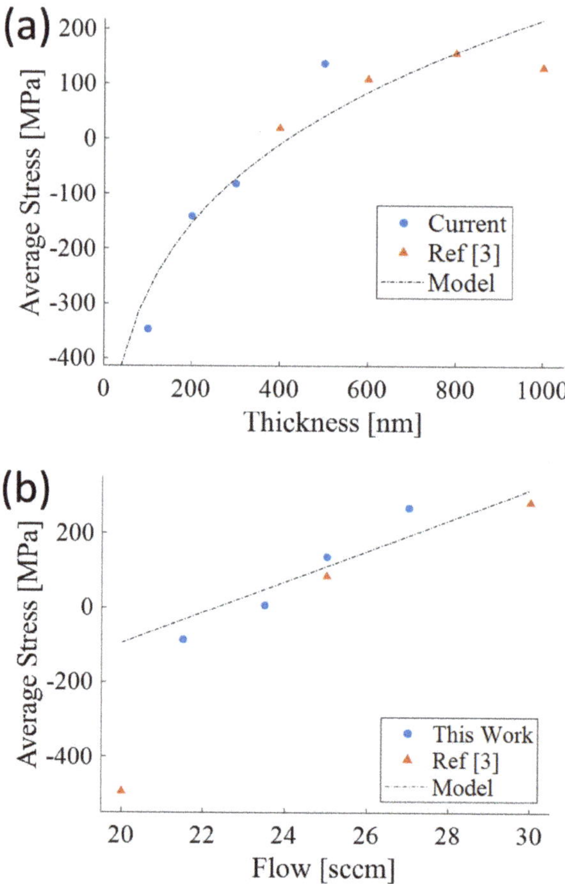

Figure 2. Average stress plots of PVD deposited $Al_{0.68}Sc_{0.32}N$ films with (**a**) Average stress versus thickness plot at a constant 25 sccm N_2 flow and (**b**) Average stress versus flow for 500 nm $Al_{0.68}Sc_{0.32}N$ with pure N_2 flow from 20–30 sccm [3].

4.2. Out-of-Plane Cantilever Deflection in Uncompensated $Al_{1-x}Sc_xN$ Materials

Low average stress (membranes) and low through-thickness stress gradient (cantilevers) are critical for realizing high yield MEMS structures with low out-of-plane bending. The multilayer stress and stress gradient compensation approach reported in this paper can be utilized to simultaneously achieve low average stress and low through-thickness stress gradient. Table 2 provides a summary of the various flow conditions and the resulting stress and out-of-plane cantilever deflections. While low average stress is achievable in all films, only a multi-layer $Al_{1-x}Sc_xN$ achieves the low through-thickness stress gradient required to realize cantilevers with low out-of-plane bending. The compressive-to-tensile stress gradient through the $Al_{1-x}Sc_xN$ film thickness results in a high degree of out-of-plane bending in uncompensated $Al_{1-x}Sc_xN$ films. A Z-axis microscope is used to measure the tip deflection with the anchor as the z = 0 reference. Figure 3 provides SEM images of the high out-of-plane bending in $Al_{0.68}Sc_{0.32}N$ cantilevers deposited with a constant flow of 25 sccm.

Table 2. Cantilever tip deflection for 500 nm $Al_{0.68}Sc_{0.32}N$ with and without stress gradient compensation.

		# of Layers		
		Single	Double	Quintuple
N_2 Flow [sccm]		25	30/20	30/27.5/25/23.5/20
Average Stress [MPa]		137.4	349.6	78.6
Seed Layer		Yes	Yes	No
Wafer Location	**Device Position**	**Out-of-Plane Deflection [µm]**		
Center	1	115.2 +/− 1.2	−3.2 +/− 0.1	−5.8 +/− 0.4
	2	108.6 +/− 4.2	−0.7 +/−0.2	−7.6 +/− 0.4
	3	117.1 +/− 0.9	0.0 +/− 0.1	−4.0 +/−0.4
North	1	50.1 +/− 0.1	0.9 +/− 0.3	7.4 +/− 0.4
	2	121.0 +/− 0.7	−1.4 +/− 0.2	−7.7 +/− 0.3
	3	145.9 +/− 0.2	−2.8 +/− 0.8	−20.0 +/− 0.1
Northeast	1	125.1 +/− 0.3	−7.0 +/− 0.2	−15.6 +/− 0.4
	2	49.3 +/− 0.1	−4.0 +/− 0.1	−21.1 +/− 0.6
	3	120.6 +/− 0.4	0.1 +/− 0.1	−0.9 +/− 0.3
East	1	128.9 +/− 0.9	0.9 +/− 0.2	−20.9 +/− 1.9
	2	126.9 +/− 42	6.2 +/− 0.2	6.4 +/− 0.2
	3	60.3 +/− 0.6	6.6 +/− 0.1	22.0 +/− 0.2

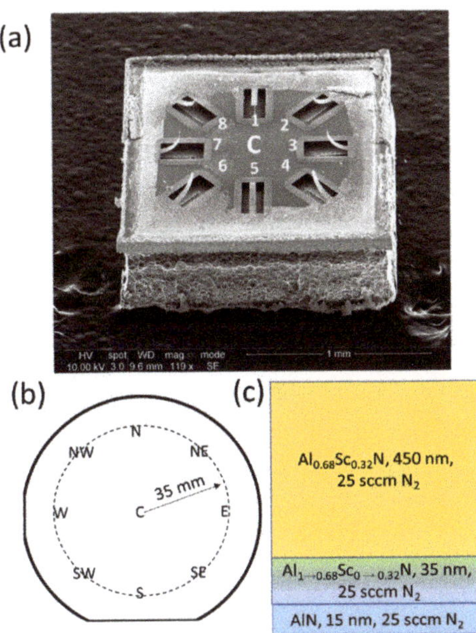

Figure 3. Graphics for a 500 nm PVD sputter-deposited $Al_{0.68}Sc_{0.32}N$ film where the N_2 process gas flow is held constant through the entire deposition, (**a**) 45-degree SEM image of center die (C) fabricated from a 100 mm wafer. Each die has 8 released structures which are labeled. Note the high out-of-plane deflection that is clearly visible in the uncompensated structures. (**b**) Schematic of locations on the 100 mm wafer where a die is pulled 35 mm from the center for imaging and measuring out-of-plane displacements. One die was pull from the north (N), northeast (NE), east (E), southeast (SE), south (S), southwest (SW), west (W), and northwest (NW) locations of the wafer (**c**) Stack-up of film with constant flow composed of a seed layer, gradient seed layer ($Al_{1\rightarrow0.68}Sc_{0\rightarrow0.32}N$) [3] and $Al_{0.68}Sc_{0.32}N$ layer.

Figure 4 shows top view SEM images of cantilevers fabricated from $Al_{0.68}Sc_{0.32}N$ films from across a 100 mm wafer demonstrating that the residual stress gradient within the film not only induces bending but also generates twisting and rotations within released structures. The center of the wafer produces structures with minimal twisting while the edge of the wafer produces significant twisting depending on the location of the die. The twisting is due to the interaction of the through-thickness stress gradient with the radial variation of the average stress across the 100 mm diameter wafer and is consistent with previous studies [3] which exhibit more compressive average stresses at the wafer edge and more tensile average stresses near the wafer center. If no stress compensation is established, depending on the performance requirements for a MEMS device, the location of the die on the wafer and the orientation of the released structures can lead to differences in out-of-plane bending and twisting.

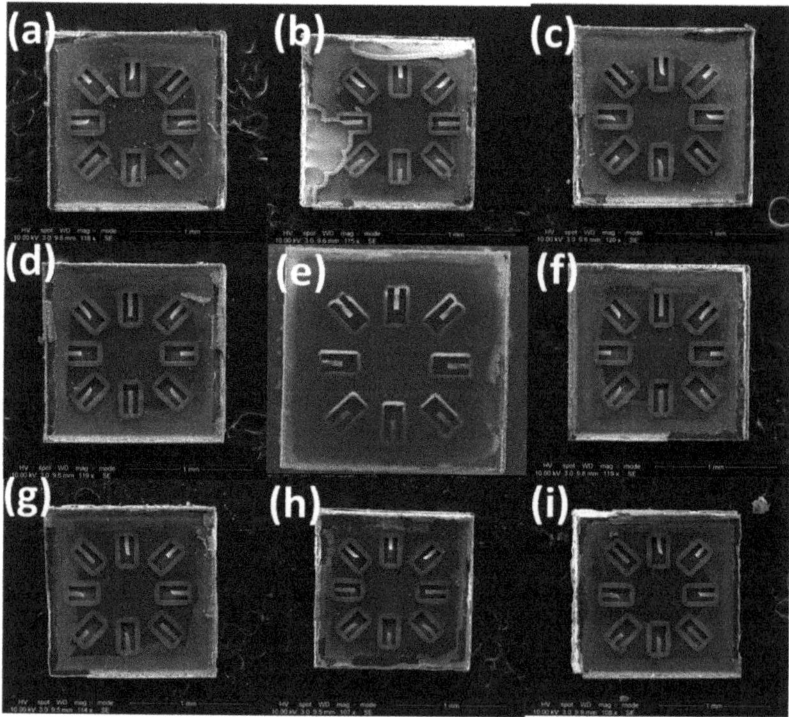

Figure 4. **(a–i)** SEM images of cantilevers formed from 500 nm thick PVD $Al_{0.68}Sc_{0.32}N$ films deposited under a constant N_2 flow of 25 sccm. Each die was pulled from the north (N), northeast (NE), east (E), southeast (SE), south (S), southwest (SW), west (W), and northwest (NW) locations of the wafer shown in Figure 3b.

4.3. Stress Gradient Compensated $Al_{1-x}Sc_xN$ Films and Cantilevers

The through-thickness stress gradient is the primary source of out-of-plane bending in released $Al_{1-x}Sc_xN$ structures. To find the additional flow needed to compensate the stress gradient, Equation (3) is used to estimate the local stress. At 25 sccm N_2 flow, the local stress within the first 15 nm after the seed layer has been deposited is approximated to be −424 MPa using Equation (3). At 500 nm, the local stress is 276 MPa. A compensating 424 MPa in the initial layers, −276 MPa at the top of the film, and the appropriate opposing stress gradient is required to compensate for the through-thickness stress gradient. Using Figure 2b, at 20 and 30 sccm N_2 flow, a 500 nm $Al_{1-x}Sc_xN$ film will yield an average stress of −450 and 317 MPa, respectively. We utilized Equation (3) to design a 2-layer stack deposited at 30 sccm (lower) and 20 sccm (upper) to compensate for the through-thickness stress gradient. Figure 5 provides an SEM image of a 2-layer $Al_{1-x}Sc_xN$ material where the N_2 flow is varied between layers to suppress the stress gradient and the resulting out-of-plane bending in cantilevers. Here, 30 sccm N_2 flow is utilized during the deposition of the AlN seed, $Al_{1-x}Sc_xN$ gradient layer, and the lower 225 nm of the bulk film while 20 sccm N_2 flow is utilized when depositing the upper 225 nm of the film stack. The approach successfully compensates for the through-thickness stress gradient and reduces the out-of-plane cantilever bending in the center of the wafer from 109 μm for the uncompensated materials to less than 3 μm for cantilevers realized in the stress gradient compensated 2-layer material.

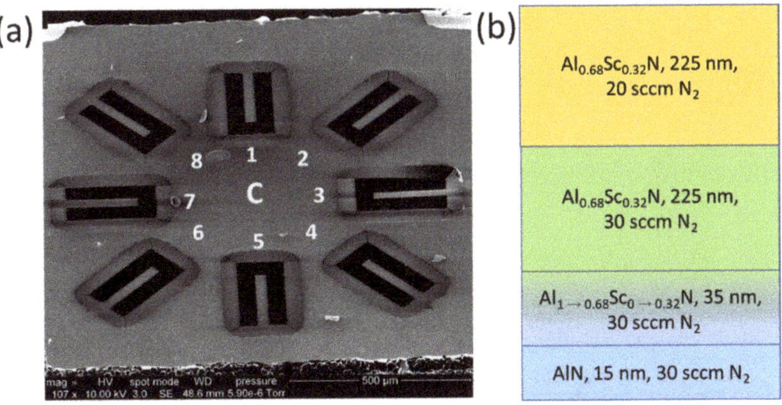

Figure 5. Graphics for multi-layer 500 nm PVD sputter-deposited $Al_{0.68}Sc_{0.32}N$ film where the N_2 process gas flow is changed between two layers with 30 sccm utilized on the bottom and 25 sccm on the top layer, (**a**) 45-degree SEM image of the center die (C) fabricated from a 100 mm wafer. Each die has 8 released structures. (**b**) Stack-up of 2-layer film composed of a seed and gradient layer ($Al_{1\rightarrow0.68}Sc_{0\rightarrow0.32}N$) [3] to suppress AOGs and two equal thickness layers with different N_2 process gas flows designed to compensate for the native through-thickness stress gradient.

Figure 6b displays a 5-layer $Al_{0.68}Sc_{0.32}N$ stack used to compensate for the through-thickness stress gradient. Since lower flows produce more compressive films while higher flows produce more tensile films, a layer stack is utilized where for each consecutive 100 nm layer, an additional 2.5 sccm of flow provides a tensile-to-compressive transition to cancel the original compressive-to-tensile stress gradient through the thickness. For the 5-layer material the N_2 flow is varied over the range from 30 to 20 sccm to yield a low average stress in addition to a low through-thickness stress gradient. After release, the cantilevers remain consistently flat, as shown in Figures 6a and 7, especially when compared to the cantilevers formed in the uncompensated films. The cantilevers in Figures 4–7 use the same naming conventions depicted in Figure 3a,b. In Figure 6a the maximum tip deflection is approximately 5.8 +/− 0.4 μm, −7.6 +/− 0.4 μm, and −4.0 +/− 0.4 μm when measured from the 1, 2, and 3 positions, respectively. The 1, 2, and 3 structures exhibited the same behavior and deflection as those directly across from them, namely cantilevers 5, 6, and 7. Table 2 compares the average wafer stress and cantilever tip deflection for the uncompensated and stress gradient compensated $Al_{0.68}Sc_{0.32}N$ materials. The out-of-plane cantilever displacement for the 5-layer, low stress material at position 2 is reduced by more than 14-fold for the center die and 19-fold for the east die 35 mm from the wafer center. Overall, substantial reductions in out-of-plane tip displacement are observed for all cantilevers fabricated in the stress gradient compensated films. The 5-layer film does not have an AlN/gradient seed layer and possess a higher tip deflection than the 2-layer film with seed layer. This is because the gradient in the N_2 process gas flow was designed using the data from Figure 2 where all the films have the seed layer. Table 3 confirms the precise control of the tip deflection by varying the range of N_2 gas flow for a 2-layer material stack. The range of gas flows controls the stress gradient while the mean N_2 gas flow controls the average stress within the film. In Table 3, a 2.5 sccm increase of the range from 27.5–22.5 sccm to 30–22.5 sccm reduces the deflection 2-fold.

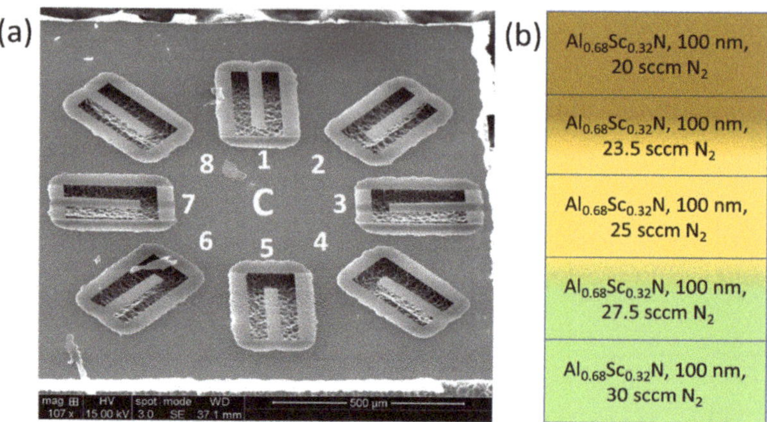

Figure 6. Graphics for multi-layer 500 nm PVD sputter-deposited $Al_{0.68}Sc_{0.32}N$ film, (**a**) 45-degree SEM image of the center die (C) fabricated from a 100 mm wafer. (**b**) Stack-up with five equal thickness layers with different N_2 process gas flows to compensate for the through-thickness stress gradient.

Figure 7. Top view SEM images of cantilevers formed from a 5-layer, 500 nm total thickness, PVD deposited $Al_{0.68}Sc_{0.32}N$ where the process gas flow for each layer is linearly changed from 30 to 20 sccm

through the five layers. Each die was pull from the north (N), northeast (NE), east (E), southeast (SE), south (S), southwest (SW), west (W), and northwest (NW) locations of the wafer shown in Figure 3b.

Table 3. Cantilever tip deflection for a die near the center of a 500 nm $Al_{0.68}Sc_{0.32}N$ film with AlN and Sc gradient ($Al_{1\rightarrow0.68}Sc_{0\rightarrow0.32}N$) seed layers.

N_2 Flow [sccm]	25	27.5/22.5	30/22.5	30/20
Seed Layer	Yes	Yes	Yes	Yes
Device Position	**Out-of-Plane Deflection [μm]**			
1	115.2 +/− 1.2	60.8 +/− 0.1	24.1 +/− 0.1	−3.2 +/− 0.1
2	108.6 +/− 4.2	55.0 +/− 0.1	28.2 +/− 1.2	−0.7 +/− 0.2
3	117.1 +/− 0.9	51.4 +/− 0.3	30.5 +/− 0.4	−0.0 +/−0.1

4.4. Discussion of Stress Gradient Cancellation Trends

This work provides, for the first-time, methods to individually control the stress and stress gradient in $Al_{1-x}Sc_xN$ films while maintaining film quality. Figure 8 shows the out-of-plane displacement along the length of the cantilever for the uncompensated and 2-layer compensated $Al_{0.68}Sc_{0.32}N$ materials while Table 3 summarizes the out-of-plane tip displacement. The compensated cantilever tip bending confirms that the radius of curvature can be controlled in the released $Al_{1-x}Sc_xN$ structures. The multilayer gas gradient method can be utilized to simultaneously and independently control both the average stress, via the average N_2 flow, and through-thickness stress gradient, via the through-thickness variation of the N_2 flow, in $Al_{1-x}Sc_xN$ thin films. Previously reported methods to control $Al_{1-x}Sc_xN$ average film stress do not provide control of the through-thickness stress gradients within the film. While the previous RF substrate bias method reported by Knisley [14] achieved independent control of stress and through-thickness stress gradient in AlN films and demonstrated reduced radius of curvature and tip displacement in AlN cantilevers, use of an RF substrate bias is less suitable for $Al_{1-x}Sc_xN$. Addition of an RF substrate bias results in more compressive film stress and is a good technique for stress control of AlN where the films are highly tensile without an RF substrate bias [14]. $Al_{0.68}Sc_{0.32}N$, by contrast, is highly compressive when deposited at the low process pressures that suppress formation of anomalously oriented grains (AOGs) without using an RF substrate bias [3]. Thus, addition of an RF substrate bias to an $Al_{0.68}Sc_{0.32}N$ growth will require even higher process pressures to achieve near-neutral average film stress, and under such process conditions a large number of AOGs would be expected.

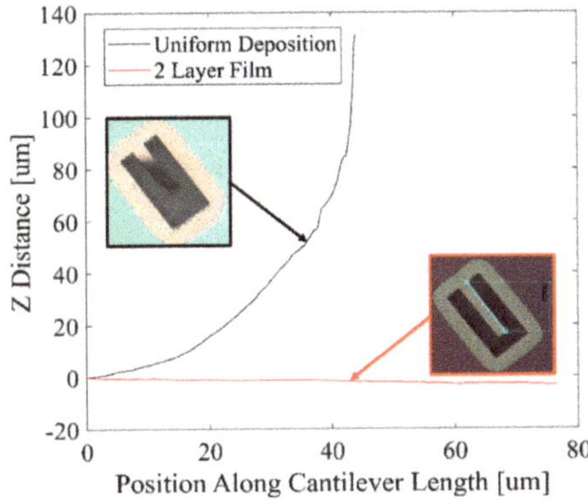

Figure 8. Z deflection for 500 nm PVD sputter-deposited $Al_{0.68}Sc_{0.32}N$ films where the N_2 process gas flow was applied at constant flow throughout the deposition (black) and a 2-layer 30/20 sccm N_2 flow stack (red).

5. Conclusions

This study reports methods to fabricate $Al_{0.68}Sc_{0.32}N$ films and cantilevers with low average stress and with low through-thickness stress gradients. $Al_{1-x}Sc_xN$ films were optimized to control stress for N_2 flows between 20 to 30 sccm. The average stress within the films ranged from 78.6 MPa to 349.6 MPa. The out-of-plane tip deflection for 100 μm long cantilevers fabricated in 500 nm thick $Al_{0.68}Sc_{0.32}N$ films was reduced from >109 μm for films without stress gradient compensation to less than 3 μm and 8 μm for 2- and 5-layer compensated film stacks for dies studied in the wafer center. The resulting deposition parameters provide methods to control stress and through-thickness stress gradients in highly Sc alloyed AlN materials and are promising for next-generation MEMS devices.

Author Contributions: Conceptualization, R.B. and R.H.O.III; methodology, R.B.; validation, R.B., K.K. and M.D; formal analysis, R.B., K.K. and M.D; investigation, R.B. and K.K.; resources, R.H.O.III; data curation, R.B., K.K. and M.D.; writing—original draft preparation, R.B.; writing—review and editing, R.B., M.D. and R.H.O.III; visualization, R.B.; supervision, R.H.O.III; project administration, R.H.O.III; funding acquisition, R.H.O.III. All authors have read and agreed to the published version of the manuscript.

Funding: This work was funded in part by the NSF CAREER Award (1944248) and in part The Defense Advanced Research Projects Agency (DARPA) Small Business Innovation Research (SBIR) under award HR0011-21-9-0004. This work was carried out in part at the Singh Center for Nanotechnology at the University of Pennsylvania, a member of the National Nanotechnology Coordinated Infrastructure (NNCI) network, which is supported by the National Science Foundation (Grant No. HR0011-21-9-0004). This research was, in part, funded by the U.S. Government. The views and conclusions contained in this document are those of the authors and should not be interpreted as representing the official policies, either expressed or implied, of the U.S. Government.

Data Availability Statement: The data that support the findings of this study are available from the corresponding author upon reasonable request.

Conflicts of Interest: The authors declare no conflict of interest.

References

1. Esteves, G.; Berg, M.; Wrasman, K.D.; Henry, M.D.; Griffin, B.A.; Douglas, E.A. CMOS compatible metal stacks for suppression of secondary grains in Sc0.125Al0.875N. *J. Vac. Sci. Technol. A Vac. Surf. Film.* **2019**, *37*, 021511. [CrossRef]
2. Olsson, R.H.; Tang, Z.; D'Agati, M. Doping of aluminum nitride and the impact on thin film piezoelectric and ferroelectric device performance. In Proceedings of the 2020 IEEE Custom Integrated Circuits Conference (CICC), Boston, MA, USA, 22–25 March 2020; IEEE: New York, NY, USA, 2020; pp. 1–6.
3. Beaucejour, R.; Roebisch, V.; Kochhar, A.; Moe, C.; Hodge, D.; Olsson, R.H., III. Controlling Residual Stress and Suppression of Anomalous Grains in Aluminum Scandium Nitride Films Grown Directly on Silicon. *J. Microelectromech. Syst.* **2022**, 1–8. Available online: https://ieeexplore.ieee.org/document/9770129 (accessed on 1 June 2022). [CrossRef]
4. Hong, Y.; Sui, L.; Zhang, M.; Shi, G. Theoretical analysis and experimental study of the effect of the neutral plane of a composite piezoelectric cantilever. *Energy Convers. Manag.* **2018**, *171*, 1020–1029. [CrossRef]
5. Wang, Q.M.; Yang, Z.; Li, F.; Smolinski, P. Analysis of thin film piezoelectric microaccelerometer using analytical and finite element modeling. *Sens. Actuators Phys.* **2004**, *113*, 1–11. [CrossRef]
6. Akiyama, M.; Kamohara, T.; Kano, T.; Teshigahara, A.; Takeuchi, Y.; Kawahara, N. Enhancement of piezoelectric response in scandium aluminum nitride alloy thin films prepared by dual reactive cosputtering. *Adv. Mater.* **2009**, *21*, 593–596. [CrossRef]
7. Reusch, M.; Cherneva, S.; Lu, Y.; Žukauskaitė, A.; Kirste, L.; Holc, K.; Datcheva, M.; Stoychev, D.; Levdeve, V.; Ambacher, O. Microstructure and mechanical properties of stress-tailored piezoelectric AlN thin films for electro-acoustic devices. *Appl. Surf. Sci.* **2017**, *407*, 307–314. [CrossRef]
8. Park, M.; Hao, Z.; Dargis, R.; Clark, A.; Ansari, A. Epitaxial aluminum scandium nitride super high frequency acoustic resonators. *J. Microelectromech. Syst.* **2020**, *29*, 490–498. [CrossRef]
9. Fichtner, S.; Reimer, T.; Chemnitz, S.; Lofink, F.; Wagner, B. Stress controlled pulsed direct current co-sputtered $Al_{1-x}Sc_xN$ as piezoelectric phase for micromechanical sensor applications. *APL Mater.* **2015**, *3*, 116102. [CrossRef]
10. Rossnagel, S.M. Thin film deposition with physical vapor deposition and related technologies. *J. Vac. Sci. Technol. A Vac. Surf. Film.* **2003**, *21*, S74–S87. [CrossRef]
11. Wang, J.; Zheng, Y.; Ansari, A. Ferroelectric Aluminum Scandium Nitride Thin Film Bulk Acoustic Resonators with Polarization-Dependent Operating States. *Phys. Status Solidi (RRL)–Rapid Res. Lett.* **2021**, *15*, 2100034. [CrossRef]
12. Thornton, J.A.; Hoffman, D.W. Stress-related effects in thin films. *Thin Solid Film.* **1989**, *171*, 5–31. [CrossRef]

13. Panjan, P.; Drnovšek, A.; Gselman, P.; Čekada, M.; Panjan, M. Review of growth defects in thin films prepared by PVD techniques. *Coatings* **2020**, *10*, 447. [CrossRef]
14. Knisely, K.E.; Hunt, B.; Troelsen, B.; Douglas, E.; Griffin, B.A.; Stevens, J.E. Method for controlling stress gradients in PVD aluminum nitride. *J. Micromech. Microeng.* **2018**, *28*, 115009. [CrossRef]
15. Henry, M.D.; Young, T.R.; Douglas, E.A.; Griffin, B.A. Reactive sputter deposition of piezoelectric $Sc_{0.12}Al_{0.88}N$ for contour mode resonators. *J. Vac. Sci. Technol. B Nanotechnol. Microelectron. Mater. Process. Meas. Phenom.* **2018**, *36*, 03E104. [CrossRef]
16. Pulskamp, J.S.; Wickenden, A.; Polcawich, R.; Piekarski, B.; Dubey, M.; Smith, G. Mitigation of residual film stress deformation in multilayer microelectromechanical systems cantilever devices. *J. Vac. Sci. Technol. B Microelectron. Nanometer Struct. Process. Meas. Phenom.* **2003**, *21*, 2482–2486. [CrossRef]
17. Sedky, S.; Howe, R.T.; King, T. Pulsed-laser annealing, a low-thermal-budget technique for eliminating stress gradient in poly-SiGe MEMS structures. *J. Microelectromech. Syst.* **2004**, *13*, 669–675. [CrossRef]
18. Zhu, W.Z.; Assylbekova, M.; McGruer, N.E. Limitations on MEMS design resulting from random stress gradient variations in sputtered thin films. *J. Micromech. Microeng.* **2021**, *31*, 045004. [CrossRef]
19. Mulloni, V.; Giacomozzi, F.; Margesin, B. Controlling stress and stress gradient during the release process in gold suspended micro-structures. *Sens. Actuators A Phys.* **2010**, *162*, 93–99. [CrossRef]
20. Shao, S.; Luo, Z.; Lu, Y.; Mazzalai, A.; Tosi, C.; Wu, T. High Quality Co-Sputtering AlScN Thin Films for Piezoelectric Lamb-Wave Resonators. *J. Microelectromech. Syst.* **2022**, *31*, 328–337. [CrossRef]
21. Dubois, M.A.; Muralt, P. Stress and piezoelectric properties of aluminum nitride thin films deposited onto metal electrodes by pulsed direct current reactive sputtering. *J. Appl. Phys.* **2001**, *89*, 6389–6395. [CrossRef]
22. Kusaka, K.; Taniguchi, D.; Hanabusa, T.; Tominaga, K. Effect of input power on crystal orientation and residual stress in AlN film deposited by dc sputtering. *Vacuum* **2000**, *59*, 806–813. [CrossRef]
23. Ng, D.K.T.; Zhang, T.; Siow, L.Y.; Xu, L.; Ho, C.P.; Cai, H.; Lee, L.Y.T.; Zhang, Q.; Singh, N. A functional CMOS compatible MEMS pyroelectric detector using 12%-doped scandium aluminum nitride. *Appl. Phys. Lett.* **2020**, *117*, 183506. [CrossRef]
24. Su, J.; Niekiel, F.; Fichtner, S.; Thormaehlen, L.; Kirchhof, C.; Meyners, D.; Quandt, E.; Wagner, B.; Lofink, F. AlScN-based MEMS magnetoelectric sensor. *Appl. Phys. Lett.* **2020**, *117*, 132903. [CrossRef]
25. Olsson, R.H.; Wojciechowski, K.E.; Baker, M.S.; Tuck, M.R.; Fleming, J.G. Post-CMOS-Compatible Aluminum Nitride Resonant MEMS Accelerometers. *J. Microelectromech. Syst.* **2009**, *18*, 671–678. [CrossRef]
26. Khanna, V.K. *Flexible Electronics*; IOP Publishing Limited: Bristol, UK, 2019; Volume 2.
27. Konno, A.; Sumisaka, M.; Teshigahara, A.; Kano, K.; Hashimo, K.Y.; Hirano, H.; Esashi, M.; Kodota, M.; Tanaka, S. ScAlN Lamb wave resonator in GHz range released by XeF_2 etching. In Proceedings of the 2013 IEEE International Ultrasonics Symposium (IUS), Prague, Czech Republic, 21–25 July 2013; IEEE: New York, NY, USA, 2013; pp. 1378–1381.

Article

Static High Voltage Actuation of Piezoelectric AlN and AlScN Based Scanning Micromirrors

Chris Stoeckel [1,2,*], Katja Meinel [2], Marcel Melzer [2], Agnė Žukauskaitė [3], Sven Zimmermann [1,2], Roman Forke [1], Karla Hiller [1,2] and Harald Kuhn [1,2]

1 Fraunhofer Institute for Electronic Nano Systems ENAS, 09126 Chemnitz, Germany; sven.zimmermann@zfm.tu-chemnitz.de (S.Z.); roman.forke@enas.fraunhofer.de (R.F.); karla.hiller@zfm.tu-chemnitz.de (K.H.); harald.kuhn@enas.fraunhofer.de (H.K.)

2 Center for Microtechnologies, Chemnitz University of Technology, 09111 Chemnitz, Germany; katja.meinel@zfm.tu-chemnitz.de (K.M.); marcel.melzer@zfm.tu-chemnitz.de (M.M.)

3 Fraunhofer Institute for Organic Electronics, Electron Beam and Plasma Technology FEP, 01277 Dresden, Germany; agne.zukauskaite@fep.fraunhofer.de

* Correspondence: chris.stoeckel@enas.fraunhofer.de

Abstract: Piezoelectric micromirrors with aluminum nitride (AlN) and aluminum scandium nitride ($Al_{0.68}Sc_{0.32}N$) are presented and compared regarding their static deflection. Two chip designs with 2×3 mm^2 (Design 1) and 4×6 mm^2 (Design 2) footprint with 600 nm AlN or 2000 nm $Al_{0.68}Sc_{0.32}N$ as piezoelectric transducer material are investigated. The chip with Design 1 and $Al_{0.68}Sc_{0.32}N$ has a resonance frequency of 1.8 kHz and a static scan angle of 38.4° at 400 V DC was measured. Design 2 has its resonance at 2.1 kHz. The maximum static scan angle is 55.6° at 220 V DC, which is the maximum deflection measurable with the experimental setup. The static deflection per electric field is increased by a factor of 10, due to the optimization of the design and the research and development of high-performance piezoelectric transducer materials with large piezoelectric coefficient and high electrical breakthrough voltage.

Keywords: AlN; AlScN; aluminum nitride; aluminum scandium nitride; micromirror; microscanner; piezoelectric

Citation: Stoeckel, C.; Meinel, K.; Melzer, M.; Žukauskaitė, A.; Zimmermann, S.; Forke, R.; Hiller, K.; Kuhn, H. Static High Voltage Actuation of Piezoelectric AlN and AlScN Based Scanning Micromirrors. *Micromachines* **2022**, *13*, 625. https://doi.org/10.3390/mi13040625

Academic Editor: Huikai Xie

Received: 8 March 2022
Accepted: 7 April 2022
Published: 15 April 2022

Publisher's Note: MDPI stays neutral with regard to jurisdictional claims in published maps and institutional affiliations.

1. Introduction

Micromirrors as scanning devices are reported intensively in literature with different electromechanical transducer principles. They are mostly classified into the electrostatic, electrothermal, electromagnetic, and piezoelectric micromirrors [1–5]. The piezoelectric transducer principle offers the advantages of high deflections at moderate excitation voltages and high dynamic ranges. Furthermore, a high degree of miniaturization and the monolithic integration of actuators and sensor elements is possible. In addition to the commonly used transducer material, lead zirconate titanate (PZT), piezoelectric AlN, and AlScN thin films can alternatively be used as piezoelectric transducers for actuation. Since 2018, several AlN and AlScN-based micromirrors have been presented. Shao et al. [6] presented the first AlN-based micromirror. The microsystem with a 0.2×0.2 mm^2 mirror plate area and L-shaped bending actuators reached a resonant scan angle of 4° at 5 V and 63.3 kHz. Since then, further publications on AlN-based micromirrors have followed. Since June 2019, our preliminary work [7,8] includes resonantly operated 1D micromirrors with a 600 nm AlN film, a mirror plate length of 0.8 mm, and a chip size of 2×3 mm^2. Large scan angles of up to 137.9° at 20 V and 3.4 kHz were reached in air. In October 2020, two 2D micromirror designs with a footprint of 2×2 mm^2 and mirror plate diameter of 0.7 mm were developed to realize Lissajous and spiral scan trajectories [9]. For the Lissajous scanning design, a scan angle of 92.4° at 12,060 Hz and 123.9° at 13,145 Hz was reached at 50 V for the x- and y-axis, respectively. The spiral scanning design reached a scan angle of 91.2° at 13,834 Hz and 50 V. In 2021, a 2D circular-scanning AlN-based mircomirror with a large aperture of 7 mm for laser material processing was published by Senger et al. [10]. In air, a scan angle of 5° is reached at 40 V and 1265 Hz. Due to the application, no large deflection angles were specified. In order to achieve higher deflections with larger mirror apertures, vacuum packaging is often used in literature. A wobbling mode AlN-scanner for automotive applications was published in October 2019 by Pensala et al. [11]. The microsystem with an aperture of 4 mm and 6.75 × 6.75 × 2 mm^3 chip size reached a scan angle of 30° at 1 V and 1.6 kHz by the implementation of a vacuum package. In 2020, Senger et al. [12] also presented a vacuum-packaged AlN-based micromirror with a 5.5 mm aperture. A Lissajous scan pattern with 50° × 20° scan angle was realized.

The previously mentioned micromirrors are exclusively driven in resonance to achieve sufficiently large tilt angles. Resonance frequency deviations, caused by variations of the ambient conditions like mechanical vibration and temperature change or heating due to light losses during laser irradiation, lead to a change of the tilt angle and, finally, result in errors in image formation and reconstruction [4,13]. Therefore, a static or quasi-static working mode has many advantages in regards of the electronics and drivers. In March 2019, Gu-Stoppel et al. presented an AlScN-based quasi-static micromirror with mirror plate diameter of 0.8 mm [14]. The mirror plate is mounted onto a pillar, which is deflected by four actuators hidden beneath. A high static scan angle of 50° at 150 V_{DC} was achieved by this novel construction. The challenge with this concept is a complex manufacturing process, which includes different wafer bonding processes for the micromirror assembly.

In this work, the static deflection and high voltage performance of the Design 1 MOEMS in [8] is investigated. Additionally, a technology is developed using a 2 μm $Al_{0.68}Sc_{0.32}N$ with high thickness as transducer material for a direct comparison of the performance of the MOEMS with higher piezoelectric coefficient. Furthermore, the design is optimized for a chip size with twice the length and width of MOEMS (Design 2) to identify the performance gain for different chip footprints and further increase the deflection. By reducing the silicon spring width in relation to previous designs, the stiffness is decreased, targeting a high deflection per voltage.

2. Design

In Figure 1 the schematics of the fabricated Al(Sc)N micromirrors with 2 × 3 mm^2 and 4 × 6 mm^2 footprint are shown. In Table 1 the mirrors parameter are shown. The mirror plate is connected with two actuators by four L-shaped springs. The design and FEA of the 2 × 3 mm^2 MOEMS is shown by Meinel et al. [8]. The MOEMS design and FEA process for optimization of the leverage effect is described by Meinel et al. [7] previously. In this publication, the static deflections of the MOEMS are calculated analytically. The parameters for the analytical calculation are given in Table 1. The actuators are described in a quarter symmetrical model of the MOEMS (Figure 2). The piezoelectric unimorphs are divided into two separate actuators in parallel. The actuator has the length l and the width w. Both actuators have a free displacement ξ_0 and a blocking force F_0.

$$\xi_0 = -3 \cdot \frac{d_{31} \cdot l^2}{t_p^2} \cdot \frac{AB \cdot (B+1)}{D} \cdot V \tag{1}$$

$$F_0 = -\frac{3}{4} \cdot t_p \cdot E_p \cdot d_{31} \cdot \frac{AB \cdot (B+1)}{AB+1} \cdot \left(\frac{w_1}{l_1} + \frac{w_2}{l_2} \right) \cdot n \cdot V \tag{2}$$

$$A = \frac{E_S}{E_p}; \; B = \frac{t_S}{t_p}; \; D = A^2 \cdot B^4 + 2A \cdot \left(2B + 3B^2 + 2B^3\right) + 1 \tag{3}$$

The free displacement is defined by the deflection of the longest actuator in the system. The blocking force is a sum of all four actuators n of the system. The actuator force in relation to the displacement at the leverage arm ξ_p can be described as $F_p(\xi_p)$:

$$F_p(\xi_p) = -\frac{F_0}{\xi_0} \cdot \xi_p + F_0 \tag{4}$$

The microsystem has a resonance frequency as a result of the systems stiffness C and mass m. The mass is approximated as the mirror plate mass only. The stiffness is calculated by the resonance frequency f of the system. The force in relation to the stiffness and the deflection at the center of the mass ξ_m can be described as $F_c(\xi_m)$.

$$F_c(\xi_m) = C \cdot \xi_m = 4\pi^2 \cdot f^2 \cdot m \cdot \xi_m \tag{5}$$

Table 1. Comparison of the parameter of the micromirror designs and transducer materials.

	Symbol [1]	Design 1 AlN	Design 1 AlScN	Design 2 AlScN
General parameter				
Piezoelectric layer		AlN	$Al_{0.68}Sc_{0.32}N$	$Al_{0.68}Sc_{0.32}N$
Thickness of piezoelectric layer [nm]	t_p	600	2000	2000
Electric field at 1 V [MV/m]		1.67	0.5	0.5
Silicon side wall shift [μm]		~0.85	~1.35	~1.35
PVD parameter				
Nitrogen concentration [%]		100	100	100
Pressure [Pa]		0.7	0.36	0.36
Substrate temperature [°C]		350	300	300
DC power [W]		2120	625 (Al) + 375 (Sc)	625 (Al) + 375 (Sc)
Geometrical parameter				
Spring width (mask) [μm]	a	5	5	7
Spring width (fabricated) [μm]	a'	~3.3	~2.3	~4.3
Lever arm distance to center [μm]	b_p	40	40	80
Distance to center of area [μm]	b_m	200	200	187.5
Lever arm distance to actuator [μm]	c	20	20	70
Spring length [μm]	d	255	255	1030
Mirror plate length/diameter [μm]	e	800	800	1000
Actuator width (quarter model) [μm]	w_1	140	140	1000
	w_2	440	440	505
Actuator length (quarter model) [μm]	l_1	760	760	2020
	l_2	360	360	1710
Parameter for analytical calculation				
Number of actuators in the model	n	4	4	4
Thickness silicon substrate [μm]	t_s	21 ± 10%	21 ± 10%	21 ± 10%
PE charge coefficient [pm/V]	d_{31}	−2 ± 10%	−5 ± 10%	−5 ± 10%
Resonance frequency [kHz]	f	3.5 ± 10%	2 ± 10%	2 ± 10%
E-Modulus of PE transducer [GPa]	E_p	108	108	108
E-Modulus of silicon [GPa]	E_s	63.9	63.9	63.9
Actuation voltage [V]	V	1.0	1.0	1.0
Mass of the mirror plate [ng]	m	31 ± 10%	31 ± 10%	38 ± 10%

[1] Symbolism according to Figures 1 and 2.

The relation between ξ_m and ξ_p is defined by the leverage arm distance to the center b_p and the distance to the area center b_m:

$$\xi_m = \frac{b_m}{b_p} \cdot \xi_p \tag{6}$$

The momentum of the actuators and the systems stiffness are in an equilibrium:

$$M_p = M_m \tag{7}$$

$$F_p \cdot b_p = F_c \cdot b_m \tag{8}$$

$$\left(-\frac{F_0}{\xi_0} \cdot \xi_p + F_0 \right) \cdot b_p = C \cdot \frac{b_m}{b_p} \cdot \xi_p \cdot b_m \tag{9}$$

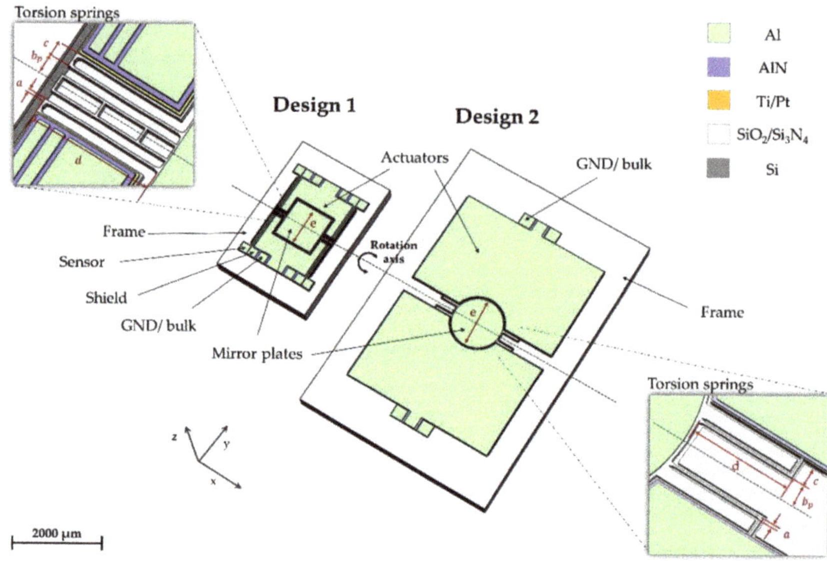

Figure 1. Schematic of the presented micromirror designs.

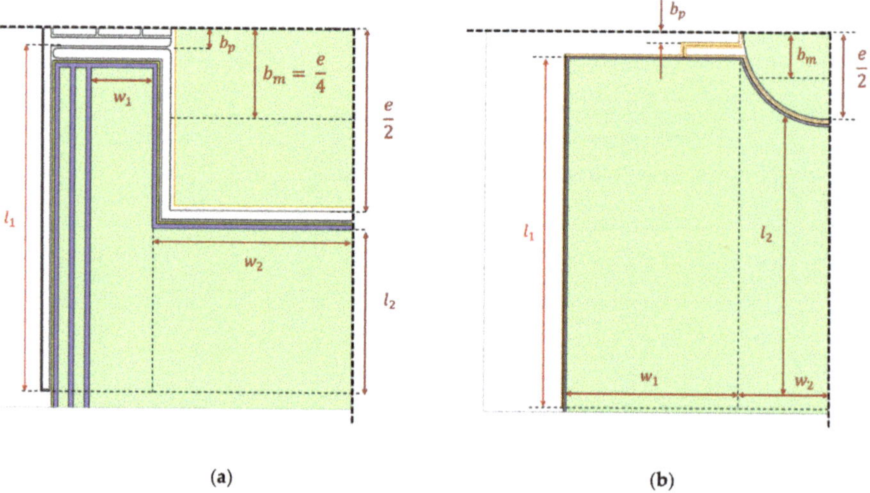

(a) (b)

Figure 2. Quarter symmetrical model of the MOEMS: (**a**) Design 1; and (**b**) Design 2.

The deflection at the leverage arm position in relation to the leverage arm distance to the center of the MOEMS can be calculated with this equilibrium of the momentum:

$$\xi_p(b_p) = \frac{F_0}{\frac{F_0}{\xi_0} - \frac{b_m^2}{b_p^2} \cdot C} \tag{10}$$

The deflection at the edge of the mirror plate ξ_e in relation to the lever arm position b_p is given by Equation (11).

$$\xi_e(b_p) = \frac{F_0}{\frac{F_0}{\xi_0} + \frac{b_m^2}{b_p^2} \cdot C} \cdot \frac{b_m}{b_p} \cdot \frac{\frac{e}{2}}{b_m} = \frac{F_0}{\frac{F_0}{\xi_0} + \frac{b_m^2}{b_p^2} \cdot C} \cdot \frac{e}{2b_p}. \tag{11}$$

It should be noticed, that a LDV-based measurement of the deflection will not be done at the exact edge of the mirror plate, due to irregular reflections of the light. The measurement of the mirrors deflection in this paper is done in a distance to the mirror edge of approximately 50 µm. Additionally, the analytic equations do not include losses of the elastic energy, or mechanical stress and strain in the torsion springs.

In Figure 3 the mirror plate deflection is shown in relation to the lever arm distance to the center. An optimum lever arm length can be identified, depending on the systems mass and stiffness as well as the blocking force and free deflection of the actuators. The parameter of silicon height, piezoelectric charge coefficient, and resonance frequency are estimated within a 10% limit of variation. For Design 1 with AlN as transducer and a 40 µm lever arm distance the calculated deflection is in the range of 65 nm/V to 116 nm/V. Using $Al_{0.68}Sc_{0.32}N$ and Design 1 increases the deflection up to 143 nm/V to 241 nm/V. Design 2 has a lever arm distance to the center of 80 µm. For this design and $Al_{0.68}Sc_{0.32}N$ a deflection of 563 nm/V to 959 nm/V is calculated.

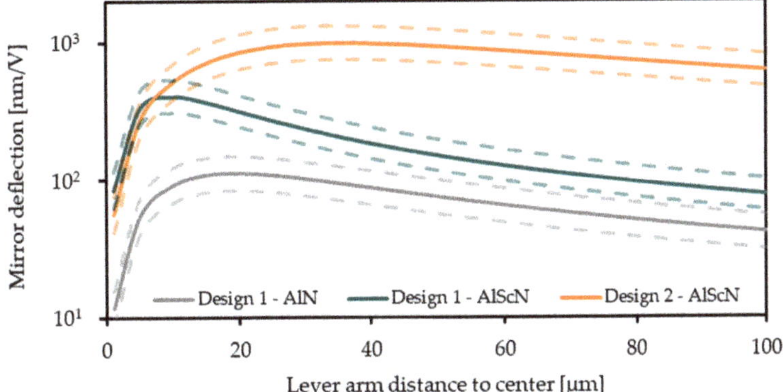

Figure 3. The static mirror plate deflection (mechanical tilt angle) per voltage in relation to the lever arm distance to the center is given for different designs and transducer materials. The parameter substrate height, piezoelectric charge coefficient, and resonance frequency are estimated with a 10% limit of variation. Therefore, an upper and lower limit of the approximated variation to the mechanical tilt angle is given.

3. Fabrication

The wafers with AlN and $Al_{0.68}Sc_{0.32}N$ piezoelectric layers are processed with an identical process flow and process parameters, except for the deposition and etching of the piezoelectric material. Figure 4 illustrates the device fabrication process flow. The microsystem fabrication is based on 150 mm SOI technology with 575 µm thick handle wafer and 20 µm device-silicon thickness. First, a 1 µm thermal oxide is grown by oxidation. LPCVD silicon nitride with a layer thickness of 100 nm is used as an isolation layer due to its selectivity against HF wet etching processes. The piezoelectric layer stack starts with a seed layer of 100 nm platinum. For a better adhesion with the substrate, a 20 nm titanium film is used. The Pt is deposited in <111> orientation to minimize the elastic energy to a <0002> Al(Sc)N crystal. A 600 nm AlN or 2000 nm $Al_{0.68}Sc_{0.32}N$, respectively, and a 100 nm PECVD SiO_2 layer are deposited as piezoelectric material. Table 1 summarizes the PVD deposition conditions for the AlN and $Al_{0.68}Sc_{0.32}N$ layer. DC magnetron process with an Al target (double ring magnetron target 120 mm and 123–236 mm, purity 5N5) in 100% nitrogen atmosphere is used for AlN. In the case of $Al_{0.68}Sc_{0.32}N$, co-sputtering from 5N5 Al and 4N pure Sc targets in pulsed DC mode in 100% nitrogen atmosphere is used at combined power of 1000 W. AlScN growth optimization, as well as structural and compositional analysis are discussed elsewhere [15,16].

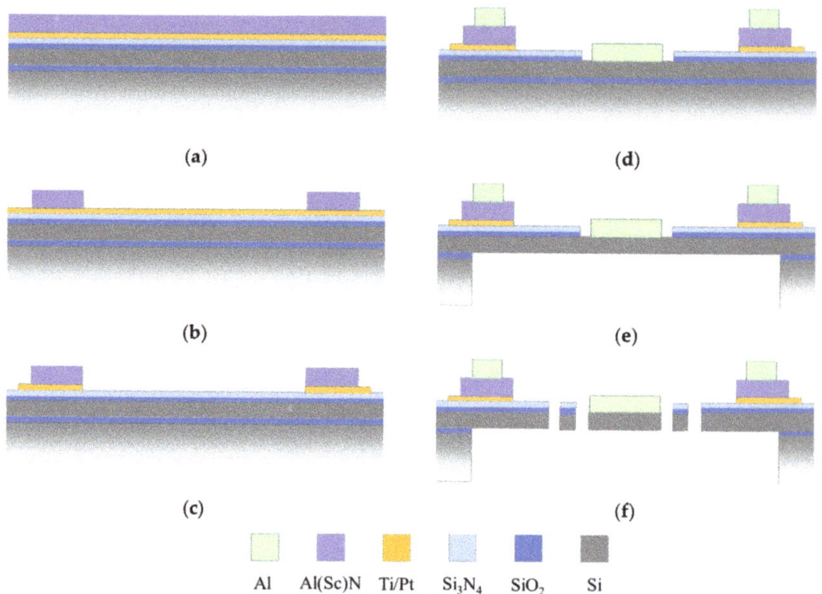

Figure 4. Fabrication process flow: (**a**) Initial layer stack; (**b**) AlN wet etching; (**c**) Pt structuring; (**d**) Dry etching of silicon nitride and wet etching of silicon oxide, aluminum deposition; (**e**) backside structuring by DRIE; and (**f**) dry etching of device silicon.

The adhesion of photoresists during the AlN wet etch process is not sufficient. Therefore, a 100 nm SiO_2 is the hard mask material for the patterning process. AlN and $Al_{0.68}Sc_{0.32}N$ wet etching is done with 85% phosphoric acid solution (H_3PO_4) at 80 °C (Figure 4b). The etch rate of AlN is 1.4 nm/s. The $Al_{0.68}Sc_{0.32}N$ has a etch rate of 6.7 nm/s. Test wafer with $Al_{0.86}Sc_{0.14}N$ have etch rates of 4.2 nm/s. This indicates a correlation of higher etch rates and higher Scandium ratios in the piezoelectric transducer.

The platinum and titanium are structured via tungsten hard mask by a dry etch process (Figure 4c) which is monitored with an optical emission spectrometer. By analyzing the species in the plasma, an etch stop can be defined as soon as the Ti/Pt is etched and the dry etching of the silicon nitride starts. In Figure 4d, the silicon nitride is patterned by RIE and the silicon oxide is wet etched. This enables an aluminum deposition on a smooth silicon surface. The 800 nm aluminum layer serves as reflective layer on the mirror plate and as upper electrode for excitation of the piezoelectric actuators. After wet etching of the aluminum layer, the handle wafer silicon is structured by DRIE using the buried SiO_2 of the initial SOI wafer as etch stop (Figure 4e).

By variation of the exposure parameters in the lithography as well as the DRIE parameters, a side wall shift of the silicon springs can be done to reduce the systems stiffness. This process can be done by using different resists, exposure times, or another DRIE process recipe. For the wafer with AlN this side wall shift is about 0.85 µm at each sidewall. This results in a change of the spring with from a = 5 µm to a′ = 3.3 µm. For the wafer with $Al_{0.68}Sc_{0.32}N$ a side wall shift of 1.35 µm is used. The lower stiffness should result in higher deflection per voltage and further increase the MOEMS performance compared to systems with high resonance frequency.

4. Measurement Setup

By using a Polytec MSA 400 Laser-Doppler-Vibrometer (LDV) with OFV 5000 Controller, the frequency-response-functions (FRF) of the MOEMS are recorded. A chirp signal with an amplitude of ± 1 V is applied to the top electrode of one actuator. The opposite actuator is driven with a 180° phase shifted signal with the same frequency and amplitude. So, the actuators work in antiphase mode. The amplitude of deflection is measured at the mirror plate edge. By measuring the resonance frequency, the stiffness of the MOEMS can be identified indirectly. This allows to compare static performance values for similar mechanical parameters of the MOEMS.

For measuring higher tilt angles, a high-deflection setup was introduced in [8,9]. Mechanical tilt angles up to approximately 15° can be measured. A laser beam is projected onto the mirror at a 45° angle. The mirror reflects it on an adjustable screen with a metric scale, which is also attached at a 45° angle to the mirror. The components like the laser mount and screen are fixed on a ring, adapted to the prober station (see Figure 5).

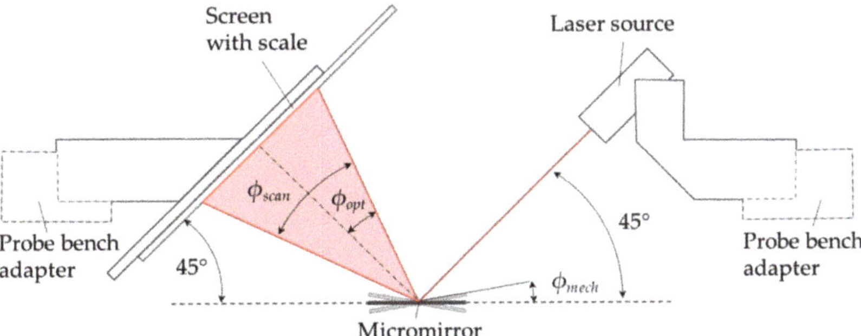

Figure 5. Schematic of the experimental setup [9].

The transversal piezoelectric coefficient defines the piezoelectric crystal deformation in result of an electric field. If samples with different piezoelectric material thicknesses and piezoelectric coefficients are used, the electric voltage as parameter for actuation is not sufficient to interpret the system performance. Therefore, the results are additionally documented in relation to the electric field in MV/m.

5. Results and Discussion

5.1. Frequency Response and Small Signal Actuation

In Figure 6a,b the FRFs are depicted in logarithmic scaling. The amplitude is measured for a range of six decades. Due to the high deflection in resonance, the LDV sensor sensitivity is low. Therefore, there is a significant noise for low amplitudes. The measurement values are given in Table 2. A motion scan image of exemplary micromirrors of Design 1 and Design 2 in torsional mode is shown in Figure 7.

An analytical calculation is shown to describe the static deflection of the 1D MOEMS. For a MOEMS with AlN and Design 1 a static deflection of 65 nm/V up to 116 nm/V is calculated. The measured deflection is 61.1 nm/V. One reason for the smaller deflection in the manufactured system can be the loss of elastic energy in the torsion spring with a high stiffness, which is not modeled by the analytical formulas. The measured MOEMS deflection with Design 1 and $Al_{0.68}Sc_{0.32}N$ is 157.6 nm/V. The measured deflection is within the calculated deflection range of 143 nm/V to 241 nm/V. For Design 2 with $Al_{0.68}Sc_{0.32}N$ as piezoelectric transducer, the modeled deflection of 563 nm/V to 959 nm/V matches with the measurement result of 667.3 nm/V.

For Design 1, the AlN MOEMS has a resonance frequency of 3444 Hz. For the same design, the resonance frequency of the $Al_{0.68}Sc_{0.32}N$ MOEMS is 1819 Hz. The side wall shift of the silicon results in a decrease of the resonance frequency and, therefore, in a lower stiffness of the system. The resonant mechanical tilt angle of the AlN based MOEMS (Design 1) is 2.8° at 1 MV/m. For the $Al_{0.68}Sc_{0.32}N$ MOEMS an angle of 11.9° at 1 MV/m is measured. The deflection in relation to the electric field is increased by a factor of 4.

(a)

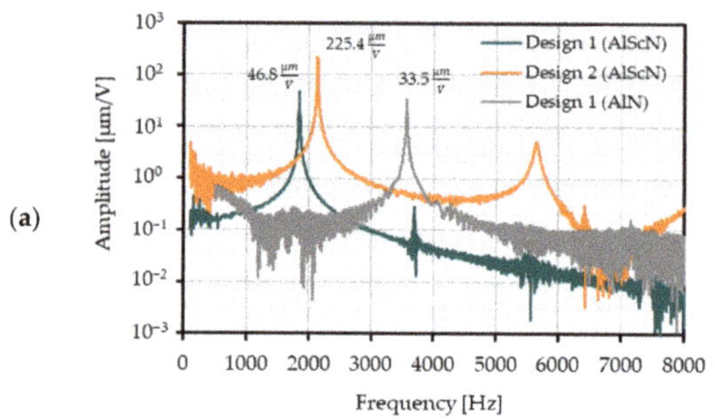

(b)

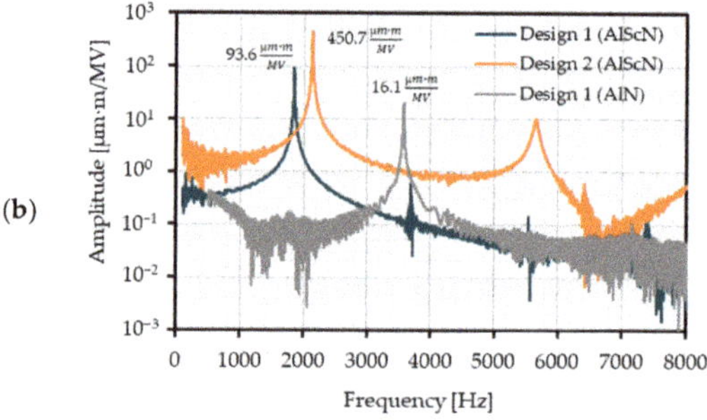

Figure 6. Frequency resonance functions of the presented micromirrors: (**a**) Mirror deflection per voltage; and (**b**) Mirror deflection per electric field.

Table 2. LDV measurement results for the resonant and static deflections of the MOEMS Designs with different transducer materials.

Parameters [1]	**Design 1**	**Design 1**	**Design 2**
	600 nm AlN	**2000 nm $Al_{0.68}Sc_{0.32}N$**	**2000 nm $Al_{0.68}Sc_{0.32}N$**
(Quasi)static parameters			
Mirror deflection (nm)			
At 1 V	61.1	157.6	667.3
At 1 MV/m	36.7	315.2	1334.6
Mech. tilt angle (m°)			
At 1 V	8.8	22.6	76.5
At 1 MV/m	5.3	45.1	152.9
Resonant parameters			
Resonance frequency (Hz)	3444	1819	2121
Mirror deflection (μm)			
At 1 V	33.1	41.4	152.9
At 1 MV/m	19.9	82.7	305.7
Resonant mech. tilt angle (°)			
At 1 V	4.7	5.9	17.8
At 1 MV/m	2.8	11.9	35.6

[1] Parameters are medium values over five samples.

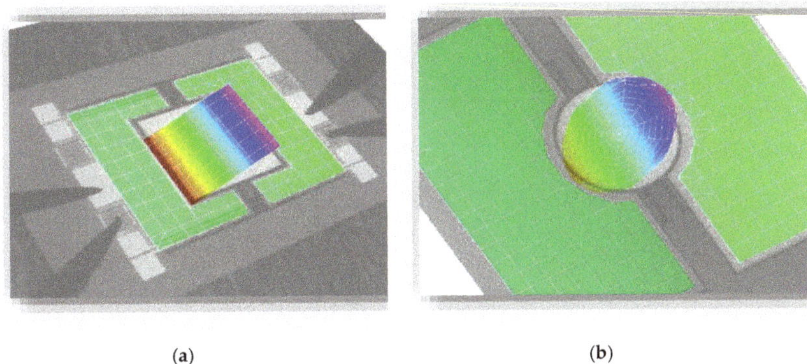

(a) (b)

Figure 7. Motion scan images of exemplary micromirrors recorded by Laser-Doppler-Vibrometry: (**a**) Design 1; and (**b**) Design 2.

Due to the larger chip area and actuator length and width of Design 2, the deflection is increased to 35.6° per MV/m in resonance. The resonance frequency for the MOEMS with Design 2 is 2121 Hz. In total, the resonant deflection from a MOEMS with AlN and Design 1 compared to $Al_{0.68}Sc_{0.32}N$ with Design 2 increased from 2.8° to 35.6° per 1 MV/m. This is a factor of 12.

5.2. Static High Voltage Actuation

In this section, the scanning characteristics of three selected micromirror samples in torsional mode for voltages of up to 400 V are shown. 400 V is the maximum voltage of the power supply in the setup. Due to limitations in the measurement setup, deflections up to 15° mechanical deflection can be measured. In Table 3 and Figure 8 the static mechanical deflections of the systems are shown.

Table 3. Comparison of the static mechanical tilt angles of the several micromirror designs and piezoelectric transducer technologies.

Static Parameters	Design 1	Design 1	Design 2
	600 nm AlN	2000 nm $Al_{0.68}Sc_{0.32}N$	2000 nm $Al_{0.68}Sc_{0.32}N$
Mech. tilt angle (°)			
At 100 V	2.1	2.1	6.3
At 200 V	4.1	4.3	12.5
At 400 V	—	9.6	— [1]
At 50 MV/m	0.6	2.1	6.3
At 100 MV/m	1.2	4.3	12.5
At 200 MV/m	2.5	9.6	— [1]
Maximum mech. tilt angle (°)	4.1 (at 200 V)	9.6 (at 400 V)	13.9 [1] (at 220 V)

[1] Limit of measurement setup.

Electric breakthroughs are observed for AlN chips at voltages higher than 200 V (Figure 9). Up to 400 V actuation voltage is used for the samples with $Al_{0.68}Sc_{0.32}N$. The reason is the high electric field for the 600 nm thin AlN layers compared to the 2 μm thick $Al_{0.68}Sc_{0.32}N$.

For Design 1 the maximum deflection is 9.6° at 400 V with $Al_{0.68}Sc_{0.32}N$ as piezoelectric transducer. AlN-based chips show deflections up to 4.1° at 200 V. By comparing the deflection of the MOEMS in relation of the electric field, the use of 2 μm $Al_{0.68}Sc_{0.32}N$ and lower system stiffness increased the static deflection by a factor of approximately 4. It can be assumed that the high piezoelectric coefficient of the $Al_{0.68}Sc_{0.32}N$ is one major reason for the larger deflection. In addition, the lower stiffness of the $Al_{0.68}Sc_{0.32}N$ based MOEMS influences the absolute and relative deflection and needs further investigations.

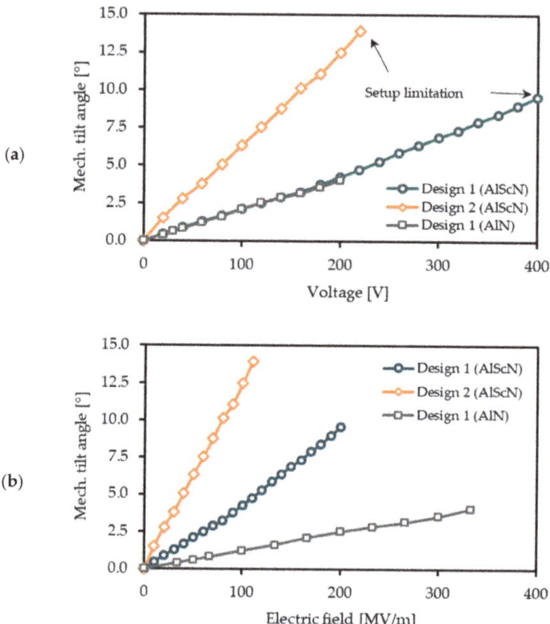

Figure 8. Performance of the presented micromirror designs at static high voltage actuation: (**a**) Mechanical tilt angle versus voltage; and (**b**) Mechanical tilt angle versus electrical field.

Figure 9. Photography of an exemplary micromirror of Design 1 with electric breakthroughs at 220 V. The electric contact of the MOEMS is done with micro needles on a probe station.

Design 2 enabled deflections of up to 13.9° at 220 V, which is the limit of the measurement setup. The deflection per electric field of Design 2 is increased by a factor of 3 in relation to $Al_{0.68}Sc_{0.32}N$-MOEMS with Design 1. Design 2 with $Al_{0.68}Sc_{0.32}N$ has more than ten times of the deflection per electric field compared to the previously reported AlN based MOEMS with Design 1. The scan angle can be defined by four times the mechanical tilt angle. Therefore, scan angles up to 55.6° for static displaced $Al_{0.68}Sc_{0.32}N$ MOEMS are shown. Figure 10 shows a photography of a static deflected MOEMS of Design 2 with $Al_{0.68}Sc_{0.32}N$ at 200 V.

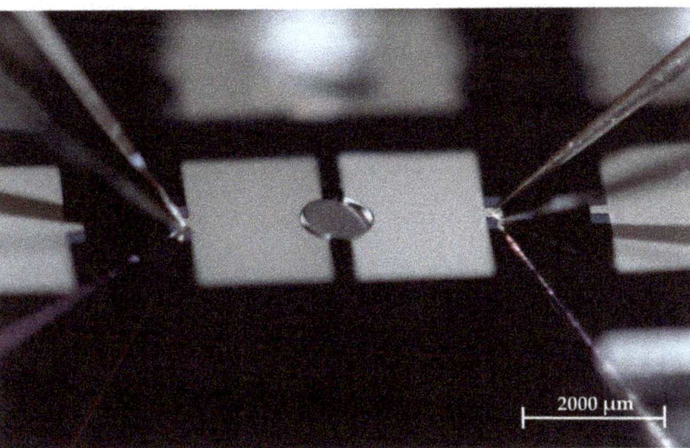

Figure 10. Photography of an exemplary micromirror of Design 2 in static operation (12.5°, 200 V). Captured by a single lens reflex (SLR) camera (Canon EOS 600D) and macro lens.

In Figure 8 the linearity for deflections < 15° can be seen. Relevant non-linear effects are not observed for the static deflections. Therefore, stress-stiffening effects have minor relevance for both MOEMS designs and static deflections < 15°.

For high electric fields the AlN and $Al_{0.68}Sc_{0.32}N$ shows electric breakthroughs. In Figure 9 a chip is shown after a breakthrough. Optically, lightning discharges were observed spontaneously. If a lightning appears at one position of the chip, an avalanche effect starts immediately and multiple areas of the chip show electric breakthroughs. The positions of the breakthroughs are random. Therefore, imperfections of the AlN and $Al_{0.68}Sc_{0.32}N$ growth could be the reason for the breakthrough. Nevertheless, a very electric field of up to 200 MV/m is applied to the piezoelectric layer, which indicates a high quality of the crystal growth and structure.

Future measurement setups need power supplies with voltages higher 400 V DC and a larger optical bank for the documentation of lager scan angles.

Figure 11 shows a photography of the Design 1 MOEMS with $2 \times 3\ mm^2$ footprint and the Design 2 MOEMS with $6 \times 8\ mm^2$ chip size in comparison.

Figure 11. Photography of the Design 1 and Design 2 MOEMS with piezoelectric $Al_{0.68}Sc_{0.32}N$ actuators on a copper sulfate crystal as a backdrop.

6. Conclusions

The AlN and $Al_{0.68}Sc_{0.32}N$ is was developed for the use at high electric fields up to 200 MV/m to increase the maximum electric energy the system can transform into a deflection. Additionally, the use of $Al_{0.68}Sc_{0.32}N$, in comparison to AlN, increased the static deflection per electric field by a factor of 3.5. This shows the impact of AlScN based transducer materials for piezoelectric microsystems. By using 2 µm thick $Al_{0.68}Sc_{0.32}N$ with high electric breakdown voltage, the maximum actuation voltage was increased up to 400 V.

In Table 4 a comparison of the MOEMS Designs 1 and 2 with AlN and AlScN and the micromirror of Gu-Stoppel et al. [14] is presented. A figure of merit (FOM) is shown as a product of mirror diameter, respectively, mirror length, and scan angle. Another figure of merit considers the influence of the stiffness of the MOEMS by including the resonance frequency [1]. The presented MOEMS Design 1 and 2 with $Al_{0.68}Sc_{0.32}N$ as transducer material have very high values for the FOMs. The FOM for the Design 2 MOEMS with $Al_{0.68}Sc_{0.32}N$ is FOM = $\theta \cdot e \cdot f_{res}$ = 116.8 m·°·Hz and therefore 3.2 times higher than the reference in literature. The reason therefore can be a 2 µm thick $Al_{0.68}Sc_{0.32}N$ with high piezoelectric coefficients and the use of high voltages as well as optimized design parameter with a leverage effect. However, Gu-Stoppel et al. [14] were able to manufacture a 2D MOEMS on a very small footprint using vertical silicon integration technologies.

Table 4. Comparison of (quasi-)static driven micromirrors based on piezoelectric AlN and AlScN of current literature and this work.

Specification	Unit	Ref. [14]	This Work		
			Design 1	Design 1	Design 2
Transducer material		AlScN	AlN	AlScN	AlScN
Material thickness	nm	1000	600	2000	2000
Mirror plate length (e)	mm	0.8	0.8	0.8	1.0
Chip size	mm^2	1.4 × 1.4 [2]	2 × 3	2 × 3	4 × 6
Res. frequency (f_{res})	kHz	0.9	3.4	1.8	2.1
Drive voltage	V_{DC}	150	200	400	220
Scan angle (θ)	°	50	8.4	38.4	55.6 [1]
FOM: $\theta \cdot e$	mm · °	40.0	6.7	30.7	55.6
FOM: $\theta \cdot e \cdot f_{res}$	m · ° · Hz	36.0	22.8	55.3	116.8

[1] Limited by measurement setup. [2] Estimated chip size with frame.

In summary, two different MOEMS designs with AlN and $Al_{0.68}Sc_{0.32}N$ as piezoelectric actuator materials are compared. AlN and $Al_{0.68}Sc_{0.32}N$ driven MOEMS with static scan angles up to 55.6° were fabricated. The chip performances for different designs and transducer materials with focus on the static actuation were compared. The use of $Al_{0.68}Sc_{0.32}N$, larger actuators, softer springs, the increased thickness of the transducer, and a material with high electrical breakdown voltages enabled the increase of the performance. The resonant deflection per electric field increased by a factor of 12. The static deflection per electric field increases more than 10 times due to the optimization in design and transducer material. The development of high-performance transducer materials and optimized MOEMS designs will allow miniaturized and robust micro optics with large static scan angles.

Author Contributions: Conceptualization, C.S.; Data curation, K.M.; Formal analysis, C.S. and K.M.; Investigation, C.S., K.M., M.M. and A.Ž.; Methodology, C.S. and R.F.; Project administration, C.S., R.F., S.Z. and K.H.; Supervision, C.S.; Writing—original draft, C.S., K.M., M.M. and A.Ž.; Writing—review and editing, R.F., S.Z. and K.H.; Funding acquisition, C.S. and H.K. All authors have read and agreed to the published version of the manuscript.

Funding: This research was funded by the European Union (EU-ESF), the Sächsische Aufbaubank SAB, the free state of Saxony, the Fraunhofer Gesellschaft FhG and the Deutsche Forschungsgemeinschaft DFG ("E-PISA", FK: 100310500, "Wafer-Level Sensorstruktur zur Bestimmung von

Ionenenergie- und Ionenwinkelverteilungsfunktionen in Niederdruckplasmen", AOBJ: 636895). The publication of this article was funded by the Fraunhofer Gesellschaft.

Conflicts of Interest: The authors declare no conflict of interest. The funders had no role in the design of the study; in the collection, analyses, or interpretation of data; in the writing of the manuscript; or in the decision to publish the results.

References

1. Holmström, S.T.S.; Baran, U. MEMS Laser Scanners: A Review. *J. Microelectromech. Syst.* **2014**, *23*, 259–275. [CrossRef]
2. Patterson, P.R.; Hah, D. Scanning micromirrors: An overview. *Proc. SPIE* **2004**, *5604*, 195–207. [CrossRef]
3. Xie, H. Editorial for the Special Issue on MEMS Mirrors. *Micromachines* **2018**, *9*, 99. [CrossRef] [PubMed]
4. Specht, H. MEMS-Laser-Display-System: Analyse, Implementierung und Testverfahrenentwicklung. Ph.D. Thesis, Chemnitz Universitiy of Technology, Chemnitz, Germany, 2011.
5. Schenk, H. Ein Neuartiger Mikroaktor zur Ein-Und Zweidimensionalen Ablenkung von Licht. Ph.D. Thesis, Gerhard-Mercator-Universität-Gesamthochschule-Duisburg, Duisburg, Germany, 2000.
6. Shao, J.; Li, Q. AlN based piezoelectric micromirror. *Opt. Lett.* **2018**, *43*, 987–990. [CrossRef] [PubMed]
7. Meinel, K.; Stoeckel, C.; Melzer, M. Piezoelectric scanning micromirror with large scan angle based on thin film aluminum nitride. In Proceedings of the 20th International Conference on Solid-State Sensors, Actuators and Microsystems and Eurosensors XXXIII (Transducers & Eurosensors XXXIII), Berlin, Germany, 23–27 June 2019.
8. Meinel, K.; Stoeckel, C. Piezoelectric scanning micromirror with built-in sensors based on thin film aluminum nitride. *IEEE Sens. J.* **2020**, *21*, 9682–9689. [CrossRef]
9. Meinel, K.; Melzer, M. 2D Scanning Micromirror with Large Scan Angle and Monolithically Integrated Angle. *Sensors* **2020**, *20*, 6599. [CrossRef] [PubMed]
10. Senger, F.; Gu-Stoppel, S. A 2D circular-scanning piezoelectric MEMS mirror for laser material processing. *Proc. SPIE* **2021**, *11697*, 1169704. [CrossRef]
11. Pensala, T.; Kyynäräinen, J. Wobbling mode AlN-piezo-mems mirror enabling 360-degree field of view LIDAR for automotive applications. In Proceedings of 2019 IEEE International Ultrasonics Symposium (IUS), Glasgow, Scotland, 6–9 October 2019.
12. Senger, F.; Albers, J. A bi-axial vacuum-packaged piezoelectric MEMS mirror for smart headlights. *Proc. SPIE* **2020**, *11293*, 1129305. [CrossRef]
13. Kobayashi, T.; Maeda, R. Piezoelectric Optical Micro Scanner with Built-in Torsion Sensors. *Jpn. J. Appl. Phys.* **2007**, *46*, 2781–2784. [CrossRef]
14. Gu-Stoppel, S.; Lisec, T. A highly linear piezoelectric quasi-static MEMS mirror with mechanical tilt angles of larger than 10°. *Proc. SPIE* **2019**, *10931*, 1093102. [CrossRef]
15. Lu, Y.; Reusch, M. Surface Morphology and Microstructure of Pulsed DC Magnetron Sputtered Piezoelectric AlN and AlScN Thin Films. *Phys. Status Solidi A* **2018**, *215*, 1700559. [CrossRef]
16. Lu, Y.; Reusch, M. Elastic modulus and coefficient of thermal expansion of piezoelectric Al1−xScxN (up to x = 0.41) thin films. *APL Mater.* **2018**, *6*, 076105. [CrossRef]

Article

SAW Resonators and Filters Based on $Sc_{0.43}Al_{0.57}N$ on Single Crystal and Polycrystalline Diamond

Miguel Sinusia Lozano [1,2], Laura Fernández-García [3,4], David López-Romero [5], Oliver A. Williams [6] and Gonzalo F. Iriarte [3,*]

1 Institute for Optoelectronic Systems and Microtechnology, Universidad Politécnica de Madrid, Avenida Complutense, 30, 28040 Madrid, Spain
2 Nanophotonics Technology Center, Universitat Politècnica de València, Camino de Vera, s/n Edificio 8F | Planta 2ª, 46022 Valencia, Spain; msinloz@ntc.upv.es
3 Departamento Ciencia de Materiales, Escuela Técnica Superior de Ingenieros de Caminos, Canales y Puertos, Universidad Politécnica de Madrid, Ciudad Universitaria, Calle del Profesor Aranguren 3, 28040 Madrid, Spain
4 Departamento de Sensores y Sistemas de Ultrasonidos, Instituto de Tecnologías Físicas y de la Información Leonardo Torres Quevedo—ITEFI, CSIC, Calle Serrano 144, 28006 Madrid, Spain; laura.fernandez@csic.es
5 Instituto de Micro y Nanotecnología, IMN-CNM, CSIC Isaac Newton, 8, Tres Cantos, 28760 Madrid, Spain; david.lopezromero@csic.es
6 School of Physics and Astronomy, Cardiff University, Cardiff CF24 3AA, UK; williamso@cardiff.ac.uk
* Correspondence: gonzalo.fuentes@upm.es; Tel.: +34-910674349

Abstract: The massive data transfer rates of nowadays mobile communication technologies demand devices not only with outstanding electric performances but with example stability in a wide range of conditions. Surface acoustic wave (SAW) devices provide a high Q-factor and properties inherent to the employed materials: thermal and chemical stability or low propagation losses. SAW resonators and filters based on $Sc_{0.43}Al_{0.57}N$ synthetized by reactive magnetron sputtering on single crystal and polycrystalline diamond substrates were fabricated and evaluated. Our SAW resonators showed high electromechanical coupling coefficients for Rayleigh and Sezawa modes, propagating at 1.2 GHz and 2.3 GHz, respectively. Finally, SAW filters were fabricated on $Sc_{0.43}Al_{0.57}N$/diamond heterostructures, with working frequencies above 4.7 GHz and ~200 MHz bandwidths, confirming that these devices are promising candidates in developing 5G technology.

Keywords: SAW devices; piezoelectricity; ScAlN thin film; diamond thin film; 5G technology; electromechanical coupling coefficient k2; Q-factor

Citation: Sinusia Lozano, M.; Fernández-García, L.; López-Romero, D.; Williams, O.A.; Iriarte, G.F. SAW Resonators and Filters Based on $Sc_{0.43}Al_{0.57}N$ on Single Crystal and Polycrystalline Diamond. *Micromachines* **2022**, *13*, 1061. https://doi.org/10.3390/mi13071061

Academic Editor: Agnė Žukauskaitė

Received: 6 June 2022
Accepted: 29 June 2022
Published: 30 June 2022

1. Introduction

Surface acoustic wave (SAW) devices leverage the piezoelectric effect to generate and detect electroacoustic signals [1,2]. These devices are fabricated with cutting-edge clean room technologies, which significantly shrinks their unitary costs. The possibility of controlling phonons and monitoring their interaction with electrons, photons, or magnetic spins have attracted the attention of the research community. Their applications have increased from the delay lines of the early 1960s to the nowadays state-of-the-art sensors or the emerging quantum technologies [3,4]. Within the scope of mobile telecommunication, the operating frequencies of these devices have been steadily increasing from 450 MHz to 6 GHz. Furthermore, SAW devices based on polycrystalline piezoelectric thin films are a cost-effective MEMS solution for the 5G technological constraints due to their outstanding capabilities such as power handling or thermal stability [5,6].

Among polycrystalline piezoelectric thin films, AlN has been extensively studied for its high SAW propagation velocity due to the stiffness of the compound, as well as its thermal and chemical stability [7,8]. However, its relatively low piezoelectric constant d_{33} restrains its applicability where large electromechanical coupling coefficients (k^2) are required. In this regard, the introduction of Sc atoms into the wurtzite AlN structure increases the piezoelectric response of the thin film [9]. The maximum increase is reported to occur at an Sc concentration of 43%. However, experimentally, the synthesis of this compound has been challenging for the competitive synthesis of the rock−salt phase (non-piezoelectric) of ScAlN, which is more energetically favorable at Sc concentrations above 55% [10–12]. Furthermore, the inclusion of Sc does not only increase the piezoelectric response of the compound but reduces its elastic constant and alters the optoelectronic properties, such as the bandgap [13,14].

In the case of loss-less materials, the electromechanical coupling coefficient (k^2) is a measure of the conversion efficiency between the mechanical and electrical energies and vice versa, and it is directly related to the piezoelectric response of the thin film [15]. Therefore, the ScAlN compound and its increase of the piezoelectric constant with the Sc concentration is a promising material to fabricate high-frequency SAW devices with outstanding performances.

In this work, SAW devices were fabricated with $Sc_{0.43}Al_{0.57}N$/diamond-based heterostructures. Highly c-axis oriented $Sc_{0.43}Al_{0.57}N$ thin films were synthesized on polycrystalline and single crystal diamond substrates. The electroacoustic properties of these layered structures were assessed. From the 1-port resonators, the effective electromechanical coupling coefficients (K^2_{eff}) and effective propagation velocities were extracted. Finally, SAW filters working at frequencies above 4.70 GHz are presented.

2. Materials and Methods

The synthesis of the piezoelectric thin film was carried out using a home-built reactive magnetron sputtering system. The target alloy ($Sc_{0.6}Al_{0.4}$) was placed 45 mm from the substrate. The process was carried out without intentional heating, and the temperature was monitored using a K-type thermocouple located below the substrate holder. The magnetron was powered using a pulsed DC generator (ENI RPG50), with a pulse width set to 1616 ns.

The polycrystalline diamond (PCD) substrates were synthesized by microwave plasma chemical vapor deposition (MPCVD) on a 500-µm thick Si (001) supporting layer. The single crystal diamond (111) (SCD) was purchased (EDP Corporation, Osaka, Japan) for evaluating its influence on the SAW propagation characteristics.

Prior to the ScAlN synthesis process, the substrates were cleaned using the following two-solvent method. First, the substrates were rinsed in acetone at 60 °C, followed by a sonication bath in methanol, both for 5 min. Afterwards, the samples were blown dry with nitrogen (N_2) and introduced into the load-lock chamber.

The synthesis chamber was conditioned for three minutes in a pure Argon (Ar) atmosphere (30 sccm), with a discharge power of 500 W and a process pressure of 1.33 Pa. Nitrogen was then introduced, the gas admixture ratio ($N_2/(N_2 + Ar)$) was adjusted to 25%, and the process pressure and discharge power were set to 1.33 Pa and 500 W, respectively, for 3 min. Finally, the process pressure was adjusted to 0.40 Pa. After 3 min, the plasma was turned off, and the substrates were transferred from the load-lock chamber. Afterwards, with the shutter closed, the plasma was ignited again using the synthesis conditions (0.40 Pa, 500 W, 25% gas admixture ratio), and after two minutes, the shutter was moved away, beginning the synthesis process.

The degree of c-axis orientation of the synthesized piezoelectric thin film was assessed via 0002 ω-θ scans using X-ray diffraction (XRD, Phillips X-Pert Pro MRD diffractometer), with Cu Kα_1 radiation (λ = 1.54059 Å), 45 kV, and 40 mA.

The interdigital transducers (IDT) were fabricated using a standard lift-off process. They were patterned with an e-beam lithography system (Crestec CABL-9500C, Hachioji, Japan). Due to the insulating behavior of the ScAlN thin film and the underlying polycrystalline diamond substrates, an organic anti-static layer (Espacer 300Z, Showa Denko K.K, Tokyo, Japan) was spun on top of the e-beam resist to avoid charge accumulation.

The fabricated 1-port resonators comprise a 2000-nm $Sc_{0.43}Al_{0.57}N$ thin film and 110-nm thick (700 nm width) Au IDT (λ = 2.8 μm). The filters were fabricated using an 850-nm thick $Sc_{0.43}Al_{0.57}N$ thin film and 300-nm wide Au IDT (λ = 1.2 μm). In this case, the Au thickness of the filters IDT were different: 130 nm and 65 nm thick for the PCD and SCD, respectively. A 5-nm thick Cr layer was employed as an adhesion layer. In all devices, a 0.5 metallization ratio was employed (Figure 1). An SEM inspection of the fabricated SAW filters can be found in (Figure S1) The 1-port resonators had 60 finger pairs and two grounded reflectors with 60 finger pairs with the designed IDT wavelength. On the other hand, the π-type SAW filters comprised 3 1-port resonators.

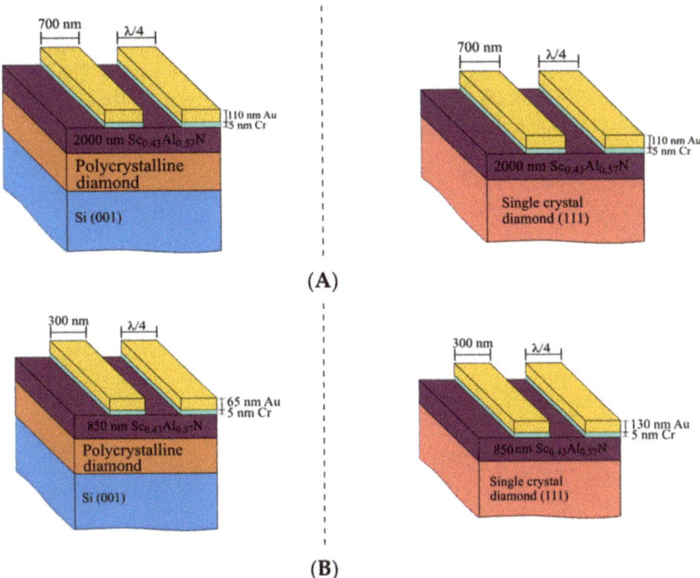

Figure 1. $Sc_{0.43}Al_{0.57}N$ heterostructures of the fabricated devices with polycrystalline and single crystal diamond substrates. (**A**) The 1-port SAW resonators and (**B**) SAW filters.

The electrical characterization of the devices was carried out using a vector network analyzer (Agilent N5230 A, Santa Clara, CA, USA), with standard, 300-μm pitch, ground-source-ground (GSG) probes (Picoprobe 40A; C style adaptor, GGB Industries, Inc, Naples, FL, USA). A standard short, open, load, through (SOLT), 50 Ω, one-port calibration was employed. In the case of the 1-port resonators, the measurement resolution was set to 16,001 points in the 0.50 GHz to 6.00 GHz frequency range, whereas the measurement of the fabricated SAW filters was carried out with a resolution of 16,001 points in the 1.00 GHz to 10 GHz frequency range. In both cases, the output power was set to 0 dBm. Afterwards, the data processing and analysis was carried out using scikit-rf, an open-source Python package [16].

3. Results

The devices presented in this work targeted a ratio between piezoelectric thin film thickness and an IDT wavelength (d/λ) of 0.71, which is reported to provide the maximum K^2_{eff} of the Sezawa mode for diamond-based heterostructures [17]. Additionally, in order to efficiently propagate the Sezawa mode at frequencies above 4 GHz, the designed IDT wavelength of the filters (1200 nm) was reduced compared to that of the resonators (2800 nm). In these devices, the resonance frequency was determined by the IDT wavelength and the propagation velocity of the wave through the layered structure [18]. Therefore, reducing the thickness of the piezoelectric thin film forces the wave to propagate predominantly through the diamond layer, leveraging its faster propagation velocity.

The single crystal diamond datasheet reported a surface roughness R_a value below 2 nm and the AFM analyses carried out on the polycrystalline diamond substrate reported R_{RMS} values below 2 nm. Regardless of the substrate, the full width at half maximum (FWHM) below 3° of the XRD 0002 ω-θ scans confirmed that the polycrystalline ScAlN thin films are highly c-axis oriented. The RBS analysis reported an Sc 43% concentration within the thin film (Figure S2).

3.1. Resonators

The reflection coefficient of the polycrystalline diamond resonators showed that several resonance frequency modes are generated together with second and third order reflections (Figure 2A). Among these, the reflection coefficient of the Rayleigh mode (1.20 GHz) outstands (~−40 dB). The Sezawa mode (2.06 GHz) and the second order Rayleigh mode (2.30 GHz) propagated with reflection coefficients below −15 dB.

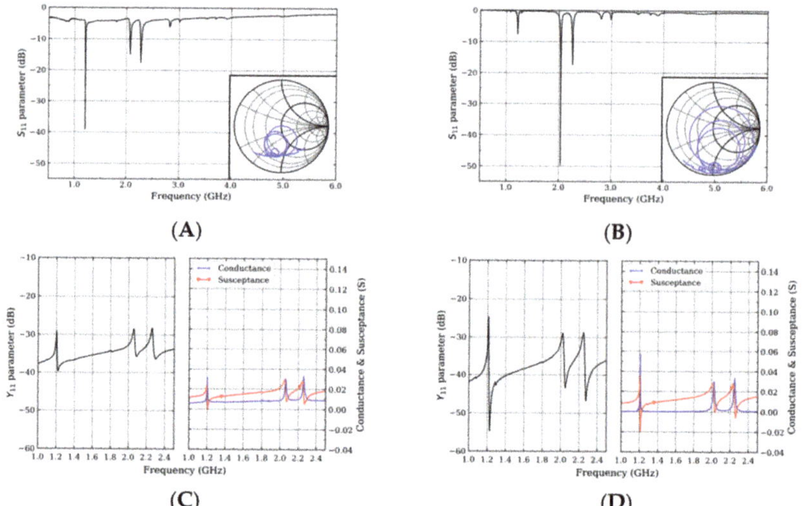

Figure 2. Measured reflection coefficient (inset: Smith chart) and admittance characteristics of the one-port resonators fabricated with the $Sc_{0.43}Al_{0.57}N$/polycrystalline (**A**,**C**) and single crystal diamond (**B**,**D**) heterostructures.

The devices using the SCD substrate reported lower insertion losses (Figure 2B). Furthermore, in these devices, the reflection coefficient of the Sezawa mode (2.03 GHz) below −45 dB indicated the efficient excitation of this mode. The Rayleigh mode (1.21 GHz) propagated with a reflection coefficient below −5 dB, whereas its second order reflection mode (2.25 GHz) propagated with a reflection coefficient below −15 dB. In both heterostructures, higher reflection modes propagated above 2.50 GHz.

From the admittance characteristics (Figure 2C,D), the series (f_s) and parallel (f_p) frequencies can be extracted and then the effective propagation velocity (v_{eff}) (Equation (1)) can be calculated, as can the effective electromechanical coupling coefficient (K^2_{eff}) (Equation (2)), where λ is the designed IDT wavelength (Table 1) [19,20].

$$v_{eff} = \frac{f_p - f_s}{2}\lambda \tag{1}$$

$$K^2_{eff} = \frac{\pi^2}{8}\frac{f_p^2 - f_s^2}{f_s^2} \tag{2}$$

A figure of merit (Table 2), usually employed to compare the performance of SAW devices, multiplied the Bode Q-factor (Equation (3)) by the electromechanical coupling coefficient. The Q-factor (Equation (4)) is defined as the ratio between the series or parallel resonance frequency and the −3-dB width of the frequency [21–23].

$$FOM_{s,p} = K^2_{eff} * Q_{s,p} \tag{3}$$

$$Q_{s,p} = \frac{f_{s,p}}{\Delta f_{-3dB(s,p)}} \tag{4}$$

Table 1. Effective velocity (v_{eff}) and effective electromechanical coupling coefficient (K^2_{eff}) from the series and parallel resonance frequencies.

	Single Crystal Diamond				Polycrystalline Diamond			
Mode	f_s (GHz)	f_p (GHz)	v_{eff} (m/s)	K^2_{eff} (%)	f_s (GHz)	f_p (GHz)	v_{eff} (m/s)	K^2_{eff} (%)
Rayleigh	1.20	1.22	3425	3.46	1.20	1.22	3402	3.19
Sezawa	2.02	2.05	5725	3.72	2.06	2.09	5830	3.65
2nd Rayleigh	2.25	2.28	6361	3.04	2.27	2.30	6407	4.24

Table 2. Quality factors and figure of merit (FOM) values from the series and parallel resonance frequencies.

	Single Crystal Diamond				Polycrystalline Diamond			
Mode	Q_s	Q_p	FOMs	FOMp	Q_s	Q_p	FOMs	FOMp
Rayleigh	251	187	8.69	6.49	103	4	3.28	0.14
Sezawa	52	69	1.91	2.58	10	10	0.376	0.372
2nd Rayleigh	67	132	2.03	4.02	62	8	2.63	0.336

3.2. Filters

The transmission coefficient of the SAW filter fabricated using the PCD heterostructure showed that the Rayleigh and Sezawa modes propagated with their center frequency (f_c) at 2.66 GHz and 4.88 GHz, respectively (Figure 3A). Both modes presented a maximum gain above −6 dB, whereas their cut−off frequencies were below −12 dB. The −3-dB bandwidth of the Rayleigh mode was above 100 MHz, whereas the Sezawa mode bandwidth was above 180 MHz (Table 3).

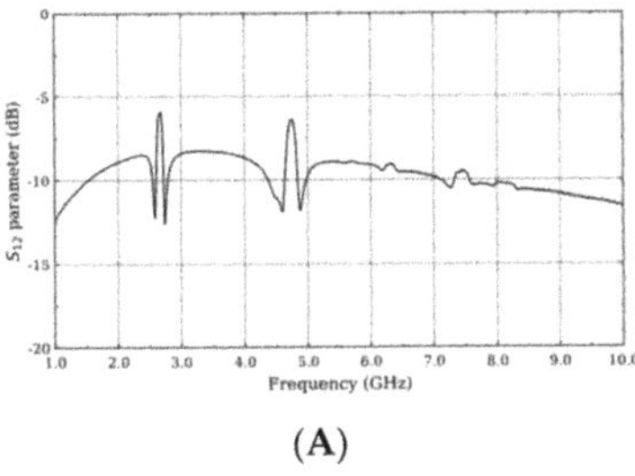

(A)

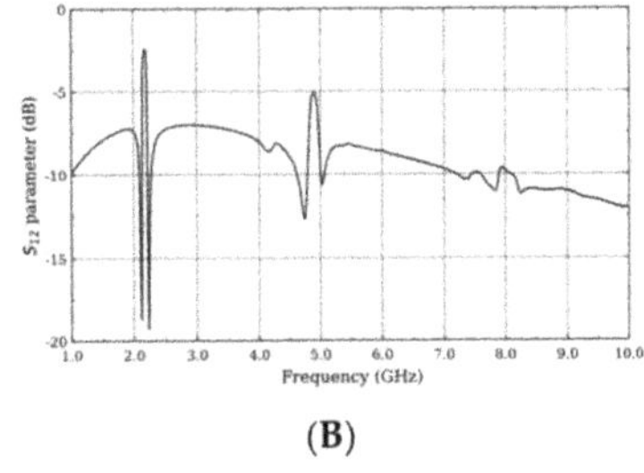

(B)

Figure 3. Transmission coefficient of a ladder-type SAW filter fabricated on $Sc_{0.43}Al_{0.57}N$/PCD (**A**) and $Sc_{0.43}Al_{0.57}N$/SCD (**B**) heterostructures.

On the other hand, the Rayleigh mode generated in the SAW filter fabricated using the SCD substrate had a band pass maximum of −2.5 dB (Figure 3B). The center frequency of this mode was 2.17 GHz. whereas the Sezawa mode f_c was at 4.74 GHz, and its band pass maximum was −5 dB. The cut-off of both the lower (f_1) and higher (f_2) frequencies in the Rayleigh mode were below −18 dB, whereas those corresponding to the Sezawa mode were below −10 dB.

Table 3. Lower (f_1) and higher (f_2) cut-off frequencies, center frequencies (f_c), and bandwidth in the Rayleigh and Sezawa modes for SCD and PCD filters.

	Single Crystalline Diamond				**Polycrystalline Diamond**			
Mode	**f_1 (GHz)**	**f_2 (GHz)**	**f_c (GHz)**	**−3 dB Bandwidth (MHz)**	**f_1 (GHz)**	**f_2 (GHz)**	**f_c (GHz)**	**−3 dB Bandwidth (MHz)**
Rayleigh	2.14	2.21	2.17	72	2.61	2.72	2.66	107
Sezawa	4.81	4.99	4.90	181	4.64	4.83	4.74	189

4. Discussion

In layered structures, the effective propagation velocity of a SAW (Equation (5)) depends on the elastic modulus (E) and density (ρ) of the particular thin films, which through these the wave propagates [24]. In this regard, both types of diamond substrates have similar stiffness, as the resonance frequency of the SAW modes in both resonators is comparable. Furthermore, when comparing the resonance frequencies with our previous works ($Sc_{0.26}Al_{0.74}N$/PCD with Pt electrodes), the softening of the piezoelectric thin film with the Sc concentration was apparent, as the resonance frequency of both Rayleigh and Sezawa modes was reduced [25].

$$v_{eff} = \sqrt{\frac{E}{\rho}} \tag{5}$$

However, the single crystal diamond substrate shifted the conductance baseline of the device towards ~0 S (Figure 2D), which showed how the scattering of the SAW wave due to defects such as grain boundaries or voids deteriorated the electrical response of the devices. This is in agreement with previous works, where the grain size of the polycrystalline diamond substrates increased the insertion losses of the devices [26–28].

Focusing on the piezoelectric thin film, particularly in its defects and how they degrade the electrical response of the devices [29], most of the defects in the thin films accumulated within the proximity of layer interfaces, mainly due to lattice mismatches and strain relaxation mechanisms, which usually require heat treatments to be minimized. However, the propagation characteristics of our devices validated the characterization via XRD analysis and the high quality $Sc_{0.43}Al_{0.57}N$ thin films which, as previously mentioned, were synthesized with no intentional heating of the substrate, on both diamond substrates.

The electromechanical coupling coefficients lay between those reported previously in the literature (Table 4). However, there have been several approaches that can be undertaken to increase this value, such as adequate device design or the selection of the substrate or the electrode metal [17,30].

On the other hand, the structure with embedded electrodes (ScAlN/IDT/diamond) has been reported to considerably increase the K^2_{eff} coefficient [31]. Additionally, in order to boost the SAW propagation, a-plane (instead of c-plane) ScAlN thin films have been recently employed [32].

The filter bandwidth was, without some other circuit elements, constrained by the effective electromechanical coupling coefficient (K^2_{eff}), as it ultimately constrained the separation between the series and parallel resonance frequencies.

The −3-dB bandwidth of the Sezawa mode in both the layered structures was above 180 MHz at the center frequencies of 4.70 GHz and 4.90 GHz for the PCD and SCD structures, respectively.

Table 4. Comparison of resonance frequencies, compound composition, and electromechanical coupling coefficient.

ScAlN Composition	Synthesis Technique	Target	Substrate	Resonance Frequency (GHz)	Electrode Metal	K^2_{eff} (%)	Q
$Sc_{0.11}Al_{0.89}N$ [29]	MBE	-	Si	3.6	Ti/Au	3.7	146
$Sc_{0.27}Al_{0.73}N$ [17]	RF	$Sc_{0.40}Al_{0.60}$	PCD	2.5–3.5	Al/Cr	5.5–4.5	396–227
$Sc_{0.27}Al_{0.73}N$ [30]	DC	ScAlN alloy	Si	0.2–0.3	Ti/Au	2	100
$Sc_{0.26}Al_{0.74}N$ [33]	Pulsed DC	$Sc_{0.40}Al_{0.60}$	Si	R 1.4	Pt	0.5	140
$Sc_{0.26}Al_{0.74}N$ [18]	Pulsed DC	$Sc_{0.40}Al_{0.60}$	PCD	R 1.5–S 2.6	Pt	2.8	R167–S180
$Sc_{0.43}Al_{0.57}N$ [34]	RF	Dual	SCD	3.75	Cu	6.1	520
			SCD	2.9	Cu	3.8	-
$Sc_{0.23}Al_{0.77}N$ [32]	Pulsed DC	Dual (Al + Sc targets)	Sapphire	1.9–1.7	Pt	1.3–2.4	659–538
This work	Pulsed DC	$Sc_{0.60}Al_{0.40}$	PCD&SCD	R 1.2–S 2.03	Cr/Au	3.2–3.7	R 250–S ~50

Higher bandwidth for both Rayleigh and Sezawa propagation modes were obtained in our $Sc_{0.43}Al_{0.57}N$-based heterostructures compared to AlN and ZnO based filters (Table 5). Furthermore, the insertion losses of our devices were similar to these polycrystalline piezoelectric thin films. However, the insertion losses of the devices based on single crystal piezoelectric thin films were lower, showing the importance of the device design and the quality of the polycrystalline thin film in the electrical response of the filters.

Table 5. Comparison of center frequencies, −3-dB bandwidth, insertion loss (IL), and the Q factor of the SAW filters with different substrates, piezoelectric thin films, and electrode metals.

Reference	Substrate	Piezoelectric Thin Film	Electrode Metal	Center Frequency (GHz)	−3 dB Bandwidth (MHz)	IL (dB)	Q
[35]	SiO_2/Si	AlN	Pt	4.47	30	−40	149
[36]	SiO_2/Si	$LiTaO_3$	Al	3.5	205	−1	17
[37]	PCD/Si	SiO_2/ZnO	Al	2.488	3	−5	700
[38]	PCD/Si	ZnO	Al	2.9	15	−20	193
This work	SCD/Si	$Sc_{0.43}Al_{0.57}N$	Cr/Au	R 2.17–S 4.90	R 72–S 181	R − 2.5–S − 5	R 30–S 27
	PCD/Si	$Sc_{0.43}Al_{0.57}N$	Cr/Au	R 2.66–S 4.74	R 107–S 189	R − 6–S − 6	R 25–S 25

5. Conclusions

SAW devices have been fabricated using $Sc_{0.43}Al_{0.57}N$ as a piezoelectric thin film, synthetized over polycrystalline and single crystalline diamond substrates. The admittance characteristics confirm that $Sc_{0.43}Al_{0.57}N$ thin films can be employed in the fabrication of SAW devices with low insertion losses above 2 GHz. The higher insertion loss and attenuation of the SAW in the PCD structure revealed the damaging influence of the electrical performance of the scattering in the grain boundaries, as well as the defects within the crystal structure.

The $Sc_{0.43}Al_{0.57}N$ thin films employed in these devices were synthesized by reactive magnetron sputtering without the intentional heating of the substrate. The energy required to minimize the defects within the thin film is provided by the plasma conditions. The Rayleigh and Sezawa mode K^2_{eff} values substantially increased with the devices comprising the $Sc_{0.43}Al_{0.57}N$ thin films, as compared to those values obtained for the modes propagating within the $Sc_{0.27}Al_{0.73}N$ thin films shown in our previous works [18,39]. These two propagating modes are efficiently excited where the reflection coefficient of the Sezawa mode propagating in the SCD heterostructure outstands, with an attenuation close to −50 dB. Additionally, SAW filters with -3-dB bandwidth above 180 MHz have been fabricated at 4.7 GHz resonance frequencies with insertion losses below -5 dB with a SCD based device, revealing their potential application to 5G technology.

Furthermore, the devices presented in this work showed promising electrical performances for sensing applications where the AlN and ScAlN thermal, chemical, or high stiffness of the compound were exploited. Therefore, the applicability of these devices will not only be constrained by 5G technology, but these results reveal the potential of the versatile ScAlN compound for the next-generation of SAW devices.

Supplementary Materials: The following supporting information can be downloaded at: https://www.mdpi.com/article/10.3390/mi13071061/s1, Figure S1: SEM inspection of the SAW filter IDT and ground reflector; Figure S2: RBS spectra of a $Sc_{0.43}Al_{0.57}N$ thin film [40].

Author Contributions: Conceptualization, M.S.L. and G.F.I.; methodology, M.S.L.; software, M.S.L.; validation, M.S.L. and D.L.-R.; formal analysis, M.S.L. and D.L.-R.; investigation, M.S.L. and D.L.-R.; resources, G.F.I. and O.A.W.; data curation, M.S.L.; writing—original draft preparation, M.S.L. and L.F.-G.; writing—review and editing, M.S.L., L.F.-G. and G.F.I.; visualization, M.S.L., D.L.-R. and L.F.-G.; supervision, G.F.I.; project administration, G.F.I.; funding acquisition, G.F.I. All authors have read and agreed to the published version of the manuscript.

Funding: This research received no external funding.

Conflicts of Interest: The authors declare no conflict of interest.

References

1. Farnell, G.; Adler, E. Elastic Wave Propagation in Thin Layers. In *Physical Acoustics*; Academic Press: New York, NY, USA, 1972; Volume 9, pp. 35–127. [CrossRef]
2. Campbell, C.; Burgess, J.C. Surface Acoustic Wave Devices and Their Signal Processing Applications. *J. Acoust. Soc. Am.* **1991**, *89*, 1479–1480. [CrossRef]
3. Morgan, D.P. History of SAW devices. In Proceedings of the 1998 IEEE International Frequency Control Symposium (Cat. No.98CH36165), Pasadena, CA, USA, 29 May 1998; pp. 439–460. [CrossRef]
4. Delsing, P.; Cleland, A.N.; Schuetz, M.J.; Knörzer, J.; Giedke, G.; Cirac, J.I.; Srinivasan, K.; Wu, M.; Balram, K.C.; Bäuerle, C.; et al. The 2019 surface acoustic waves roadmap. *J. Phys. D Appl. Phys.* **2019**, *52*, 353001. [CrossRef]
5. Hashimoto, K. Advances in RF SAW devices: What are demanded? In Proceedings of the 2016 European Frequency and Time Forum (EFTF), York, UK, 4–7 April 2016; pp. 1–4. [CrossRef]
6. Wang, W.; Wu, H. High selectivity SAW DMS filter with in-between shorted-gratings. In Proceedings of the 2010 IEEE International Ultrasonics Symposium, San Diego, CA, USA, 11–14 October 2010; no. 10774073. pp. 1263–1266. [CrossRef]
7. Ababneh, A.; Schmid, U.; Hernando, J.; Sánchez-Rojas, J.L.; Seidel, H. The influence of sputter deposition parameters on piezoelectric and mechanical properties of AlN thin films. *Mater. Sci. Eng. B* **2010**, *172*, 253–258. [CrossRef]
8. Duquenne, C.; Besland, M.-P.; Tessier, P.-Y.; Gautron, E.; Scudeller, Y.; Averty, D. Thermal conductivity of aluminium nitride thin films prepared by reactive magnetron sputtering. *J. Phys. D Appl. Phys.* **2012**, *45*, 015301. [CrossRef]
9. Akiyama, M.; Kamohara, T.; Kano, K.; Teshigahara, A.; Takeuchi, Y.; Kawahara, N. Enhancement of Piezoelectric Response in Scandium Aluminum Nitride Alloy Thin Films Prepared by Dual Reactive Cosputtering. *Adv. Mater.* **2009**, *21*, 593–596. [CrossRef]
10. Tasnádi, F.; Alling, B.; Höglund, C.; Wingqvist, G.; Birch, J.; Hultman, L.; Abrikosov, I.A. Origin of the Anomalous Piezoelectric Response in WurtziteScxAl1−xNAlloys. *Phys. Rev. Lett.* **2010**, *104*, 137601. [CrossRef]
11. Högund, C.; Birch, J.; Alling, B.; Bareño, J.; Czigany, Z.; Persson, P.; Wingqvist, G.; Zukauskaite, A.; Hultman, L. Wurtzite structure Sc1−xAlxN solid solution films grown by reactive magnetron sputter epitaxy: Structural characterization and first-principles calculations. *J. Appl. Phys.* **2010**, *107*, 123515. [CrossRef]
12. Talley, K.R.; Millican, S.L.; Mangum, J.; Siol, S.; Musgrave, C.B.; Gorman, B.; Holder, A.M.; Zakutayev, A.; Brennecka, G.L. Implications of heterostructural alloying for enhanced piezoelectric performance of (Al,Sc)N. *Phys. Rev. Mater.* **2018**, *2*, 063802. [CrossRef]
13. Caro, M.A.; Zhang, S.; Riekkinen, T.; Ylilammi, M.; Moram, A.M.; Lopez-Acevedo, O.; Molarius, J.; Laurila, T. Piezoelectric coefficients and spontaneous polarization of ScAlN. *J. Phys. Condens. Matter* **2015**, *27*, 245901. [CrossRef] [PubMed]
14. Zhang, S.; Holec, D.; Fu, W.Y.; Humphreys, C.J.; Moram, M.A. Tunable optoelectronic and ferroelectric properties in Sc-based III-nitrides. *J. Appl. Phys.* **2013**, *114*, 133510. [CrossRef]
15. Berlincourt, D.A.; Curran, D.R.; Jaffe, H. Piezoelectric and piezomagnetic materials and their function in transducers. In *Physical Acoustics Principles and Methods*; Mason, W.P., Ed.; Academic Press: New York, NY, USA, 1964; pp. 170–270.
16. Arsenovic, A.; Hillairet, J.; Anderson, J.; Forsten, H.; Ries, V.; Eller, M.; Sauber, N.; Weikle, R.; Barnhart, W.; Forstmayr, F. scikit-rf: An Open Source Python Package for Microwave Network Creation, Analysis, and Calibration [Speaker's Corner]. *IEEE Microw. Mag.* **2022**, *23*, 98–105. [CrossRef]
17. Kobayashi, Y.; Tsuchiya, T.; Okazaki, M.; Asao, Y.; Hashimoto, K.; Shikata, S. High-frequency surface acoustic wave resonator with ScAlN/hetero-epitaxial diamond. *Diam. Relat. Mater.* **2021**, *111*, 108190. [CrossRef]
18. Lozano, M.S.; Chen, Z.; Williams, O.A.; Iriarte, G.F. Giant Reflection Coefficient on Sc 0.26 Al 0.74 N Polycrystalline Diamond Surface Acoustic Wave Resonators. *Phys. Status Solidi* **2019**, *216*, 1900360. [CrossRef]
19. Liu, H.; Zhang, Q.; Zhao, X.; Wang, F.; Chen, M.; Li, B.; Fu, S.; Wang, W. Highly coupled leaky surface acoustic wave on hetero acoustic layer structures based on ScAlN thin films with a c-axis tilt angle. *Jpn. J. Appl. Phys.* **2021**, *60*, 031002. [CrossRef]

20. Lu, R.; Li, M.-H.; Yang, Y.; Manzaneque, T.; Gong, S. Accurate Extraction of Large Electromechanical Coupling in Piezoelectric MEMS Resonators. *J. Microelectromech. Syst.* **2019**, *28*, 209–218. [CrossRef]
21. Feld, D.A.; Parker, R.; Ruby, R.; Bradley, P.; Dong, S. After 60 years: A new formula for computing quality factor is warranted. In Proceedings of the 2008 IEEE Ultrasonics Symposium, Beijing, China, 2–5 November 2008; pp. 431–436. [CrossRef]
22. Ruby, R. 11E-2 Review and Comparison of Bulk Acoustic Wave FBAR, SMR Technology. In Proceedings of the 2007 IEEE Ultrasonics Symposium Proceedings, New York, NY, USA, 28–31 October 2007; pp. 1029–1040. [CrossRef]
23. Zhang, S.; Li, F.; Jiang, X.; Kim, J.; Luo, J.; Geng, X. Advantages and challenges of relaxor-PbTiO3 ferroelectric crystals for electroacoustic transducers—A review. *Prog. Mater. Sci.* **2015**, *68*, 1–66. [CrossRef]
24. Hashimoto, K. *Surface Acoustic Wave Devices in Telecommunications*; Springer: Berlin/Heidelberg, Germany, 2000. [CrossRef]
25. Wu, D.; Chen, Y.; Manna, S.; Talley, K.; Zakutayev, A.; Brennecka, G.L.; Ciobanu, C.V.; Constantine, P.; Packard, C.E. Characterization of Elastic Modulus Across the (Al1–xScx)N System Using DFT and Substrate-Effect-Corrected Nanoindentation. *IEEE Trans. Ultrason. Ferroelectr. Freq. Control* **2018**, *65*, 2167–2175. [CrossRef]
26. Fujii, S.; Odawara, T.; Yamada, H.; Omori, T.; Hashimoto, K.-Y.; Torii, H.; Umezawa, H.; Shikata, S. Low propagation loss in a one-port SAW resonator fabricated on single-crystal diamond for super-high-frequency applications. *IEEE Trans. Ultrason. Ferroelectr. Freq. Control* **2013**, *60*, 986–992. [CrossRef]
27. Fujii, S.; Shikata, S.; Uemura, T.; Nakahata, H.; Harima, H. Effect of crystalline quality of diamond film to the propagation loss of surface acoustic wave devices. *IEEE Trans. Ultrason. Ferroelectr. Freq. Control* **2005**, *52*, 1817–1822. [CrossRef]
28. Elmazria, O.; Bénédic, F.; El Hakiki, M.; Moubchir, H.; Assouar, M.; Silva, F.; Assouar, B. Nanocrystalline diamond films for surface acoustic wave devices. *Diam. Relat. Mater.* **2006**, *15*, 193–198. [CrossRef]
29. Hao, Z.; Park, M.; Kim, D.G.; Clark, A.; Dargis, R.; Zhu, H.; Ansari, A. Single Crystalline ScAlN Surface Acoustic Wave Resonators with Large Figure of Merit ($Q \times k$). In Proceedings of the IEEE MTT-S International Microwave Symposium Digest, Boston, MA, USA, 7 June 2019; pp. 786–789. [CrossRef]
30. Wang, W.; Mayrhofer, P.M.; He, X.; Gillinger, M.; Ye, Z.; Wang, X.; Bittner, A.; Schmid, U.; Luo, J. High performance AlScN thin film based surface acoustic wave devices with large electromechanical coupling coefficient. *Appl. Phys. Lett.* **2014**, *105*, 133502. [CrossRef]
31. Zhang, Q.; Han, T.; Chen, J.; Wang, W.; Hashimoto, K. Enhanced coupling factor of surface acoustic wave devices employing ScAlN/diamond layered structure with embedded electrodes. *Diam. Relat. Mater.* **2015**, *58*, 31–34. [CrossRef]
32. Ding, A.; Kirste, L.; Lu, Y.; Driad, R.; Kurz, N.; Lebedev, V.; Christoph, T.; Feil, N.M.; Lozar, R.; Metzger, T.; et al. Enhanced electromechanical coupling in SAW resonators based on sputtered non-polar Al0.77Sc0.23N $11\bar{2}0$ thin films. *Appl. Phys. Lett.* **2020**, *116*, 101903. [CrossRef]
33. Lozano, M.S.; Pérez-Campos, A.; Reusch, M.; Kirste, L.; Fuchs, T.; Zukauskaite, A.; Chen, Z.; Iriarte, G.F. Piezoelectric characterization of Sc0.26Al0.74N layers on Si (001) substrates. *Mater. Res. Express* **2018**, *5*, 036407. [CrossRef]
34. Hashimoto, K.-Y.; Fujii, T.; Sato, S.; Omori, T.; Ahn, C.; Teshigahara, A.; Kano, K.; Umezawa, H.; Shikata, S.-I. High Q surface acoustic wave resonators in 2-3 GHz range using ScAlN/single crystalline diamond structure. In Proceedings of the 2012 IEEE International Ultrasonics Symposium, Dresden, Germany, 7–10 October 2012; pp. 1–4. [CrossRef]
35. Wang, F.; Xiao, F.; Song, D.; Qian, L.; Feng, Y.; Fu, B.; Dong, K.; Li, C.; Zhang, K. Research of micro area piezoelectric properties of AlN films and fabrication of high frequency SAW devices. *Microelectron. Eng.* **2018**, *199*, 63–68. [CrossRef]
36. Kimura, T.; Omura, M.; Kishimoto, Y.; Hashimoto, K. Comparative Study of Acoustic Wave Devices Using Thin Piezoelectric Plates in the 3–5-GHz Range. *IEEE Trans. Microw. Theory Tech.* **2019**, *67*, 915–921. [CrossRef]
37. Uemura, T.; Fujii, S.; Kitabayashi, H.; Itakura, K.; Hachigo, A.; Nakahata, H.; Shikata, S.-I.; Ishibashi, K.; Imai, T. Low-Loss Diamond Surface Acoustic Wave Devices Using Small-Grain Poly-Crystalline Diamond. *Jpn. J. Appl. Phys.* **2002**, *41*, 3476–3479. [CrossRef]
38. Higaki, K.; Nakahata, H.; Kitabayashi, H.; Fujii, S.; Tanabe, K.; Seki, Y.; Shikata, S. High power durability of diamond surface acoustic wave filter. *IEEE Trans. Ultrason. Ferroelectr. Freq. Control* **1997**, *44*, 1395–1400. [CrossRef]
39. Lozano, M.S.; Chen, Z.; Williams, A.O.; Iriarte, G.F. Temperature characteristics of SAW resonators on Sc0.26Al0.74N/polycrystalline diamond heterostructures. *Smart Mater. Struct.* **2018**, *27*, 075015. [CrossRef]
40. Majer, M. "SIMNRA." MAX-PLANCK-INSTITUT FÜR PLASMAPHYSIK. 1997. Available online: https://home.mpcdf.mpg.de/~{}mam/index.html (accessed on 5 June 2022).

Article

Dual-Mode Scandium-Aluminum Nitride Lamb-Wave Resonators Using Reconfigurable Periodic Poling

Sushant Rassay , Dicheng Mo and Roozbeh Tabrizian *

Electrical and Computer Engineering Department, University of Florida, Gainesville, FL 32603, USA; sushantrassay@ufl.edu (S.R.); dicheng.mo@ufl.edu (D.M.)
* Correspondence: rtabrizian@ece.ufl.edu

Abstract: This paper presents the use of ferroelectric behavior in scandium–aluminum nitride ($Sc_xAl_{1-x}N$) to create dual-mode Lamb-wave resonators for the realization of intrinsically configurable radio-frequency front-end systems. An integrated array of intrinsically switchable dual-mode Lamb-wave resonators with frequencies covering the 0.45–3 GHz spectrum. The resonators are created in ferroelectric scandium–aluminum nitride ($Sc_{0.28}Al_{0.72}N$) film and rely on period poling for intrinsic configuration between Lamb modes with highly different wavelengths and frequencies. A comprehensive analytical model is presented, formulating intrinsically switchable dual-mode operation and providing closed-form derivation of electromechanical coupling (k_t^2) in the two resonance modes as a function of electrode dimensions and scandium content. Fabricated resonator prototypes show k_t^2s as high as 4.95%, when operating in the first modes over 0.45–1.6 GHz, 2.23% when operating in the second mode of operation over 0.8–3 GHz, and series quality factors (Q_s) over 300–800. Benefiting from lithographical frequency tailorability and intrinsic switchability that alleviate the need for external multiplexers, and large k_t^2 and Q, dual-mode $Sc_{0.28}Al_{0.72}N$ Lamb-wave resonators are promising candidates to realize single-chip multi-band reconfigurable spectral processors for radio-frequency front-ends of modern wireless systems.

Keywords: ferroelectric; scandium–aluminum nitride; Lamb-wave resonators; complementary switchable

Citation: Rassay, S.; Mo, D.; Tabrizian, R. Dual-Mode Scandium-Aluminum Nitride Lamb-Wave Resonators Using Reconfigurable Periodic Poling. *Micromachines* **2022**, *13*, 1003. https://doi.org/10.3390/mi13071003

Academic Editor: Agnė Žukauskaitė

Received: 18 May 2022
Accepted: 24 June 2022
Published: 26 June 2022

1. Introduction

Scandium–aluminum nitride ($Sc_xAl_{1-x}N$) has recently emerged as a transforming piezoelectric material for creation of high-performance electroacoustic resonators and filters. Benefiting from the large piezoelectric coefficients that only increase with scandium content [1], $Sc_xAl_{1-x}N$ enables the radical enhancement of electromechanical coupling (k_t^2) in electroacoustic resonators. This facilitates the creation of radio-frequency (RF) filters with significantly lower loss and higher bandwidth compared to AlN counterparts [2–5]. Besides the k_t^2 enhancement, the recent discovery of ferroelectricity in $Sc_xAl_{1-x}N$ [6] has initiated extensive research efforts for the creation of configurable RF components, such as varactors, and tunable and switchable resonators and filters [7–12]. These components are of particular interest for emerging wireless communication systems that require multi-band adaptive operation over a wide frequency spectrum [13,14].

Currently, the RF front-end of wireless systems rely on a large set of AlN thickness-extensional bulk acoustic wave (BAW) filters that are arrayed at the board level using external multiplexers, to enable spectral processing over the 0.4 GHz to 6 GHz spectrum [13,15,16]. The frequency of BAW filters is defined by the thickness of the metal–piezoelectric–metal stack and cannot be tailored with lithography. This imposes the need for a large number of separately packaged filter chips to address the numerous bands for different wireless applications and protocols.

Lamb-wave AlN resonators have been extensively explored as an alternative to BAW, as they provide lithographical frequency scalability and enable the integration of multi-band filters on a single chip [17]. However, the lower k_t^2 of Lamb-wave resonators compared to their BAW counterparts, and the resulting limitation in maximum attainable filter bandwidth has set a barrier for their adoption in RFFE. The lower k_t^2 of Lamb-wave resonators is due to the smaller transverse piezoelectric coefficient (e_{31}) compared to longitudinal (e_{33}) in AlN films.

However, the k_t^2 shortcoming in Lamb-wave resonators can be resolved considering the substantial increase in e_{31} with sufficiently high Sc doping that enables the realization of $Sc_xAl_{1-x}N$ Lamb-wave resonators with k_t^2 on par with or exceeding AlN BAW resonators [2,3,5]. Further, the ferroelectricity in $Sc_xAl_{1-x}N$ provides new opportunities for the intrinsic and on-chip reconfiguration of Lamb-wave resonators, to further reduce the number of filters and external switches, and their corresponding load on RFFE footprint, power consumption, and latency [18]. Another application of ferroelectricity is the use of polarization engineering to tailor excitable resonance modes for the performance optimization of electroacoustic resonators and filters. Polarization engineering has been previously demonstrated in lithium niobate electroacoustic devices for improving the response of an acoustically coupled filter [19], in the extreme frequency scaling of a resonator by enabling excitation of higher harmonics [20], and in the creation of acoustic stop-bands in waveguides [21]. In these efforts, polarization tailoring is applied as a part of the fabrication process. This approach does not allow the use of on-chip polarization tuning for the dynamic reconfiguration of device operation that is highly desirable for adaptive spectral processing applications.

In this work, we demonstrate high-k_t^2 dual-mode intrinsically switchable $Sc_{0.28}Al_{0.72}N$ Lamb-wave resonators. Intrinsic switchability and dual-mode operation are realized by the periodic poling of $Sc_{0.28}Al_{0.72}N$, using pulsed switching, to enable the selective excitation of Lamb modes with different wavelengths and frequencies. Dual-mode Lamb-wave resonators with frequencies covering the entire ultra-high-frequency regime are implemented in the same batch, and their intrinsic switchability and dual-mode operation are analytically formulated and experimentally verified.

2. Concept

Lamb-wave resonators are created from cascading unit-cells with patterned interdigitated transducers (IDT) (Figure 1a).

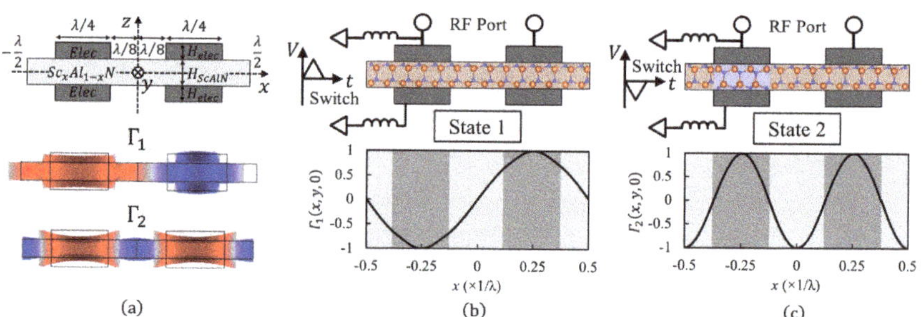

Figure 1. (**a**) Cross−sectional schematic of unit-cell in the Lamb-wave resonators, with strain mode−shapes when operating in Γ_1 and Γ_2 modes. (**b**) Operation State 1 (unified polarization): polarization under all the IDTs is in same direction, enabling the high-k_t^2 excitation of Γ_1 while Γ_2 mode is turned off. (**c**) Operation State 2 (alternating polarization): polarization under consecutive IDTs is in the opposite direction, enabling the high-k_t^2 excitation of Γ_2 while Γ_1 mode is turned off.

Considering the resonator as a waveguide extended in the x-axis direction, the mechanical resonance modes correspond to the eigenmodes of the unit cell, when a periodic boundary condition is applied:

$$\Gamma_i(x,y,z) = \Gamma_i(x+\lambda,y,z) \tag{1}$$

Here, $\Gamma_i(x,y,z)$ is the strain mode–shape function, and λ is the unit-cell length in x-axis direction. Figure 1a shows the COMSOL-simulated deformation mode–shape for two different eigenmodes, corresponding to the zeroth-order symmetric Lamb waves (i.e., S_0) propagating in the x-axis direction, that optimally match the lateral electric-field excitation scheme using top-surface IDTs. The simulation is performed for a unit-cell with $Sc_xAl_{1-x}N$ thickness of 200 nm, molybdenum (Mo) bottom electrode and IDTs with a thickness of 100 nm, and λ of 6 mm. These modes benefit from the efficient electromechanical excitation enabled by the large e_{31} in $Sc_xAl_{1-x}N$. This can be formulated using the excited electric displacement in the two modes ($D_{S_0,i}$, $i = 1,2$) as:

$$D_{S_0,i}(x) = e_{31,eff}\Gamma_{S_0,i} \tag{2}$$

Here, $e_{31,eff}$ is the effective transverse piezoelectric coefficient that is linearly proportional to the normalized instantaneous polarization $P_{inst}(x)$, due to the ferroelectric characteristic in $Sc_{0.28}Al_{0.72}N$ and is formulated as:

$$e_{31,eff} = e_{31}P_{inst}(x) \tag{3}$$

$P_{inst}(x)$ can be spatially tuned between 1 (i.e., nitrogen polar) or -1 (i.e., metal polar) by the application of proper polarization-switching pulses to $Sc_{0.28}Al_{0.72}N$, between each IDT finger and the bottom electrode. The motional charge per unit length of the y-axis ($Q_{m,i}$) excited between the two IDT fingers in the unit-cell is derived from:

$$Q_{m,i} = \frac{1}{2}\sum_{j=1,2}\left((-1)^j \int_{x_{elec,j}}^{x_{elec,j}+\lambda/4} D_{S_0,i}(x)dx\right) \tag{4}$$

Assuming a similar acoustic velocity in metal electrodes and $Sc_{0.28}Al_{0.72}N$, and infinite dimension of the waveguide in y-axis direction, the S_0 mode–shapes are:

$$\begin{aligned} \Gamma_1(x,y,z) &\cong \sin\left(\tfrac{2\pi x}{\lambda}\right), \\ \Gamma_2(x,y,z) &\cong \cos\left(\tfrac{4\pi x}{\lambda}\right) \end{aligned} \tag{5}$$

The coupling coefficient ($k_{t,i}^2$) of the ith S_0 mode ($i = 1,2$) is formulated from mechanical and electrical energies as [22]:

$$k_{t,i}^2 \cong \frac{\frac{1}{2}\frac{Q_{m,i}^2}{C_0}}{\frac{1}{2}\frac{Q_{m,i}^2}{C_0} + \int_0^{H_{ScAlN}}\int_0^{\lambda}\frac{c_{11}}{2}\left(\Gamma_{S_0,i}\right)^2 dxdz + 4\int_0^{H_{elec}}\int_{\frac{\lambda}{4}-\frac{W_f}{2}}^{\frac{\lambda}{4}+\frac{W_f}{2}}\frac{E_{elec}}{2}\left(\Gamma_{S_0,i}\right)^2 dxdz} \tag{6}$$

Here, H_{ScAlN} and H_{elec} are the thickness of the piezoelectric film and electrodes, respectively. c_{11} and E_{elec} are, respectively, the elastic constants of the piezoelectric film and electrodes in the wave-propagation direction. W_f is the IDT finger width. C_0 is the capacitance between the two IDT fingers, per unit length in the y-axis direction, and is approximated as:

$$C_0 = \frac{\epsilon_{33}W_f}{2H} \tag{7}$$

Here, ϵ_{33} is the piezoelectric film permittivity. Replacing Equations (4) and (5) in Equation (6), $k_{t,i=1,2}^2$ is simplified to:

$$k_{t,i}^2 = \left(\frac{P_{inst,1} - (-1)^i P_{inst,2}}{2}\right)^2 \frac{\frac{4}{\pi^2}K_{31}^2\alpha_i}{\beta_i + \frac{4}{\pi^2}K_{31}^2\alpha_i} \tag{8}$$

Here, α_i is a scaling factor corresponding to the relative width of IDT fingers to unit-cell length, calculated as:

$$\alpha_i = \frac{\lambda}{i^2 W_f}\left(\sin\left(\frac{i\pi W_f}{\lambda}\right)\right)^2 \tag{9}$$

β_i is a scaling factor representing the relative energy distribution in overall unit cell and the piezoelectric layer, calculated as:

$$\beta_i = \frac{c_{11}H_{ScAlN} + \frac{2\alpha_i W_f}{\lambda}E_{elec}H_{elec}}{c_{11}H_{ScAlN}} \tag{10}$$

$P_{inst,j}$ $(j = 1,2)$ are the net polarization of $Sc_{0.28}Al_{0.72}N$ under the two IDT fingers in the unit-cell, and K_{31}^2 is the transverse piezoelectric coupling constant formulated as:

$$K_{31}^2 = \frac{e_{31}^2}{\epsilon_{33}c_{11}} \tag{11}$$

Considering Equation (8), two complementary polarization states exist where either the first (i.e., Γ_1) or the second (i.e., Γ_2) S_0 mode has the maximum k_t^2. In State 1, wherein the polarizations under both IDT fingers are unified (i.e., all in the same direction: $P_{inst,j}$ $(j = 1,2) = \pm 1$), Γ_1 is excited with the maximum k_t^2, while Γ_2 is switched off (i.e., $k_{t,2}^2 = 0$). In State 2, wherein the polarizations under IDT fingers are periodically alternating (i.e., in opposite directions: $P_{inst,1} = -P_{inst,2} = \pm 1$), Γ_2 is excited with the maximum k_t^2, while Γ_1 is switched off (i.e., $k_{t,1}^2 = 0$). Figure 1b,c shows the complementary operation states corresponding to different polarization configurations and the x-axis strain mode–shape function for the active mode.

The complementary operation enables intrinsic switching of the resonator between fundamental and second harmonics of Lamb modes, with a frequency ratio near 2. Considering Equations (8)–(10), the relative magnitude of k_t^2 for these modes depends on the electrode finger width. Figure 2a shows the normalized k_t^2 of Γ_1 and Γ_2 modes, across different finger widths, extracted using the presented analytical model. It is evident that the k_t^2 of Γ_1 mode is always higher than that of Γ_2. When using the dual-mode resonator to implement a dual-band bandpass filter, the lower k_t^2 of mode Γ_2 translates into a lower fractional bandwidth. However, the absolute bandwidth of the filter remain nearly the same in either operation modes, considering the higher frequency of Γ_2 mode. As modern wireless networking protocol applies similar channel bandwidth at different center frequencies (e.g., 40 MHz in both 2.4 GHz and 5 GHz in IEEE 802.11 n), a halved k_t^2 of Γ_2 at a frequency that is nearly twice Γ_1 enables the realization of a dual-band filter with the same absolute bandwidth in both operation states. Figure 2b shows the maximum k_t^2 achievable in mode Γ_1, for resonators created from $Sc_xAl_{1-x}N$ films with Sc content over 0% to 40%. This plot is derived using Equation (8) and for different thicknesses of the metal electrode relative to $Sc_xAl_{1-x}N$ film. It is evident that, for $Sc_xAl_{1-x}N$ films exceeding 30% Sc content, assuming 0.1 relative thickness of electrodes, the k_t^2 of both modes exceeds the 6% typical value in AlN BAW resonators. It should also be noted that, considering the very large polarization-switching fields in $Sc_xAl_{1-x}N$, thinner films are desirable to enable a configuration between the two modes with reasonable voltages. Therefore, opting for 0.1 relative thickness of metal films may result in excessive electrode loss. In this work, the resonators are implemented in ~200 nm $Sc_{0.28}Al_{0.72}N$ films, and ~100 nm Mo electrodes (i.e., 0.5 relative electrode thickness) are used.

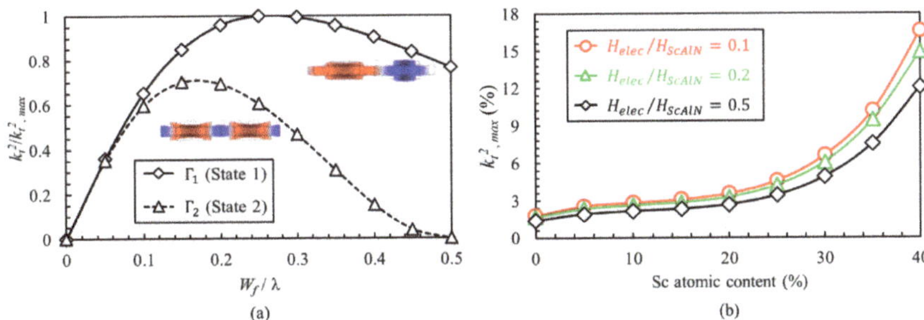

Figure 2. (**a**) Normalized k_t^2 of Γ_1 and Γ_2 modes across different electrode finger widths and for arbitrary Sc content and electrode thickness. (**b**) Maximum achievable k_t^2 of Γ_1 mode for different Sc contents and electrode thicknesses.

3. Fabrication Process

Figure 3 shows the fabrication process for the creation of the dual-mode $Sc_{0.28}Al_{0.72}N$ Lamb-wave resonators. The process consists of the DC sputtering of a 100 nm Mo layer atop a 30 nm AlN film that serves as a seed for (110)-textured growth of Mo film [23]. The Mo layer is then patterned using a boron trichloride (BCl_3) gas-based recipe in a reactive-ion-etching inductively coupled plasma (RIE-ICP) system. Prior to etching, a tapered photoresist mask was developed via the proximity exposure method. The resulting tapered photoresist profile thus enabled the formation of tapered sidewalls in the Mo layer, which promoted the crack-free growth of the subsequent $Sc_{0.28}Al_{0.72}N$ layer. Following this, a highly crystalline *c*-axis-oriented 200 nm $Sc_{0.28}Al_{0.72}N$ layer was deposited using reactive magnetron sputtering from segmented scandium–aluminum targets [24]. Finally, atop this, a ~120 nm thick Mo layer was deposited, to serve as the top electrode for the resonator.

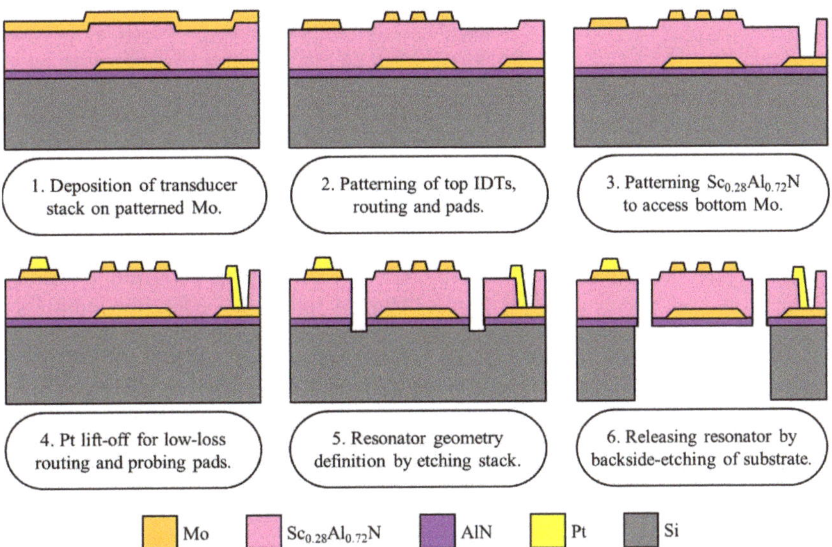

Figure 3. Fabrication process flow of $Sc_{0.28}Al_{0.72}N$ Lamb-wave resonators.

After the deposition of the transducer stack, the top Mo layer is patterned using SF_6 in the RIE-ICP system to form IDTs. Next, to access the bottom electrode, $Sc_{0.28}Al_{0.72}N$ is patterned using timed Cl_2 dry-etch in a RIE-ICP system. This is followed by the deposition of 500 nm thick Cr/Pt through lift-off, to create low-loss lines and probing pads. After metallization, the lateral geometry of the resonator is formed by etching trenches using a high-power Cl_2/BCl_3-based recipe in the RIE-ICP system, wherein low-frequency PECVD SiO_2 is used as a hard-mask for the etching process. Finally, the device is released from the backside of the silicon substrate by deep reactive ion etching (DRIE). Figure 4a shows the SEM image of the Lamb-wave resonator and highlights the patterned IDTs. Figure 4b shows the cross-sectional SEM image of the resonator stack, highlighting the thickness of the constituent layers.

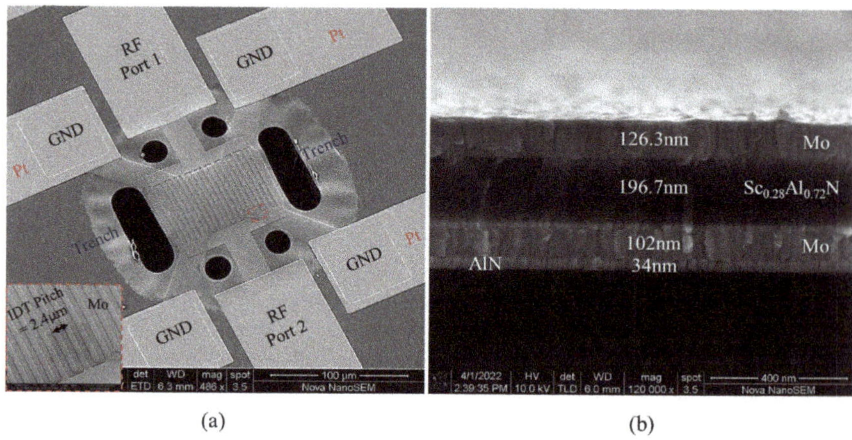

(a) (b)

Figure 4. (**a**) SEM image of $Sc_{0.28}Al_{0.72}N$ Lamb−wave resonator. The inset shows the IDT with a 2.4 μm pitch size. (**b**) Cross−sectional SEM image of the resonator, detailing constituent-layer thicknesses.

4. Characterization

Dual-mode $Sc_{0.28}Al_{0.72}N$ Lamb-wave resonators with different IDT pitch sizes were measured to identify their admittance and switching behavior. Ferroelectric polarization hysteresis loop measurement and the periodic poling of the $Sc_{0.28}Al_{0.72}N$ were performed using a Radiant PiezoMEMS ferroelectric tester. The resonators' RF performance was measured using a Keysight N5222A PNA vector network analyzer, along with a short-open-load-through (SOLT) calibration procedure enabled by a CS-5 calibration kit from GGB Industries INC.

4.1. Ferroelectric Characterization

To identify the switching voltage, polarization hysteresis loops were measured using 100 kHz bipolar triangular pulses with 125 V amplitude. Figure 5a shows the hysteresis loop measured at an IDT port and is compared with the loop measured for a 100 μm × 100 μm capacitor. The slight degradation of the loop measured at the IDT port, defined by a lower remanent polarization and higher coercive field, may correspond to the nonuniform distribution of the electric field at excessive edges of IDTs. A coercive voltage of 114 V is extracted for the IDT port, identifying the required voltage for the periodic poling of $Sc_{0.28}Al_{0.72}N$ to switch resonator operation between Γ_1 and Γ_2 modes. Figure 5b shows the instantaneous current measured at the IDT port, upon the application of a 45 kHz negative positive-up-negative-down (PUND) pulse sequence with a 112 V amplitude. The large instantaneous current induced upon a change in the sign of deriving voltage pulse indicates the polarization inversion of $Sc_{0.28}Al_{0.72}N$ between metal- and nitrogen-polar states. A similar waveform, with slightly lower voltage of 110 V, is used for periodic poling and the intrinsic switching of the resonator between the two operation states. Opting for lower voltage enables the observation of resonator admittance evolution during the transition between the two operation states and the corresponding complementary excitation and suppression of Γ_1 and Γ_2 modes.

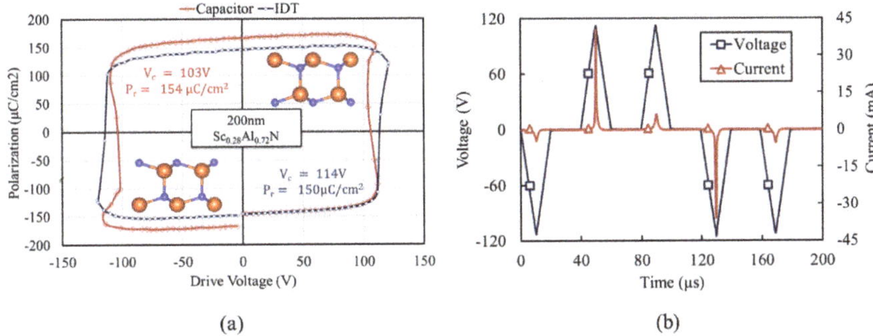

Figure 5. (**a**) Polarization−voltage (P−V) hysteresis loop measured at an IDT port and a 100 μm × 100 μm capacitor. (**b**) The measured instantaneous current at the IDT port, upon the application of a 45 kHz triangular PUND pulse sequence, highlighting polarization reversal in $Sc_{0.28}Al_{0.78}N$ film.

4.2. RF Characterization

The admittance of resonators was extracted from the measured reflection coefficient (i.e., S_{11}). One-port measurements were performed through the application of a signal between the two IDT ports while keeping the bottom electrode floating. The intrinsic switching of the resonators between the two operation states is performed by applying pulsed poling voltages between one of the IDT ports and the bottom electrode. Figure 6a–e shows the measured admittances for resonators with different IDT pitch sizes ranging over 2.4 mm to 8 mm. For each resonator, the measured admittances are shown when operating in each state. The complementary switchable dual-mode operation in Γ_1 and Γ_2 modes is evident. This is enabled by the pulsed periodic poling of $Sc_{0.28}Al_{0.72}N$ under one of the IDT ports. Figure 6f shows the frequency of the Γ_1 and Γ_2 modes for different IDT pitch sizes, highlighting the coverage of the 0.45–3 GHz spectrum.

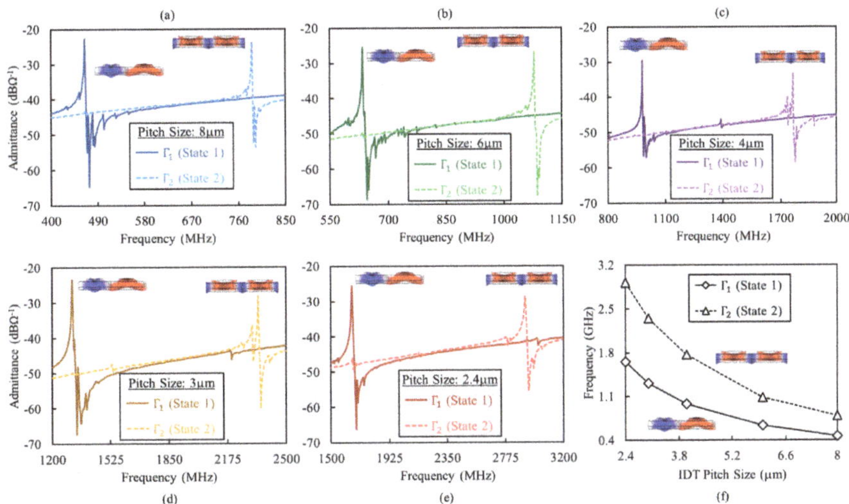

Figure 6. Measured admittance of dual−mode $Sc_{0.28}Al_{0.72}N$ Lamb−wave resonators when operating in either of the complementary switchable states defined by periodic poling procedure. The admittances are shown for resonators with (**a**) 8 mm, (**b**) 6 mm, (**c**) 4 mm, (**d**) 3 mm, and (**e**) 2.4 mm IDT pitch sizes. (**f**) The frequency of Γ_1 and Γ_2 modes for different IDT pitch sizes, highlighting the coverage of the 0.45−3 GHz spectrum.

Figure 7 shows the short-span admittance of the $Sc_{0.28}Al_{0.72}N$ resonator with 2.4 mm IDT pitch size, around Γ_1 and Γ_2 resonance frequencies. The evolution of admittance upon the application of three −110 V 45 kHz monopolar poling pulses, between one of IDT terminals and the floating bottom electrode, is evident. Starting from the pristine film with uniform polarization, the application of the first and second poling pulses to one of the IDT terminals results in periodic, yet partial, polarization switching. This translates to the gradual suppression of mode Γ_1 and the emergence of mode Γ_2. After the third pulse, when the polarization of all the domains under the corresponding IDT terminal are fully reversed, Γ_1 is fully suppressed, while Γ_2 has emerged with its maximum k_t^2. The reversibility of this procedure is verified by the application of three 45 kHz monopolar poling pulses, but with a 110 V amplitude.

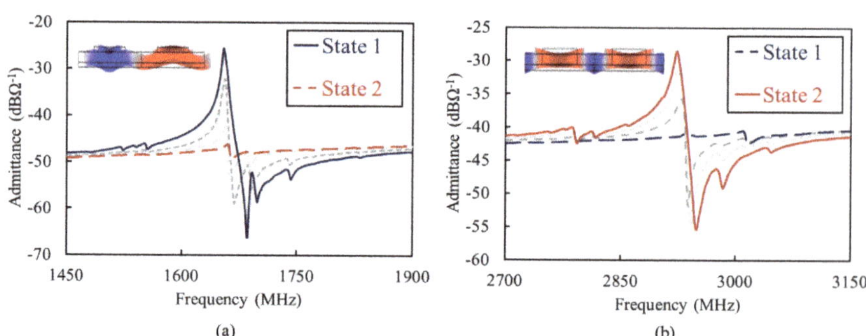

Figure 7. Short−span measured admittance of the Lamb−wave resonator with an IDT pitch size of 2.4 mm when operating in (**a**) State 1 and (**b**) State 2. The intermediate admittance plots, shown in dashed gray line, highlight the transition between the two operation states.

Figure 8 shows the measured k_t^2, Q_s, and Q_p (i.e., Q at series and parallel resonance) of resonators with different IDT pitch sizes over 2.4 mm to 8 mm, covering 0.45 GHz to 3 GHz in two operation states. For each IDT pitch size, ten resonators are measured across the wafer. The resonators k_t^2 and Q are extracted from the admittances using [25,26]:

$$k_t^2 = \frac{\pi^2}{8}\left(\frac{{f_p}^2 - {f_s}^2}{{f_s}^2}\right),\ Q = \frac{f}{2}\left|\frac{\partial \varphi_Y}{\partial f}\right| \tag{12}$$

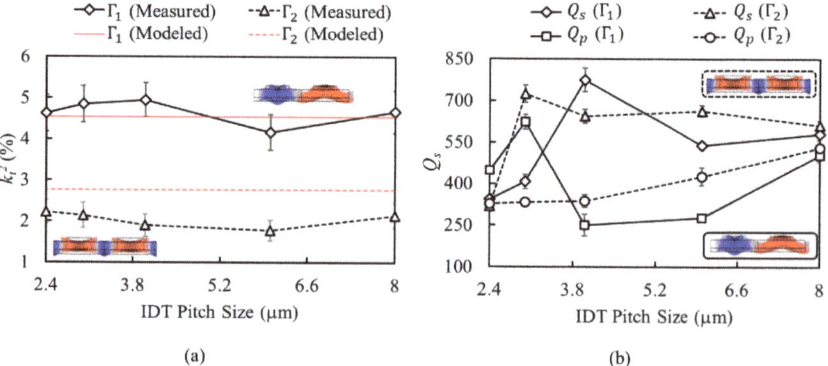

Figure 8. (**a**) Measured k_t^2 of the two modes for resonators with different IDT pitch sizes, in comparison with values extracted from analytical model. (**b**) Measured Q_s and Q_p of the two modes, for resonators with different IDT pitch sizes. For each IDT pitch size, the k_t^2, Q_s, and Q_p are the average of values measured from ten resonators across the 4-inch substrate.

Here, f_s and f_p are the frequencies of series and parallel resonance modes, and φ_Y is the admittance phase. The measured k_t^2 is compared with the values extracted from the presented analytical model.

k_t^2 mean values over 4.17% to 4.95% are measured for the resonators when operating in Γ_1 mode (i.e., State 1 configuration). This is very close to the 4.58% value extracted from the analytical model. When operating in Γ_2 mode (i.e., State 2 configuration), k_t^2 mean values over 1.78% to 2.23% are measured for the resonators. The measured k_t^2 are slightly lower than the 2.77% extracted from the analytical model. This discrepancy may be attributed to the nonuniformity of strain field across the transducer thickness, which is aggravated at higher frequency (i.e., smaller wavelength) and results in higher energy concentration in Mo electrodes. Spurious modes are generated due to the two-dimensional nature of Lamb-wave propagation in a transducer membrane. Ideally, having a structure with infinitely long IDTs enables the creation of a spurious-free S_0 resonator. In practice, the finite length of the IDTs and the mechanical boundary at the substrate-anchoring region result in the energy localization of Lamb waves with a non-zero wavenumber in the IDT length direction. These waves create spurious modes with distribution and frequency defined by IDT length and the mechanical termination of the membrane. The spurious modes can be suppressed through the proper apodization of IDTs to avoid their excitation or by reducing their coupling through charge cancelation [27].

Finally, as Figure 8b suggests, large Q variations, with no clear trend, is observed in resonators with different IDT pitch size. This may correspond to the varying length of IDTs in different resonator designs, and its influence on the energy localization and Q_s. These variations are also observed in Q_p, which may be attributed to the non-homogeneous distribution of spurious modes for different IDT dimensions. Opting for optimized IDT length and exploiting apodization techniques enable achieving consistent Q_s that are only limited by fundamental material-related energy-dissipation mechanisms [27–29]. Considering the k_t^2 and Q_s values presented in Figure 8, the maximum figure of merits (i.e., $k_t^2 \times Q_s$) of ~38 and ~16 are measured for Γ_1 and Γ_2 modes, when operating in State 1 and State 2, respectively. These values can be further improved by opting for higher Sc content in the $Sc_xAl_{1-x}N$ film or by reducing W_f to $\sim \frac{\lambda}{5}$ to achieve an optimized k_t^2 for both modes.

5. Conclusions

This paper presented a new reconfigurable $Sc_xAl_{1-x}N$ Lamb-wave resonator technology based on the use of ferroelectric behavior. Periodic polarization tuning, through interdigitated transducers (IDT), was used for complementary switching between two Lamb modes with highly different wavelengths and frequencies. A comprehensive analytical model was presented to verify the complementary switchable dual-mode operation of the resonator and to provide closed-form formulation to identify the electromechanical coupling (k_t^2) of the two modes as a function of scandium content, IDT electrode thickness, and finger width. The fabrication process for the implementation of a dual-mode Lamb-wave resonator in $Sc_{0.28}Al_{0.72}N$ film was presented. Prototypes with IDT pitch sizes over 2.4 mm to 8 mm were characterized to identify their switching behavior and RF admittance. Period poling was performed through the application of 110 V 45 kHz triangular pulses between one of the IDT ports and the bottom electrode, enabling successful complementary switching between two modes of operation. k_t^2s over 4.17–4.95%, when operating in the first modes over 0.45–1.6 GHz, and 1.78–2.23% when operating in the second mode of operation over 0.8–3 GHz, were measured. Series quality factors (Q_s) over 300–800 were extracted for resonators operating in first and second modes over 0.45–3 GHz. The presented dual-mode complementary-switchable $Sc_xAl_{1-x}N$ Lamb-wave resonator technology provides lithographical frequency scaling over the entire ultra-high-frequency GHz bands; large k_t^2 exceeding AlN BAW, once doped with sufficiently high scandium content; and intrinsic switchability to relieve the need for external multiplexers. These highlight a high potential to create single-chip multi-band spectral processors for modern wireless systems.

Author Contributions: Conceptualization, S.R., D.M. and R.T.; methodology, S.R., D.M. and R.T.; validation, S.R. and D.M.; formal analysis, S.R., D.M. and R.T.; writing—original draft preparation, S.R. and D.M.; writing—review and editing, R.T.; visualization, S.R. and D.M.; supervision, R.T.; project administration, R.T.; funding acquisition, R.T. All authors have read and agreed to the published version of the manuscript.

Funding: This work was supported in parts by the Defense Advanced Research Projects Agency (DARPA), Tunable Ferroelectric Nitrides (TUFEN) Program under Grant HR00112090049, and the National Science Foundation (NSF) through the CAREER award (Grant ECCS-1752206).

Data Availability Statement: Data available on request due to restrictions e.g., privacy or ethical.

Acknowledgments: The authors would like to thank Plasma-Therm LLC and the staff at the Nanoscale Research Facility at the University of Florida with device-fabrication support.

Conflicts of Interest: The authors declare no conflict of interest.

References

1. Akiyama, M.; Kamohara, T.; Kano, K.; Teshigahara, A.; Takeuchi, Y.; Kawahara, N. Enhancement of piezoelectric response in scandium aluminum nitride alloy thin films prepared by dual reactive cosputtering. *Adv. Mater.* **2009**, *21*, 593–596. [CrossRef] [PubMed]
2. Shao, S.; Luo, Z.; Wu, T. High Figure-of-Merit Lamb Wave Resonators Based on $Al_{0.7}Sc_{0.3}N$ Thin Film. *IEEE Electron Device Lett.* **2021**, *4*, 1378–1381. [CrossRef]
3. Esteves, G.; Young, T.R.; Tang, Z.; Yen, S.; Bauer, T.M.; Henry, M.D.; Olsson, R.H., III. $Al_{0.68}Sc_{0.32}N$ Lamb wave resonators with electromechanical coupling coefficients near 10.28%. *Appl. Phys. Lett.* **2021**, *118*, 71902. [CrossRef]
4. Zhao, X.; Kaya, O.; Pirro, M.; Assylbekova, M.; Colombo, L.; Simeoni, P.; Cassella, C. A 5.3 GHz $Al_{0.76}Sc_{0.24}N$ Two-Dimensional Resonant Rods Resonator with a Record k_t^2 of 23.9%. *arXiv* **2022**, arXiv:2202.11284.
5. Giribaldi, G.; Colombo, L.; Bersano, F.; Cassella, C.; Rinaldi, M. Investigation on the Impact of Scandium-doping on the k_t^2 of $Sc_xAl_{1-x}N$ Cross-sectional Lamé Mode Resonators. In Proceedings of the 2020 IEEE International Ultrasonics Symposium (IUS), Las Vegas, NV, USA, 7–11 September 2020.
6. Fichtner, S.; Wolff, N.; Lofink, F.; Kienle, L.; Wagner, B. AlScN: A III-V semiconductor based ferroelectric. *Appl. Phys.* **2019**, *125*, 114103. [CrossRef]
7. Wang, J.; Park, M.; Mertin, S.; Pensala, T.; Ayazi, F.; Ansari, A. A film bulk acoustic resonator based on ferroelectric aluminum scandium nitride films. *J. Microelectromech. Syst.* **2020**, *29*, 741–747. [CrossRef]
8. Rassay, S.; Mo, D.; Li, C.; Choudhary, N.; Forgey, C.; Tabrizian, R. Intrinsically switchable ferroelectric scandium aluminum nitride lamb-mode resonators. *IEEE Electron Device Lett.* **2021**, *42*, 1065–1068. [CrossRef]
9. Wang, J.; Zheng, Y.; Ansari, A. Ferroelectric Aluminum Scandium Nitride Thin Film Bulk Acoustic Resonators with Polarization-Dependent Operating States. *Phys. Status Solidi (RRL) Rapid Res. Lett.* **2021**, *15*, 2100034. [CrossRef]
10. Rassay, S.; Hakim, F.; Ramezani, M.; Tabrizian, R. Acoustically Coupled Wideband RF Filters with Bandwidth Reconfigurablity Using Ferroelectric Aluminum Scandium Nitride Film. In Proceedings of the 2020 IEEE 33rd International Conference on Micro Electro Mechanical Systems (MEMS), Vancouver, BC, Canada, 18–22 April 2020.
11. Dabas, S.; Mo, D.; Rassay, S.; Tabrizian, R. Intrinsically Tunable Laminated Ferroelectric Scandium Aluminum Nitride Extensional Resonator Based on Local Polarization Switching. In Proceedings of the 2022 IEEE 35th International Conference on Micro Electro Mechanical Systems Conference (MEMS), Tokyo, Japan, 9–13 February 2022.
12. Mo, D.; Dabas, S.; Rassay, S.; Tabrizian, R. Complementary-Switchable Dual-Mode SHF Scandium-Aluminum Nitride BAW Resonator. *arXiv* **2022**, arXiv:2205.03446.
13. Ruby, R. A snapshot in time: The future in filters for cell phones. *IEEE Microw. Mag.* **2015**, *16*, 46–59. [CrossRef]
14. Matthaiou, M.; Yurduseven, O.; Ngo, H.Q.; Morales-Jimenez, D.; Cotton, S.L.; Fusco, V.F. The road to 6G: Ten physical layer challenges for communications engineers. *IEEE Commun. Mag.* **2021**, *59*, 64–69. [CrossRef]
15. Liu, Y.; Cai, Y.; Zhang, Y.; Tovstopyat, A.; Liu, S.; Sun, C. Materials, design, and characteristics of bulk acoustic wave resonator: A review. *Micromachines* **2020**, *11*, 630. [CrossRef] [PubMed]
16. Vetury, R.; Hodge, M.D.; Shealy, J.B. High Power, Wideband Single Crystal XBAW Technology for sub-6 GHz Micro RF Filter Applications. In Proceedings of the 2018 IEEE International Ultrasonics Symposium (IUS), Kobe, Japan, 22–25 October 2018.
17. Lin, C.; Yantchev, V.; Zou, J.; Chen, Y.; Pisano, A.P. Micromachined One-Port Aluminum Nitride Lamb Wave Resonators Utilizing the Lowest-Order Symmetric Mode. *J. Microelectromech. Syst.* **2014**, *23*, 78–91. [CrossRef]
18. Mo, D.; Rassay, S.; Tabrizian, R. Intrinsically Switchable Ferroelectric Scandium Aluminum Nitride Bulk Acoustic Wave Resonators. In Proceedings of the 2021 21st International Conference on Solid-State Sensors, Actuators and Microsystems (Transducers), Orlando, FL, USA, 20–24 June 2021.
19. Wang, R.; Bhave, S.A. Etch-A-Sketch filter. In Proceedings of the 2015 Transducers—2015 18th International Conference on Solid-State Sensors, Actuators and Microsystems (TRANSDUCERS), Anchorage, AK, USA, 21–25 June 2015.

20. Lu, R.; Gong, S. A 15.8 GHz A6 Mode Resonator with Q of 720 in Complementarily Oriented Piezoelectric Lithium Niobate Thin Films. In Proceedings of the Joint Conference of the European Frequency and Time Forum and IEEE International Frequency Control Symposium (EFTF/IFCS), Virtual Conference, Gainesville, FL, USA, 7–17 July 2021.
21. Ostrovskii, I.V.; Klymko, V.A.; Nadtochii, A.B. Plate wave stop-bands in periodically poled lithium niobate. *JASA Expr. Lett.* **2009**, *125*, EL129–EL133. [CrossRef] [PubMed]
22. Rosenbaum, J.F. *Bulk Acoustic Wave Theory and Devices*; Artech House Acoustics Library: Boston, MA, USA; London, UK, 1988; pp. 125–163.
23. Felmetsger, V.; Mikhov, M.; Ramezani, M.; Tabrizian, R. Sputter Process Optimization for $Al_{0.7}Sc_{0.3}N$ Piezoelectric Films. In Proceedings of the 2019 IEEE International Ultrasonics Symposium (IUS), Glasgow, UK, 6–9 October 2019.
24. Rassay, S.; Hakim, F.; Li, C.; Forgey, C.; Choudhary, N.; Tabrizian, R. A Segmented-Target Sputtering Process for Growth of Sub-50 nm Ferroelectric Scandium–Aluminum–Nitride Films with Composition and Stress Tuning. *Phys. Status Solidi (RRL)–Rapid Res. Lett.* **2021**, *15*, 2100087. [CrossRef]
25. Lu, R.; Li, M.-H.; Yang, Y.; Gananoque, T.; Gong, S. Accurate Extraction of Large Electromechanical Coupling in Piezoelectric MEMS Resonators. *J. Microelectromech. Syst.* **2019**, *28*, 209–218. [CrossRef]
26. Feld, D.A.; Parker, R.; Ruby, R.; Bradley, P.; Dong, S. After 60 years: A new formula for computing quality factor is warranted. In Proceedings of the 2008 IEEE Ultrasonics Symposium, Beijing, China, 2–5 November 2008.
27. Zhang, H.; Liang, J.; Zhou, X.; Zhang, H.; Zhang, D.; Pang, W. Transverse Mode Spurious Resonance Suppression in Lamb Wave MEMS Resonators: Theory, Modeling, and Experiment. *IEEE Trans. Electron Devices* **2015**, *62*, 3034–3041. [CrossRef]
28. Tabrizian, R.; Rais-Zadeh, M. The effect of charge redistribution on limiting the kt2.Q product of piezoelectrically transduced resonators. In Proceedings of the 2015 Transducers—2015 18th International Conference on Solid-State Sensors, Actuators and Microsystems (TRANSDUCERS), Anchorage, AK, USA, 21–25 June 2015.
29. Segovia-Fernandez, J.; Piazza, G. Thermoelastic damping in the electrodes determines Q of AlN contour mode resonators. *J. Microelectromech. Syst.* **2017**, *26*, 550–558. [CrossRef]

MDPI
St. Alban-Anlage 66
4052 Basel
Switzerland
Tel. +41 61 683 77 34
Fax +41 61 302 89 18
www.mdpi.com

Micromachines Editorial Office
E-mail: micromachines@mdpi.com
www.mdpi.com/journal/micromachines

www.ingramcontent.com/pod-product-compliance
Lightning Source LLC
LaVergne TN
LVHW070921170726
843515LV00005B/834
* 9 7 8 3 0 3 6 5 6 3 6 7 1 *